Coastal and Marine Governance and Protection

Coastal and Marine Governance and Protection

Guest Editors

Celene B. Milanes
Camilo M. Botero
Daniel O. Suman

MDPI

Basel • Beijing • Wuhan • Barcelona • Belgrade • Novi Sad • Cluj • Manchester

Guest Editors
Celene B. Milanes
Faculty of Engineering
Universidad del Magdalena
Santa Marta
Colombia

Camilo M. Botero
School of Law
Universidad Sergio Arboleda
Santa Marta
Colombia

Daniel O. Suman
Rosenstiel School of Marine,
Atmosphere, and Earth
Science (EVR)
University of Miami
Miami, FL
USA

Editorial Office
MDPI AG
Grosspeteranlage 5
4052 Basel, Switzerland

This is a reprint of the Special Issue, published open access by the journal *Water* (ISSN 2073-4441), freely accessible at: https://www.mdpi.com/journal/water/special_issues/F5M45KEH56.

For citation purposes, cite each article independently as indicated on the article page online and as indicated below:

Lastname, A.A.; Lastname, B.B. Article Title. *Journal Name* **Year**, *Volume Number*, Page Range.

ISBN 978-3-7258-5625-1 (Hbk)
ISBN 978-3-7258-5626-8 (PDF)
https://doi.org/10.3390/books978-3-7258-5626-8

Cover image courtesy of Daniel O. Suman and Camilo Mateo Botero Saltaren

Contents

About the Editors

Celene B. Milanes

Celene B. Milanes completed her postdoctoral research in Geography at the Federal University of Pará, Brazil. Milanes holds a Ph.D. in Technical Sciences and two master's degrees: one in Integrated Coastal Zone Management and another in Conservation and Rehabilitation of Built Heritage. Her scholarly work is strongly oriented towards the United Nations Sustainable Development Goals (SDGs). Milanes has extensive expertise in land-use planning, vulnerability assessment, urban risk, and tourism. Her research further encompasses governance and public policy, coastal pollution, integrated ecosystem management, and urban resilience, with a particular emphasis on island nations and coastal cities. Her academic distinctions include the National Award of the Academy of Sciences of the Republic of Cuba and the Stephen Olsen International Chair in Integrated Coastal Management. She is currently a Research Associate at the Coiba Scientific Station (Coiba AIP) in Panama and in CEMarin, Colombia.

Camilo M. Botero

Camilo M. Botero is a renowned expert in integrated coastal management, marine law, and beach management. He has served as Technical Secretary for key national committees on Tsunamis, El Niño, Marine Pollution, and Integrated Coastal Zone Management within the Colombian Oceanic Commission. Botero is a distinguished professor and researcher at universities across Latin America, the Caribbean, and Canada, and has extensive experience in teaching and research on coastal and marine issues. His research focuses on coastal regulation and marine–coastal land-use planning. A Full Fellow of esteemed institutions like the Academy of Geographical Sciences of Colombia and the Coastal Education and Research Foundation, he possesses strong communication and language skills, including English, Italian, Portuguese, and Esperanto. He is a prolific author and researcher with over 150 publications, including books and peer-reviewed articles. He has supervised numerous graduate and doctoral theses and is a frequent lecturer and journal evaluator. Beyond his academic achievements, he is an entrepreneurial innovator in beach and coastal tourism. He possesses a unique blend of expertise, incorporating his knowledge of innovative approaches such as collaborative board game development, audiovisual production, citizen science, and the gamification of educational and tourism experiences.

Daniel O. Suman

Daniel O. Suman is a Professor of Environmental Science and Policy at the Rosenstiel School of Marine, Atmosphere, and Earth Science of the University of Miami, as well as an Adjunct Professor at the School of Law. His research and project areas focus on coastal management, adaptation to climate change, governance of marine resources and space, management of mangroves and coastal wetlands, and marine protected areas—particularly in Latin America. Suman earned a Ph.D. in Oceanography from the Scripps Institution of Oceanography (University of California, San Diego), a J.D. from the University of California, Berkeley, and an M.A., M.Ed., and Certificate of Latin American Studies from Columbia University. At the University of Miami, he has taught courses for 35 years in Environmental Law, Environmental Planning, Coastal Management, Coastal Law, and Water Resources Policy. In addition, Suman has engaged in collaborative research and training projects throughout the world.

Preface

The nine papers in this Reprint offer a diverse, yet interconnected, set of contributions to the field of coastal and marine governance, spanning case studies, methodological advances, and critical reviews. The geographical focus of the set of essays ranges from local case studies in Brazil, China, Cuba, Italy, Panama, and Portugal to global assessments. Common concerns include the fragmentation of institutional frameworks, the low enforcement of legal instruments, and the socio-ecological impacts of coastal transformation. Coastal and marine areas are undergoing significant transformations due to urbanization, industrialization, port development, climate change, pollution, over-exploitation, and other human activities, with consequent impacts on ecological integrity and traditional livelihoods. The essays collectively advocate for the urgent need to improve governance structures, integrate technological innovations, coordinate the management of watersheds, coastal areas, and adjacent marine areas, and improve participatory mechanisms that enhance inclusion of all stakeholders in decision-making processes. Collectively, the essays in the Reprint advocate for more inclusive, data-informed, and adaptive approaches that address the complex challenges faced by coastal and marine ecosystems today. The Reprint will be of interest to individuals concerned about the wise and sustainable management of coastal and marine systems, particularly practitioners and students.

Celene B. Milanes, Camilo M. Botero, and Daniel O. Suman
Guest Editors

water

MDPI

Editorial

The Multiple Challenges Faced by Coastal and Marine Governance

Camilo M. Botero [1,2,*], Daniel O. Suman [2,3] and Celene B. Milanes [2,4,5]

1 School of Law, University Sergio Arboleda, Santa Marta 470001, Colombia
2 Estación Científica (Coiba AIP), Gustavo Lara Street, Building 145B, City of Knowledge, Clayton, Panama City 0801, Panama; dsuman@earth.miami.edu (D.O.S.); cmilanes@unimagdalena.edu.co (C.B.M.)
3 Rosenstiel School of Marine, Atmospheric, and Earth Sciences & School of Law, University of Miami, Miami, FL 33149-1031, USA
4 Faculty of Engineering, and BioRed Institute, Universidad del Magdalena, Calle 29H3 No 22-01, Santa Marta 470004, Colombia
5 Center of Excellence in Marine Sciences (CEMarin), Block 1 Floor 7, Cra 4 # 22-61, Bogotá 111321, Colombia
* Correspondence: camilo.botero@usa.edu.co

1. Introduction

Coastal and marine areas are only partially integrated into governance across different geographical scales, with persistent fragmentation and sectoral approaches remaining common challenges [1]. While integrated management frameworks—such as integrated coastal zone management (ICZM) and marine spatial planning (MSP)—are widely promoted to address these issues, their implementation often struggles with institutional inertia, conflicting mandates, and limited cross-sectoral coordination, resulting in slow or incomplete integration, especially at the interface of land and sea and across administrative levels [2]. Case studies from several regions around the world reveal that planning tools and governance mechanisms frequently remain siloed, focusing on compliance with sector-specific regulations rather than fostering holistic, ecosystem-based management [3].

Efforts to align governance vertically (from local to national), horizontally (across sectors), and programmatically (from policy to implementation) are often hampered by misaligned objectives and resource constraints [4]. Additionally, the integration of traditional and local knowledge into governance processes remains limited, though participatory and deliberative approaches show promise for more inclusive and adaptive management [5,6]. Overall, while progress has been made in recognizing the need for integrated approaches, significant barriers—such as institutional complexity, lack of technical capacity, and entrenched policy frameworks—continue to impede the full integration of coastal and marine areas into governance at all levels [1,4,7].

On the other hand, the inclusion of bottom-up approaches in coastal and marine governance has gained recognition for its potential to enhance sustainability, legitimacy, and adaptability; however, its integration remains uneven and faces significant challenges. Bottom-up strategies—such as empowering local communities, participatory planning, and community-based management—can provide valuable local knowledge, foster stewardship, and improve compliance with regulations [8]. Successful examples, such as the Algoa Bay Project in South Africa, demonstrate that combining bottom-up engagement with top-down authority can identify practical pathways for integrating local knowledge into marine spatial planning, increase transparency, and enhance access to information [6].

However, in many cases, top-down approaches still dominate, with bottom-up participation often limited to consultation rather than genuine power-sharing, as seen in several

 https://doi.org/10.3390/w17152322

designations of marine protected areas [9,10]. Comparative studies in the Global South reveal that while local collective action can address certain challenges, external drivers, such as market forces and policy constraints, often limit the effectiveness of bottom-up initiatives, underscoring the need for flexible, multi-level governance that aligns local and national priorities [11,12]. Case studies also demonstrate that empowering communities necessitates sustained engagement, long-term capacity building, and acknowledgment of traditional rights [6,8]. Ultimately, the most effective governance models combine bottom-up and top-down approaches, tailoring participation to local contexts and ensuring that community voices are not only heard, but also have a real influence over decision-making [8,12,13].

2. Main Contributions of the Special Issue

The nine papers in this Special Issue offer a diverse, yet interconnected, set of contributions to the field of coastal and marine governance, spanning case studies, methodological advances, and critical reviews. While they differ in geographical focus—from local case studies in China (Contributions 1 and 6), Cuba (Contribution 4), Brazil, Panama, and Portugal (Contribution 8), and Italy (Contribution 7) to global assessments (Contributions 2, 5 and 9)—and in disciplinary approach, they converge around key themes such as the urgent need to improve governance structures, integrate technological innovation, and facilitate participatory mechanisms. Common concerns include the fragmentation of institutional frameworks, the low enforcement of legal instruments, and the socio-ecological impacts of coastal transformation. Despite methodological variation—from geospatial analysis to qualitative fieldwork and legal critique—the papers collectively advocate for more inclusive, data-informed, and adaptive approaches to address the complex challenges facing coastal and marine systems today.

Reviewing this set of papers reveals that they employ a diverse range of research types. Five articles use case studies (Contributions 1, 4, 6, 7, and 8), three analyze methodological or technological development papers (Contributions 1, 3, and 9), one presents an international legal review (Contribution 2), and three developed comprehensive reviews (Contributions 5, 8, and 9). Together, these contributions reflect a balanced integration of empirical investigation, theoretical synthesis, and applied innovation, advancing our understanding of coastal and marine governance.

The papers in this Special Issue encompass a broad geographical scope, with case studies and analyses spanning Asia (China), Europe (Italy and Portugal), and the Americas (Brazil, Panama, and Cuba), in addition to global assessments. Several papers adopt a global or multi-regional lens—particularly those on marine pollution monitoring, marine protected areas, legal frameworks, and nature-based solutions—thereby offering both localized insights and internationally relevant perspectives.

The authors employed a range of qualitative, quantitative, and mixed methods, including field surveys and stakeholder interviews (e.g., Contribution 4), document analysis (e.g., Contribution 8), and spatial planning reviews (e.g., Contribution 7). Several techniques are utilized, such as machine learning algorithms (e.g., Contribution 3), social media analysis (e.g., Contribution 1), geospatial techniques (e.g., Contribution 5), and doctrinal legal analysis (e.g., Contribution 2), to examine governance frameworks, environmental monitoring, and spatial prioritization at both national and global scales.

The papers collectively highlight critical governance challenges across scales, including institutional fragmentation, limited stakeholder inclusion, weak enforcement of international legal instruments, and insufficient integration between land and sea governance. Several studies emphasize the importance of participatory and adaptive governance models, including community engagement, co-management, digital platforms, and spatial planning, as essential pathways to strengthen coastal and marine policy implementation and enhance environmental and social resilience.

The studies reveal that coastal and marine areas are undergoing significant transformation due to urbanization, industrialization, port development, climate change, and pollution, with consequent impacts on ecological integrity, traditional livelihoods, and spatial dynamics (e.g., Contributions 1, 6, and 7). While several papers emphasize the degradation of coastal ecosystems and the fragmentation of marine governance, others propose integrated management approaches—such as marine protected areas, nature-based solutions, and marine spatial planning (e.g., Contributions 5, 7 and 8)—to support more sustainable and inclusive stewardship of coastal and marine environments.

In summary, the papers collectively recommend strengthening multi-level governance frameworks through improved legal instruments, enhanced stakeholder participation, integrated data, and spatial planning, with a particular emphasis on the role of innovative tools such as remote sensing, naval surveillance, and social media platforms. Future directions call for advancing interdisciplinary and cross-jurisdictional collaboration, embedding nature-based and community-led approaches in policy design, and scaling up scientifically grounded marine protection efforts to meet global biodiversity and sustainability targets.

3. Urgent Challenges

The challenges facing coastal and marine areas are numerous, with different priorities according to the analytical perspective. Nevertheless, global economic and political processes impact some communities and ecosystems more intensely than others. The hope sparked by the United Nations Convention on Environment and Development—UNCED, in 1992, in Rio de Janeiro, with Agenda 21 and its Chapter 17, barely exists after 30 years. The integrated coastal zone management framework, launched internationally at this convention, was almost forgotten by the end of the first decade of the new century. Instead of ICZM, marine spatial planning has taken the lead as the preferred management tool for marine areas, frequently disconnecting coastal areas from decision-making [14].

However, coastal and marine governance extends beyond integrated management approaches. Governance implies multi-sectoral approaches, multi-level decision-making, but mainly real public participation. The contributions included in this Special Issue highlight most of these concerns, with a blend of social, legal, environmental, economic, and technological perspectives. As stated earlier, the challenges to coastal and marine governance are numerous, but we, as editors of this Special Issue, would like to highlight the eight most pressing issues.

3.1. *Institutional Fragmentation*

Coastal and marine areas require new institutional frameworks that extend beyond collective bodies of already-established institutions. The governance of coastal and marine areas should be led by a single body with sufficient authority to encompass and integrate three diverse perspectives: environmental, social, and economic. Coastal and marine areas should be recognized as a governance unit, which could be called Maremtory, as a semantic evolution of the coastal and marine territory [15].

3.2. *Legal Enforcement*

Policy and regulatory landscapes in coastal and marine areas are complex, and sometimes, could be even inadequate [16]. However, the main flaws in regulations include the enforcement of the many laws and resolutions. One reason for low levels of enforcement is that regulations have been created with minimal participation from local communities, imposed on them rather than meeting their needs. A new approach to coastal and marine governance should incorporate innovative methods for establishing agreements that reflect the interests and desires of local communities, rather than those of multinational powers.

3.3. Bottom-Up Methodologies

Decision-making has historically been a top-down process. From kingdoms to republics, power has been centralized by a select few individuals who decide the fate of the majority. Nevertheless, the social achievements of the twentieth century enabled large consultation processes, reinforced by current social technologies. The new challenge is implementing new governance methodologies that integrate the needs of the majority and converting them into decisions for the general benefit.

3.4. Urbanization and Traditions

Urban areas have been the primary focus of decision-making in coastal regions, due to their significant social and economic importance in quantitative terms. As a consequence, rural and natural areas have been relegated to secondary status, resulting in several environmental and cultural impacts caused by urbanization. The challenge for coastal governance turns around the best way to conserve rich traditional heritage without indiscriminately prejudicing new developments.

3.5. Limited Stakeholder Participation and Social Inclusion

Sometimes, different governance processes exclude the participation of local coastal communities, Indigenous peoples, women, and young people from decision-making, undermining the legitimacy of coastal and marine policies and reducing compliance with regulations. Participatory governance and citizen science are essential for promoting environmental justice and ensuring that management strategies align with local needs and knowledge [17].

3.6. Insufficient Scientific Data and Monitoring

A lack of up-to-date, site-specific data on biodiversity, pollution, erosion, sea-level rise, and ecosystem services hinders evidence-based policymaking. Without robust indicators and continuous monitoring, it becomes extremely difficult to assess risks accurately or adapt management to changing environmental conditions [18].

3.7. Conflicts Between Economic Interests and Conservation Goals

Coastal zones are often sites of competing interests, including tourism, fisheries, urban development, oil and gas extraction, and conservation. These conflicting uses often prioritize short-term economic gain over long-term sustainability, resulting in habitat degradation, resource overexploitation, and social inequities. Balancing these pressures requires transparent, multi-stakeholder governance frameworks.

3.8. Climate Change and Disaster Risk Vulnerability

Rising sea levels, ocean acidification, saline intrusion, an increase in the intensity and frequency of storms, and coastal erosion intensify existing vulnerabilities [19]. Many local governance systems are reactive rather than proactive, lacking adaptive management mechanisms to prepare for future impacts. Integrating climate adaptation and disaster risk reduction into governance is urgent, particularly in small island developing states and low-lying coastal areas [20].

Conflicts of Interest: The authors declare no conflicts of interest.

References

1. Kelly, C.; Ellis, G.; Flannery, W. Unravelling Persistent Problems to Transformative Marine Governance. *Front. Mar. Sci.* **2019**, *6*, 213. [CrossRef]
2. Le Tissier, M. Unravelling the Relationship between Ecosystem-Based Management, Integrated Coastal Zone Management and Marine Spatial Planning. In *Ecosystem-Based Management, Ecosystem Services and Aquatic Biodiversity*; O'Higgins, T., Lago, M., DeWitt, T., Eds.; Springer: Cham, Switzerland, 2020. [CrossRef]
3. Lin, T.; Liu, W.; Chang, Y.; Hsiao, S. Capacity assessment of integrated coastal management for Taiwanese local government. *Mar. Policy* **2021**, *134*, 104769. [CrossRef]
4. Singh, G.G.; Cottrell, R.S.; Eddy, T.D.; Cisneros-Montemayor, A.M. Governing the Land-Sea Interface to Achieve Sustainable Coastal Development. *Front. Mar. Sci.* **2021**, *8*, 709947. [CrossRef]
5. Brooks, K.; Barclay, K.; Grafton, R.; Gollan, N. Transforming coastal and marine management: Deliberative democracy and integrated management in New South Wales, Australia. *Mar. Policy* **2020**, *139*, 104053. [CrossRef]
6. Rivers, N.; Strand, M.; Fernandes, M.; Metuge, D.; Lemahieu, A.; Nonyane, C.; Benkenstein, A.; Snow, B. Pathways to integrate Indigenous and local knowledge in ocean governance processes: Lessons from the Algoa Bay Project, South Africa. *Front. Mar. Sci.* **2023**, *9*, 1084674. [CrossRef]
7. O'Hagan, A.; Paterson, S.; Tissier, M. Addressing the tangled web of governance mechanisms for land-sea interactions: Assessing implementation challenges across scales. *Mar. Policy* **2020**, *112*, 103715. [CrossRef]
8. Gaymer, C.; Stadel, A.; Ban, N.; Cárcamo, P.; Ierna, J.; Lieberknecht, L. Merging top-down and bottom-up approaches in marine protected areas planning: Experiences from around the globe. *Aquat. Conserv.-Mar. Freshw. Ecosyst.* **2014**, *24*, 128–144. [CrossRef]
9. Cockerell, L.; Jones, P. Governance Analysis of St Anne Marine National Park, Seychelles. *Mar. Policy* **2021**, *127*, 103912. [CrossRef]
10. Jones, P. Marine protected areas in the UK: Challenges in combining top-down and bottom-up approaches to governance. *Environ. Conserv.* **2012**, *39*, 248–258. [CrossRef]
11. Andriesse, E.; Saguin, K.; Ablo, A.; Kittitornkool, J.; Kongkaew, C.; Mang'ena, J.; Onyango, P.; Owusu, V.; Yang, J. Aligning bottom-up initiatives and top-down policies? A comparative analysis of overfishing and coastal governance in Ghana, Tanzania, the Philippines, and Thailand. *J. Rural Studies* **2022**, *92*, 404–414. [CrossRef]
12. Parker, Q.; Longosoa, H.; Long, S.; Jones, P. A longitudinal governance analysis of a locally managed marine area: Ankobohobo wetland small-scale mud crab fishery, Madagascar. *Mar. Policy* **2024**, *163*, 106138. [CrossRef]
13. Yang, J.; Chang, Y.; Hsiao, S. Finding harmony in the sea: Resolving conflicts by regional marine spatial planning. *Ocean Coast. Manag.* **2024**, *254*, 107200. [CrossRef]
14. Ramieri, E.; Bocci, M.; Marković, M. Linking Integrated Coastal Zone Management to Maritime Spatial Planning: The Mediterranean Experience. In *Maritime Spatial Planning*; Palgrave Macmillan: Cham, Switzerland, 2019. [CrossRef]
15. Botero, C.M. *Maremtorio: Descubrimiento de una Colombia Invisible*; Serie Geographic Expositions N° 17; Colombian Geographic Society: Bogotá D.C., Colombia; ISBN 978-958-521-0219.
16. Taljaard, S.; van Niekerk, L.; Weerts, S.P. The legal landscape governing South Africa's coastal marine environment—Helping with the 'horrendogram'. *Ocean Coast. Manag.* **2019**, *178*, 104801. [CrossRef]
17. Pomeroy, R.; Rivera-Guieb, R. *Fishery Co-Management: A Practical Handbook*; International Development Research Centre (IDRC): Ottawa, ON, Canada, 2006.
18. United Nations Environment Programme (UNEP). *Coastal Resilience and Climate Adaptation Strategies: Guidance for Policy and Planning*; UNEP: Nairobi, Kenya, 2021. Available online: https://www.unep.org/resources/report/coastal-resilience-and-climate-adaptation-strategies (accessed on 10 July 2025).
19. Montero, O.P.; Batista, C.M. Social perception of coastal risk in the face of hurricanes in the southeastern region of Cuba. *Ocean Coast. Manag.* **2019**, *184*, 105010. [CrossRef]
20. IPCC (Intergovernmental Panel on Climate Change). *Climate Change 2022: Impacts, Adaptation and Vulnerability*; Contribution of Working Group II to the Sixth Assessment Report of the IPCC; Cambridge University Press: Cambridge, UK, 2022. [CrossRef]

water

MDPI

Article

Meaningful Multi-Stakeholder Participation via Social Media in Coastal Fishing Village Spatial Planning and Governance

Jing Wang [1,†], Ming-Ming He [1,†], Su-Hsin Lee [2] and Shu-Chen Tsai [1,*]

[1] College of Arts and Design, Jimei University, Xiamen 361021, China; wangjing@jmu.edu.cn (J.W.); 202311130014@jmu.edu.cn (M.-M.H.)
[2] Department of Geography, National Taiwan Normal University, Taipei 10644, Taiwan; shlee@ntnu.edu.tw
* Correspondence: tsaishuchen@jmu.edu.cn
† These authors contributed equally to this work.

Abstract: Due to the rapid development of China's economy, the current situation of fishing villages in the southeastern coastal areas is spatial disorder caused by changes in population composition and industrial transformation. This study analyses the differences between the clan structure and the multi-stakeholder engagement model in traditional fishing villages. The main aim is to illustrate contemporary issues that fishing villages' spaces need to deal with in governance and decision making. With the development of information technology, social media has become an important platform through which stakeholders can communicate and make decisions. The aims of this paper were as follows: (1) Identify the stakeholders involved in the governance of fishing villages; (2) explore how stakeholders participate in the planning and governance of fishing villages through social media; (3) examine the mechanisms of social media and its impact on the spatial planning of fishing villages. Through qualitative research methods such as field surveys and in-depth interviews, the following results were obtained: (1) Social media subverts the traditional fishing village governance model, and the scope of the governance subject is expanded; (2) spatial changes in fishing villages are affected by the joint influence of people, the environment, and the economy, and a social network acts as an intermediary to compensate for the deficiencies that existed in previous fishing village governance processes.

Keywords: experiential tourism; intangible cultural heritage; community engagement; social inclusion; relationship

1. Introduction

Effective governance through spatial planning has a long history of development in Western society, and there are many forms of spatial planning and governance, such as the famous De Architectura treatise, which became a principle of urban planning two thousand years ago and laid the foundation for typical patterns of development in many European cities [1]. In modern times, city planning is typically state-led or market-led, with many functional zones forming the pattern of the city [2]. In the 21st century, more emphasis is being placed on community participation in design, which will help ameliorate the social and economic issues arising from the mismanagement of physical space [3]. Consequently, over the past two decades, community participation has played an increasingly important role in urban renewal [3].

In recent years, top-down community participation led by national programs has also been implemented in developing countries, but national leadership typically focuses more on political responsibility than on community interests, resulting in low levels of satisfaction among residents and a failure to generate positive benefits [4]. Traditional urban planning can no longer meet the needs of the new era, and public participation workshops for urban renewal must be changed into cooperative workshops according to the different conditions of the community to achieve the goals effectively [5]. The single theory of community participation, therefore, cannot be applied to spatial governance in different national conditions. For example, in China, urban regeneration has been ongoing since 1978, but recent studies have shown that most Chinese cities have not considered social sustainability in the urban regeneration process, resulting in more urban sprawl, social exclusion, and environmental pollution [6]. Case studies in China have shown that state-led community participation, in the absence of democratization, tends to be a formality as more structural problems remain unresolved and thus fail to produce the desired sense of residential cohesion and local belonging [4].

However, are participatory approaches not suitable for China? Or is there a participatory approach that has not yet been discovered in China? Most of China's rural areas are frozen in time and space and have maintained traditional patterns, architectural forms, and lifestyles for hundreds of years; however, rural areas near the urban periphery are experiencing intense urbanization, mainly due to changes forced by urban expansion [7–10]. Furthermore, almost none of the fishing villages along the coast of China have been spared the impact of industrial restructuring and urbanization [11–13]. Mainly because coastal cities serve as China's primary gateways [12,14] to globalization, industrial concentration, population migration [15], and socio-economic transformation [16,17] have collectively led to either the demolition of some fishing villages or the preservation of only fragmented historical districts and buildings [18]. Without fisheries and fishermen, these villages lose their essence as fishing communities. Some fortunate ones have been transformed into coastal resorts [19–23]; others have become abandoned villages [13]. In the tide of national development, fishing villages have not only lost their population but also their unique characteristics [12,17,24]. However, fishing villages are of utmost importance in the history of human development [25].

Fishing villages are usually located in coastal waters or estuaries, where fishermen have long retained skills passed down from generation to generation and rarely disseminate their social networks outwards [25,26]. As these fishing villages are mostly clustered and have strong defenses, their governance model is clan-based [17,27,28], and within the village, public gathering spaces centered on wells, ponds, and ancestral halls constitute the spatial core of traditional fishing villages [13]. However, the decline of subsistence fisheries necessitated a transformation of both fishermen and villages, with a state-dominated institutional framework replacing the clan-based spatial governance model [17,18]. However, in practice, due to the over-reliance on administrative power and the lack of social incentives, it is difficult to stimulate fishermen's collective action to safeguard the resources and eliminate the prisoner's dilemma of competition, and the effect of governance is not as effective as it should be [29,30]. Research on "how to improve the national regulatory model through the coordination of the relationship between the state and society, governance and development" has therefore been gradually conducted. In 2012, Ostrom et al. proposed and developed a community-owned governance model (i.e., good governance of the commons through multi-party participation and democratic consultation), but this model reveals obvious limitations when faced with such a large-scale dilemma as that of the marine fisheries industry [31]. In fact, since its inception, Western governance theories such as community-owned governance have clearly conveyed the position that the state should withdraw and return power to society, but many failures in governance practices in developing countries are due precisely to the dismantling of state logic in governance practices [32].

In recent years, with the rapid socio-economic transformation of China's urban society, a variety of heterogeneous subjects, including the government, village committees, enterprises, rural talents, and villagers, are actively participating in rural spatial practices. Multi-party participation has accelerated the spatial transformation, reconstruction, and commercialization of rural areas in China from productivism to post-productivism and multifunctional modern countryside [33]. In addition, with the development of technology, the forms of participation have changed. In addition to traditional in-person participation, relevant stakeholders can use the Internet to collaborate in a more direct, real-time, and networked manner [34,35]. In previous studies, new mechanisms, forms, and modes of transformation have been discussed in detail to seek the sustainable development of fishing villages [17,18,24,36], but e-participation through social media, and thus, changing the spatial planning model of fishing villages, has not been adequately studied.

E-participation in social media has led to a more diverse public sphere constructed by stakeholders [35,37] and these digital platforms also enhance citizen empowerment and social inclusion [38,39]. The purpose of this study was to clarify through a case study who the multiple stakeholders are, what mediums they use to participate in the spatial planning and management of a fishing village, and what manifestations occur in the space. Importantly, the cases show that the value lies not in the outcome of centralized planning but in restoring organic co-development between local residents, the living environment, and the economic system.

2. Materials and Methods

2.1. Study Area

Xunpu village (XPV) in Quanzhou City, Fujian Province, is a typical fishing village in the southern coastal cities of China, located at 118°44′ E longitude and 24°51′ N latitude, covering an area of 3.8 square kilometers (Figure 1). As of 2023, Xunpu village (XPV) had 1678 households and a population of about 6355 [40]. Before 2010, the fishing village was left isolated because of the urban renewal around it, and the overall landscape was depressed, leaving only elder residents and broken houses in the fishing village. XPV is a place with a profound cultural element, including religious and cultural buildings (Mazu Temple), ancient and famous trees, water wells, traditional buildings (Oster house), and fishing port landscapes.

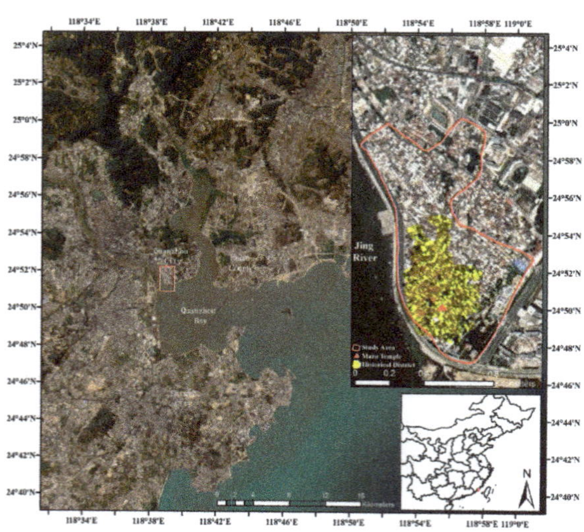

Figure 1. Study area.

XPV was the end point of the entire Quanzhou east coast tourist route [41], which is also situated near the famous UNESCO World Heritage Site: Quanzhou: Emporium of the World in Song-Yuan China [42]. It is therefore positioned as a service space for coastal tourism, and the government should ideally transform XPV into a "modern" historical and cultural district that meets the needs of modern tourism: neat, uniform, and modern. In 2013, except for the surrounding area of Mazu Temple being retained as a district, the context of the entire fishing village was almost entirely destroyed via the modern district square planning method, in accordance with the 2013 urban renewal plan [43]. Between 2013 and 2022, in a partnership between the government and developers, many green spaces were designated as sites for the planned construction of commercial and residential areas, e.g., Fenghai Road, which circles the coast, and Binhai Road, which runs from southeast to northwest surrounding XPV.

However, the COVID-19 pandemic initially delayed the demolition of XPV, and after the lockdown was lifted, the popularity of intangible cultural heritage (ICH) tourism attracted public attention. The planned demolition was then stopped, and the historic district was preserved.

2.2. Study Methods

The term Xunpu women refers to the women of Xunpu village, known for their traditional hair decoration style, clothes, and festivals, which have been given national ICH status in China [44]. As early as around 2020, the practices of Xunpu fishing women were regarded as local folk customs, and they would only appear during local festivals. Though loved by the public, these local customs enjoyed limited appreciation from an outsider's perspective. However, this phenomenon has since spread throughout China in a short period of time and has become a prevalent study issue [36,45,46].

This study adopted a qualitative research approach integrating descriptive and explanatory frameworks. First, the process of XPV spatial change, industrial transformation, and social media participation in governance is documented and described through observation and interviews; second, the causes behind the phenomenon are analyzed in depth, i.e., (1) how multi-stakeholders participate in the collaborative governance of the fishing village space via social media; (2) how the fishing village space (XPV), ICH (fisherwomen's culture), and multi-stakeholders are involved in the governance of the fishing village space; (3) the fishing village space (XPV), ICH (fisherwomen's culture), and multi-stakeholder (governance) interaction mechanism (Figure 2); and (4) the significance of social media participation for fishing village governance.

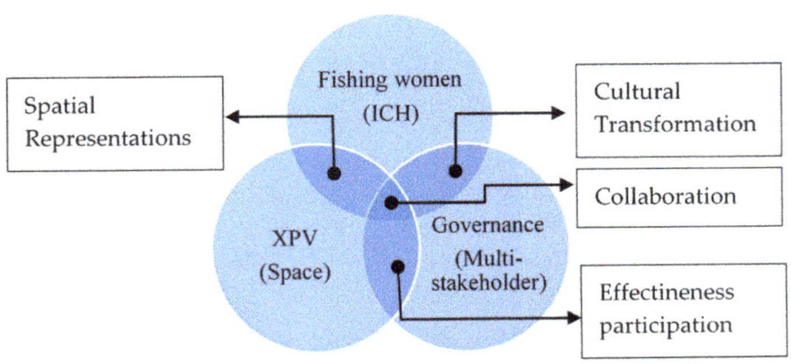

Figure 2. Study concept.

This study utilized a case study approach to longitudinally track a single case (XPV), combining observations, in-depth interviews, and physical information to obtain data. The study process was as follows: On 9 September 2023, and 4 March 2024, a field investigation was conducted in XPV. The target of the investigation was the number and spatial distribution of intangible cultural heritage experience shops (ICHESs), mainly Zanhuawei shops (including clothing, headwear, photography, and makeup) The geographical location, natural environment, location characteristics, and Zanhuawei status quo of XPV were observed and recorded.

In the process of sampling investigation, the systematic sampling method was adopted in this study. Starting from the first ICHES in the first village entrance of XPV, one ICHES was taken as a sample every four shops. Finally, relevant information on 50 ICHESs was collected, such as the opening time, type, and service. The data are shown in Table 1. The total number of shops is approximately 246. In addition, the experiences and opinions of local residents, ICH inheritors, village party secretaries, tourists, makeup artists, and photographers were obtained through in-depth interviews (11 samples in Table S1). The data from the in-depth interview was keyed in verbatim for a subsequent text analysis (Table S1). Social media platforms are the focus of this study, and this study lists relevant social platforms and analyzes their roles and functions in XPV's ICH tourism (Table S2). Case studies can provide rich and detailed descriptions of actual phenomena, especially single case studies, which are more likely to shed light on the "what", "why", and "how" [47]. There are specific patterns in the information obtained from sample surveys about the composition of ICHESs, the range of services, and the length of time in business, and in-depth interviews can reveal the underlying mechanisms and potential changes behind such complex phenomena.

Table 1. Description of sampling stores in this study.

Categorization	Operator		Shop Source		Opening Time		
	Native	Outsider	Tenancy	Own House	Before 2023	2023–2024	After 2024
Number of shops	31	19	29	21	4	39	7
Subtotal	50		50		50		

3. Results

3.1. Transformation of Fishing Village Economy

Since the Tang and Song dynasties (around 618 AD), women in the fishing villages of XPV have continued to perform various folk activities. During major festivals, XPV women would wear traditional costumes and decorate their hair with flowers into a circle (called Zanhuawei) to celebrate together. This custom has continued for thousands of years [48]. However, the Cultural Revolution put XPV's culture at risk of extinction, while the reform and opening up of the last century put the fishing village space at risk of demolition.

It was not until 2004 that the Quanzhou Municipal Government realized that the Xunpu fishing women's folk customs are important and began to protect them, such as by establishing the "Grandma Waist Drum Team" (consisting of old women from Quanzhou Bay Village wearing traditional costumes and Zanhuawei, participating in various activities); incorporating Xunpu women's customs into teaching materials; and performing traditional activities (teaching children traditional skills such as Zanhuawei and carving shells). As a result, the Xunpu women's customs were included in China's national ICH list in 2005 [49]. Subsequently, Fengze District incorporated Xunpu women's customs into the school curriculum, and local school students learned ICH skills such as Zanhuawei.

In 2013, the first ICHES opened in XPV, focusing on providing an experience of ICH services and traditional fishing women's costumes, and more ICHESs were subsequently opened in the village. In the following ten years, although the government continued to lead the promotion of XPV, the leisure industry featuring Xunpu women was only promoted within the entire Quanzhou city. It was not until 2023 that a Chinese film and television star (Zhao Liying) appeared on the cover of a magazine wearing Zanhuawei, and the photos swept across all major media platforms [40]. Influenced by the movie star effect, many Internet celebrities came to XPV to experience Zanhua (a verb meaning to decorate one's hair with flowers like Xunpu fishing women's), requesting "the same style as Zhao Liying" and uploading their selfies to social media platforms. For a time, topics such as "Be a Xunpu girl for a day" and "Zanhua in this life, be beautiful in the next life" became hot search topics on social media platforms. In addition, media platforms further promoted the development of Xunpu's tourism and Zanhua industries. ICHESs are now spreading in XPV at a rate of three a day, and almost all shops, whether shops or residential buildings, have become Zanhua places.

According to a Xiecheng (a Chinese online travel service company) big data analysis, the overall travel orders in Quanzhou during the 2024 New Year holiday increased by 276% year on year. According to third-party estimates, Quanzhou City received 2.0458 million tourists during the New Year's Day holiday in 2024, a year-on-year increase of 155.8% and an increase of 17.9% over the same period in 2019; the tourism revenue reached USD 193 million, an increase of 174.0% and 17.0% year on year [50]. In 2023, the number of domestic tourists in Fengze District reached 10.43 million, an increase of 57.9% year on year. Domestic tourism revenue reached USD 1.53 billion, up 69.3% year on year. Among the various trending topics, "Xunpu women. Zanhuawei" became popular on REDnote and other social media platforms, attracting more than 5 billion likes and receiving praise from UNESCO [51].

As of 6 April 2024, the survey findings in this study indicate that the number of shops offering Zanhuawei services (ICHESs) in XPV is approximately 246. With government support, XPV has constructed tourist service centers, shared bicycles, electric vehicles, and other supporting facilities. XPV has thus gradually transitioned from fishing to tourism. The traditional fishing industry has been on the decline, and the Zanhuawei-led cultural tourism industry has replaced the fishing industry as a new economic pillar of XPV. According to the statistical data obtained from field research, the daily turnover of XPV relying on ICHESs alone reaches USD 88,584 in the off-peak season and USD 272,568 in the peak season (Table 2).

Table 2. ICHES revenue estimation.

Post	Average Price	Quantity of Shops	Off-Peak Season Daily Average Number of People	Peak Season Daily Average Number of People	Off-Peak Season Daily Turnover/USD	Peak Season Daily Turnover/USD
Zanhua	5.54	246	65	200	88,584.6	272,568
Makeup	20.77	246	65	200	332,112.3	1,021,884
Photography	27.70	246	65	200	442,923	1,362,840
Sum					863,619.9	2,657,292

Notes: Calculation basis: As of the statistical date, there were about 246 ICHESs in XPV. According to the average price of the market, the number of tourists is between 30 and 100 people per shop per day in the off-season, and between 100 and 200 in the peak season. The number of people is averaged for calculation.

However, over time, the field survey found that by 12 October 2024, the Zanhua industry development in XPV had gradually formed an M-shaped trend, and ICHESs with good momentum had not only opened branches in XPV but also in "West Street" (the main tourist attraction in Quanzhou), thus showing a trend of expansion. Those who did not operate well have closed down their shops or are transferring their shops (Figure 3). All in all, the Xunpu women's ICH represented by Zanhuawei has greatly promoted the development of tourism in Quanzhou. As an important supplement to the fishing industry in XPV, tourism brings more opportunities to local residents.

Figure 3. Distribution map of Zanhua shops.

3.2. Role of Social Media in Stakeholder Engagement

In existing research on social media, only a few studies come close to defining or clarifying the concept of social media. However, it is evident that they all consider social media to comprise various user-driven platforms. These platforms facilitate the dissemination of information, the creation of conversations, and communication with a broader audience. Essentially, social media is a digital space created by people and serving people, and it provides a network environment that is conducive to interactions at different levels (e.g., personal, professional, commercial, marketing, political, and social) [52,53].

New media platforms serve as both participants and intermediaries in the process of fishing village governance [54]. On the one hand, social platforms have their inherent operating mechanisms. On the other hand, they provide new avenues for other entities to participate [55]. The pyramid structure (Figure 4) within social platforms comprises platform rule makers at the top tier, who are responsible for controlling algorithms and traffic flow; content creators and multi-channel network (MCN) organizations at the middle tier, who are tasked with adapting to and making use of the rules; and ordinary users at the bottom tier, who function as both consumers and data providers for the algorithms. Multiple stakeholders also occupy different positions upon entering the social platform. For instance, government agencies, social organizations, Internet celebrities, and some large-scale investors often belong to the middle tier. They possess the ability to create content and can better adapt to and even exploit the platform rules to achieve their own goals. In contrast, villagers, tourists, netizens, etc., are at the bottom tier. They are not only influenced by the top-level rules and middle-level content, but their behaviors also feed back to the social platforms, serving as providers of algorithmic data.

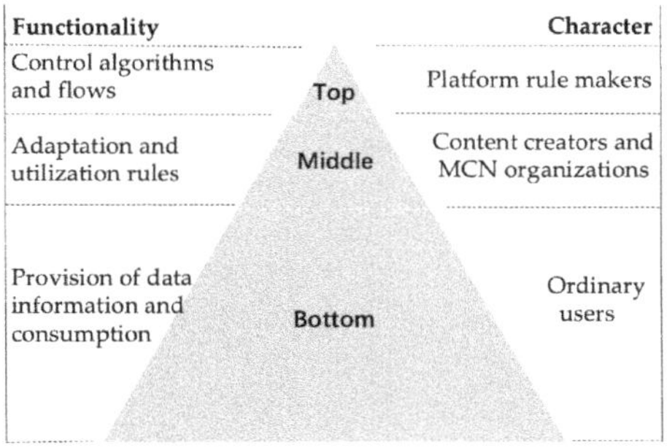

Figure 4. Social platform internal structure.

The intrinsic operational mechanism of social platforms (Figure 5) comprises four core components [55]:

1. User-generated content (UGC) serves as the foundational fuel, enabling users to create and consume multimedia content (text, images, videos, audio) that sustains platform activity. Motivations for UGC production include social engagement (identity formation through audience interaction) and economic incentives (monetization via traffic sharing and advertising).
2. Traffic allocation mechanisms act as centralized power nodes, prioritizing efficiency (promoting viral content), diversity (mitigating content monopoly), and commercial interests (advertisement integration).
3. Algorithmic recommendation systems construct user profiles based on demographic data (age, gender, location) to deliver personalized content.
4. User behavioral feedback (e.g., dwell time, likes, comments, shares, search queries) constitutes training data for algorithmic optimization.

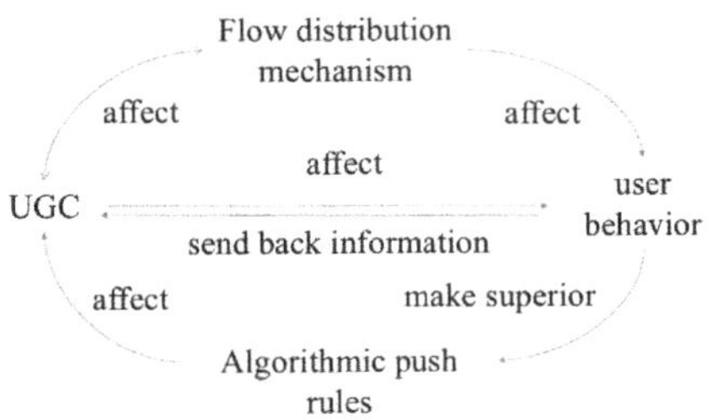

Figure 5. The inner workings of social platforms.

The thriving development of ICH tourism and the transformation of XPV can be attributed to the collaborative participation of multiple stakeholders. Through a "cultural heritage preservation + digital empowerment + regional collaboration" model, local cultural assets such as oyster-shell houses and Xunpu women's ICH are integrated to construct an "online engagement–offline experience" closed-loop ecosystem. Leveraging new media platforms (also known as TikTok and REDnote), XPV employs short videos and live broadcasts to achieve cross-platform viral dissemination of cultural IPs, fostering a differentiated cultural tourism brand. In 2023, the Quanzhou Municipal People's Government released its "Cultural Tourism +" Action Plan, which outlines an Online Travel Agency (OTA) new media operation mechanism encompassing four core modules: content production, traffic management, user engagement, and data-driven decision making. This systematic strategy enables omnichannel brand promotion. The resultant "policy guidance + market operation

+ technology-driven" framework provides a replicable model for ICH preservation and cultural tourism upgrading.

In the context of cultural–tourism integration, social media has emerged as a critical avenue for participatory engagement, with content-driven mechanisms serving as the cornerstone of user interaction. The most notable characteristic is its content-centric model, where content quality directly influences engagement metrics (likes, retweets, comments). Platforms employ a user-screening mechanism to foster niche communities through shared-interest aggregation (e.g., mutual following and content curation). In the context of cultural and tourism integration, social media has become an important way for individuals to make their voices heard. In addition, its most notable feature is the content-led model, where content is the key to attracting individual likes, retweets, and comments. Its operation mechanism is to attract and screen users through content (users with the same hobby pay attention to each other, like each other's content, etc.), and the screened users form a high-quality community. In addition, in accordance with the push rules, the traffic-oriented community constantly stimulates users to publish better quality content, with better quality content then constantly stimulating new users to join the community and ultimately forming a virtuous circle such that the scale of users in the community and the amount of high-quality content are increasingly expanding.

Taking XPV as a case study, key stakeholders (celebrities, influencers, and large-scale travel agencies) act as flow catalysts, generating viral discussions around ICH tourism. This decentralized content production model shifts authority from institutional creators to collective member contributions, enabling crowdsourced cultural dissemination. Leveraging big data algorithms, platforms enhance dissemination efficiency by delivering tailored content to target audiences based on behavioral patterns (e.g., dwell time, click-through rates). The combination of high user engagement, algorithmic personalization, and community-driven content creation constitutes a replicable framework for cultural heritage activation in the digital era.

The substantial online engagement with ICH tourism has been effectively translated into offline tourism activities through cross-sectoral collaborative efforts. For instance, the Quanzhou Municipal Bureau of Culture and Tourism's official social media accounts have launched a series of promotional videos highlighting local cultural heritage and tourist attractions. These initiatives include user-generated content campaigns encouraging netizens to document "hidden gems," and co-creation projects with influencers to develop viral photo spots and stimulate public discourse, thereby generating momentum for tourism activities. This strategy transforms the "ICH tourism fever" into a city-wide developmental opportunity. The offline implementation of online publicity materializes through three typologies of XPV's ICH experience shops: (1) Government-invited galleries: established photography institutions with proven marketing models and financial resources, attracting private investment through public–private partnerships; (2) community-run enterprises: locally operated businesses managed by villagers; (3) externally invested ventures: projects funded by non-local entities.

Notably, government-invited galleries serve as catalysts for attracting additional investors, with all stakeholders leveraging social media as a primary promotional channel to sustain XPV's ICH tourism ecosystem. Complementing these efforts, the Quanzhou government has introduced a dedicated "white bus" transit network connecting key attractions. As a taxi driver noted, "During holidays, traffic police are deployed to manage crowds, and all major Quanzhou attractions remain free of charge," reflecting a visitor-centric approach to enhancing experiential quality.

3.3. Spatial Changes in Multi-Party Governance

Located at the mouth of the Jin River, XPV is a typical marine fishing village. Surrounding the village, there is the Mazu Temple, built in 1661 during the Qing Dynasty. There are shopping malls, high schools, and elementary schools in the fishing village. The commercial and residential areas established under the urban plan are located in the northwest of the fishing village.

With the participation of multiple stakeholders, the fishing village space has undergone significant changes (Figure 6). Firstly, the function of the village space has transformed from a single-purpose production and living space into a multifunctional composite space. Satellite maps from 2006 show that the surrounding area of XPV was still not fully developed. Fenghai Road had not been completed, and on the southeast side of the village, there was a large shallow beach with numerous puddles, which might have been used by local residents for aquaculture. Seven years later, in 2013, the previously unused open space on the southeast side of the village had completely dried up, and Fenghai Road was also dry, with a commercial residential area established on the northwest side of the village. By 2013, Fenghai Road had been fully repaired.

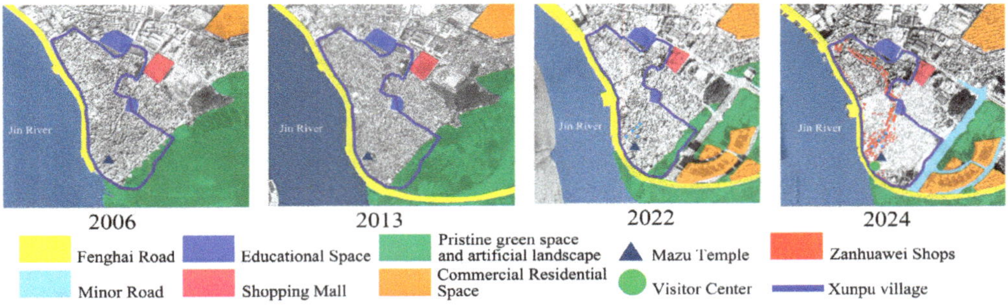

Figure 6. Urban renewal and change around XPV (2006–2024).

In 2022, another road starting from the southeast corner of XPV and extending to the northeast emerged, and the vacant land to the east of this road was developed into a commercial and residential area. The real transformation, however, took place after 2023. As ICH tours went viral on the Internet, many tourists flocked to XPV, leading to a soaring demand for experience shops. In 2024, the second year after XPV gained popularity, hundreds of experience shops and travel photography shops opened on both sides of XPV's main road. Moreover, the village established a visitor service center near the Mazu Temple, turning it into a public space shared by both local residents and tourists.

The direct and indirect participation of multiple stakeholders has brought about changes in the spatial layout of the fishing village. The government has played a pivotal role in providing planning guidance and financial support, ranging from the overall planning of the fishing village to the construction of infrastructure. From the "Quanzhou City Xunpu Folk Culture Village Protection and Rectification Plan" issued by the Quanzhou Natural Resources Bureau on 17 December 2013, and the "Fujian Provincial Traditional Village: Protection and Development Plan for Xunpu Community, Donghai Street, Fengze District, Quanzhou City (2024–2040)" released on 2 September 2024, it is not difficult to observe the evolutions in government decision making under the joint participation of multiple stakeholders.

Firstly, the planning scope has been expanded. It was initially delineated as 12 hectares in 2013 and has now been increased to 51.56 hectares. Except for the core protection zone, whose scope remains relatively unchanged, in 2024, the coastal scenery on the southwest side of Fenghai Road will be fully exploited to construct four parks and squares. The aim is to create new trendy photo spots that align with the current popular esthetic preferences, similar to those often seen on Internet celebrity check-in spots.

Tourists' needs and feedback have exerted a significant influence on the design and functionality of the fishing village space, propelling its transformation into a tourist-oriented area. Tourists shape the spatial planning of the fishing village through their tourism activities. For instance, the popular photography activities have led to a functional transformation of the traditional spaces within the fishing village.

In 2013, the core protection zone of XPV was divided into five areas: The Folk Cultural Activities Display Zone, the Religious Culture Pilgrimage Zone, the Special Architectural and Cultural Display Zone, the Art and Culture Plaza Zone, and the Fishermen's Production and Experience Zone. However, the overall planning at that time was not detailed enough and has undergone substantial changes since then (Figure 7).

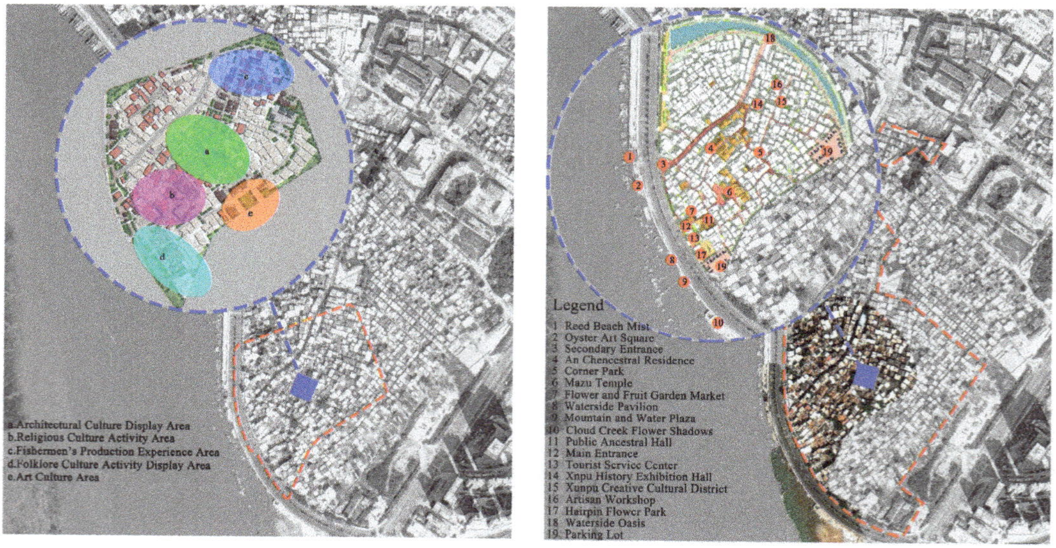

Figure 7. Spatial change with multi-stakeholder participation 2013 (**left**) and 2024 (**right**).

In contrast, the 2024 planning scheme clearly demonstrates the influence of resident participation, market regulation, and tourist feedback. For example, the Mazu Temple, ancestral halls, oyster shell houses, seascapes, and fishing boats, which were originally used by fishermen for religious rituals, worship, and daily living, have, due to their strong regional characteristics and cultural significance, become popular tourist attractions or been transformed into tourist shops and have been incorporated into the new planning. During an interview, the staff of a travel photography shop replied that "*This is a clan temple, the ancestral home of a family, but now it serves as the studio of Earl's Travel Photography Shop, where tourists come to get their makeup done and take photos*"(Table S1, F-3).

Moreover, some tourists expressed their dissatisfaction on the Internet, with comments like "Zanhuawei" and remarks such as "*There are too many people. It's a dilapidated village, and the experience is really poor*" (Table S2, line 4). These feedback remarks were promptly reflected in the transformation of the fishing village space. In October 2024, a Visitor Experience Center was established, and new commercial establishments such as milk tea shops, cultural and creative product shops, and coffee shops were added. Additionally, new commercial spaces, including flower and fruit parks, bazaars, and cultural and creative neighborhoods, have been created to enhance the overall travel experience for visitors. In response to tourists' online complaints about the lack of parking spaces, the new plan has also included two additional parking lots.

Overall, the synergistic influence of multiple stakeholders has resulted in demand-driven transformations in the function, form, and utilization of the fishing village space. Driven by the tourism boom, local residents have gradually converted the traditional production and living space into a consumption-oriented one. They do so by opening ICH experience halls to engage in market competition, creating photo-shooting scenarios with distinctive XPV features to meet tourists' demands, or selling local specialties along the streets and in front of the residences.

The photos of girls adorned with hairpins and flower bows have gone viral on short video platforms, portraying XPV as a picturesque and idyllic fishing village. Such a beautiful imagination serves as the crux for the success of the Internet celebrity economy, which entices people to flock to the tourist destination.

The involvement of capital has accelerated the commercial transformation of the fishing village space. The impact of the Internet celebrity economy has attracted many tourists to XPV to experience wearing Zanhuawei. Consequently, local residents' houses have been repurposed as ICHESs. Moreover, the Mazu Temple, the oyster-shell houses, and even the Xunpu women have become popular Internet celebrity check-in spots, and the living space has also been commercialized.

On the other hand, the cultural heritage mission of ICH is conducted by government-organized groups and schools, which contribute to the sustainable development of the fishing village space by participating in cultural activities and environmental protection activities.

4. Discussion

4.1. E-Participation Spatial Governance Model for Fishing Villages

As part of digital social governance, the digitization of rural areas is growing rapidly and has become an effective way to address rural development issues [56], but research on digital governance has focused more on the technical aspects [57], such as the creation of digitally smart villages explored in most papers in China [58]; the smart village proposal initiated by the European Union [59]; and the digital inclusion strategy implemented in the UK [60]. Although there is no lack of cases of economic transformation facilitated by online communities in China (Zibo Barbecue [61], Tianshui Spicy Hot Pot [62], Luoyang Hanfu [63], etc.), the case of fishing villages governed by multi-stakeholders through social networks is rarely discussed.

The essence of public participation lies in citizens exercising their rights to ensure their opinions are systematically incorporated into government decision making [64]. As an extension of e-government and e-democracy, e-participation is widely recognized as an online mode of public participation—a civic engagement mechanism enabling individuals and organizations to influence policy making through digital platforms. This process relies on information technology to facilitate involvement in managing public affairs, aligning with the modernization of governance [53,65]. However, the difference between e-participation and traditional civic participation is not only the difference between online and offline modes; e-participation has the characteristics of high interactivity, real-timelessness, inter-temporality, and anonymity brought by information technology. Especially with the development of information technology and the evolution of public participation methods, the governance model of fishing villages is changing.

The earliest governance of fishing villages was mainly based on clan participation and decision making. However, due to national governance and urbanization, the governance of fishing village space has been decentralized. The planning of the fishing village space has also been detached from the living and production experience of traditional fishing villages, causing the fishing village to lose its original features. The subsequent main body of governance consists of clan elders, grass-roots governments (village committees, township governments), and villagers. The governance mechanism relies on traditional customs and township rules and regulations, and spatial development depends on the development of living and production activities. The popularity of social media has gradually compensated for the previous shortcomings. Social media allows residents of fishing villages to obtain policy information, learn about resource allocation dynamics, and stay instantly informed about public affairs, reducing the distortion and lag in information transmission caused by the hierarchical system. In addition, the user-generated content (UGC) model enables marginalized groups in traditional governance, such as fishermen, women, and youth, to voice their opinions through social media, creating online public opinion pressure, which provides unlimited possibilities for public participation. Local residents in XPV can learn about fisheries policy and tourism development through WeChat public numbers, WeChat groups, and short videos.

In the era of new media, the governance pattern of fishing villages has undergone a remarkable evolution. It has transitioned from the initial stage where government departments took the lead in management, to a phase of de-governmentalization and community autonomy, and ultimately towards the establishment of a comprehensive governance community. This new model effectively mobilizes the resources of the state, the market, and civil society, fostering a more collaborative and inclusive governance environment [30]. Meanwhile, the competition for discourse power has gradually shifted from the offline public domain to online social platforms. The original structure of the discourse power system has been disrupted. The dominance of elites, the government, and traditional media in terms of discourse power has been dispersed, while public intellectuals, various public interest groups, netizens, and bloggers have emerged as forces with an increasingly growing influence in shaping public discourse [66]. This shift has made e-participation a powerful force in the governance of fishing villages. Instead of relying on traditional promotion methods for ICH and tourism development, XPV has adapted to the trends of the times. The government and civil society organizations first invite celebrities and Internet influencers to experience the ICH items and then upload the related photographs and videos on social media platforms to expand their influence. Then, they invite renowned domestic photography studios to settle in XPV. By applying their proven business models, the influence of XPV is further expanded. Finally, in the face of the influx of tourists, the government responds to the challenges brought about by the rapid development of the tourism industry by establishing tourist service centers and improving infrastructure.

With the involvement of Internet celebrities and bloggers, ICH tourism has become even more popular. On the one hand, Internet celebrities leverage the popularity of celebrities to increase their own online traffic. On the other hand, they use their influence to promote ICH tourism. Local residents have also embraced the tourism industry. Some of them have chosen to open their own ICHESs or cooperate with others. Photo studios, freelance photographers, and makeup artists have also joined the trend, and they attract tourists by posting relevant videos and photos. At the same time, tourists' online feedback and discussions have expanded the online influence of the ICH. The scope of discussion about the Zanhuawei is positively correlated with the diversity and number of users. The rapid dissemination features of the Internet have accelerated the spatial diffusion of the ICH. Although the ICHESs in XPV have reached a saturation point, their influence is spreading to Quanzhou City, Fujian Province, and even across the whole country.

4.2. The Social Network Jointly Constituted by Strong and Weak Ties

The change in strong and weak relationships has led to changes in the spatial governance model of fishing villages. The traditional governance model of fishing villages is dominated by strong relationships, but the addition of social media has brought the advantages of weak relationships to the fore, compensating for the shortcomings of strong relationships in the governance of fishing villages, and its influence has been expanding, and it has even become the key to the transformation of the economy of fishing villages.

In his 1973 article 'The Strength of Weak Ties', the American sociologist Mark Granovetter first proposed two models of relationships: strong ties and weak ties. Strong ties are characterized by a high degree of information overlap, a high degree of trust, and a small range, while weak ties are characterized by a low degree of information overlap and a wide range of communication [67]. The social relationship in traditional fishing villages is a typical "strong-tie" network. However, with the development of digital social media, the social network of the fishing village has gradually become a pattern composed of both strong ties and weak ties. Social media not only provide an online channel for strong ties (e.g., WeChat) but also magnify the advantages of weak ties (e.g., Weibo, TikTok, REDnote, etc.).

"The credit in rural society does not lie in the emphasis on contracts, but in the reliability that emerges when one is so familiar with the norms of a behavior that it becomes instinctive" [68]. Marine fishing villages, due to their geographical isolation, have formed relatively closed acquaintance-based societies dominated by strong ties rooted in blood, kinship, and fictive kinship. Clan elders and other community leaders exercise "intergenerational authority" to manage resources. Through long-term, high-frequency social interactions, villagers have developed strong interpersonal trust. Modern relationships, such as cadre–community ties and occupational networks, also contribute to these strong relational systems. As village leaders have a sense of the local culture of the fishing villages where they have lived for a long time, they will consciously undertake and fulfill their responsibilities of representing the fishing villages in communicating with government agencies and managing public affairs on behalf of the villagers [69]. The role of strong ties in promoting ICH activities is manifested in fishermen's preference for transacting and transferring resources (e.g., property, land, and technical expertise) within their strong-tie networks. For instance, in XPV, local residents have preferential access to prime and affordable housing when establishing ICHESs compared to outsiders, thus leveraging familial, neighborly, and acquaintance-based social ties.

The continuous evolution of society has spurred rural differentiation, with traditional villages transitioning from subsistence economies to market-oriented systems [70]. The interplay between state intervention—manifested through policies like fishery subsidies, environmental conservation bans, and village committees' integration of bureaucratic regulations into acquaintance-based societies—and market forces (e.g., e-commerce) has fostered the emergence of modern "weak ties". This process has reshaped social networks into a hybrid model of coexisting strong and weak ties, fundamentally altering the spatial distribution of rural populations.

This hybrid social network extends beyond offline interactions. Online community relationships, characterized as semi-familiar ties between acquaintances and strangers, exhibit both connectivity and alienation—representing quintessential weak ties [71]. These loose, fluid connections form spontaneously and scale easily. Catalyzed by the Internet, they generate aggregation and fission effects [71]. For example, by 16 January 2025, a search for "Zanhua Quanzhou" on REDnote yielded over 650,000 posts, and the most popular post garnered 160,000 likes, 17,000 comments, and 59,000 saves.

On the other hand, big data algorithms tailor content to users based on their preferences, enabling individuals with similar interests to cluster and form strong ties. These ties influence both individual and collective social behaviors through the emotional bonds they foster [72,73]. Empirical studies indicate that strong-tie networks exhibit higher emotional intimacy, shared interests, and congruent values [74], which facilitate information exchange and tailored advice seeking. For instance, individuals browsing Facebook demonstrate preferential attention to strong-tie contacts, leading to enhanced self-esteem outcomes [75]. Similarly, celebrity–fan dynamics illustrate how strong ties motivate followers to amplify exposure through repetitive social media actions (e.g., retweets, comments, likes). These behaviors, in turn, trigger aggregation and fission effects within weak-tie online communities. Huang Liyong [76], the cultural ambassador for Xunpu women's folk traditions in Fengze District, has dedicated over a decade to promoting Xunpu culture. She said the following:

> *"Although many celebrities have engaged with ICH activities over the years, this sudden surge in popularity this year caught everyone by surprise. After seeing Zhao Liying's photos, people from Beijing, Shanghai, and even overseas—such as Canada and the U.S.—have flown here to experience Zanhuawei. Some have even contacted us to purchase Zanhua materials for international shipping".*

This highlights how strong ties influence individual behavior, while weak ties, augmented by online media, generate robust information aggregation and fission effects. It is this dynamic synergy that empowers stakeholders to achieve governance of fishing villages through social media platforms.

4.3. The Relevance of Social Media and Its Operating Mechanisms to Spatial Governance in Fishing Villages

Historically, fishing village governance relied on three main actors: government agencies, cooperatives, and clan elders. Government entities employed top-down spatial planning, encompassing central policy oversight, local implementation, and resource allocation. Cooperatives operated via market-driven self-regulation, while clan leaders managed resources through kinship-based authority and traditional norms. Fishing villages exhibited integrated spatial systems blending residential areas, productive zones (fisheries, ports), and ecological habitats (marine environments) (Table 3). Traditional governance faced limitations: isolation stifled economic diversity, administrative–custom conflicts caused inefficiencies, and experience-based resource management lacked scientific rigor.

Table 3. Traditional fishing village governance patterns and their spatial characteristics.

Governing Body	Intermediary	Governance Mechanisms	Spatial Feature	
Clan elders	Offline focus	Bloodline authority, township rules, customs and practices	Living space	Village, Mazu Temple, clan hall, etc.
Cooperative		Economic self-reliance, market regulation, laws and regulations	Production space	Tidal zone, fishing harbors, fishing grounds, etc.
Government		Macro-control by central government, policy formulation by local government, administrative implementation by grass-roots government	Ecological space	Marine resources, intangible cultural resources, etc.

However, the popularization and application of social media have led to significant changes in the governance model of fishing villages. Community relationships in online environments are widely accepted in the context of real-world interpersonal alienation [71]. People who share the same needs or purposes can rapidly assemble via social networks and form a group with weak ties. Moreover, depending on diverse needs, individuals can simultaneously join multiple communities or switch between different ones, giving rise to a phenomenon known as "mobile aggregation". This mobility has greatly facilitated the dissemination of information among different groups and propelled the explosive development of weak relationships within communities. For example, live-streaming e-commerce and community group purchasing are direct products of the development of weak relationships [77]. ICHESs fully leverage weak ties through platforms like REDnote, TikTok, and Weibo to conduct fission-style information dissemination. The photos of Zanhuawei taken by tourists and shared on WeChat Moments can serve as effective promotion. Therefore, social media platforms have become crucial media for information dissemination, public opinion supervision, and resource matching.

Foreign investors develop tourism activities by injecting capital. For instance, they acquire or lease local residential houses and transform them into homestays or ICHESs, which has led to the commercialization of living spaces. Tourists and netizens have transitioned from being mere onlookers to active space shapers. They influence the image of fishing villages through online reviews, check-ins, and sharing of experiences, thereby indirectly participating in governance decisions and promoting the symbolization of the spatial aspects of fishing villages. Their online complaints and negative reviews prompt the government to improve the environment and upgrade infrastructure. Government agencies and social organizations can utilize social media to promote local culture, popularize scientific knowledge, and advocate for environmental and cultural heritage protection. Villagers can also leverage live streaming and online promotion to attract tourists to their shops to experience traditional ICH, thus securing a place for themselves amidst the influx of external capital.

When all stakeholders jointly engage in the spatial governance of fishing villages via social media, these villages continue to evolve, developing cultural spaces, capitalized spaces, consumption spaces, and even virtual community spaces (Figure 8). Social media has broken the isolation of the traditional governance model, giving rise to a new scenario where multiple entities interact and compete. Online public opinion and feedback have a reciprocal impact on the governance plans and spatial planning of fishing villages. Moreover, the spatial functions have expanded from "production–living–ecology" to "consumption–capital–culture".

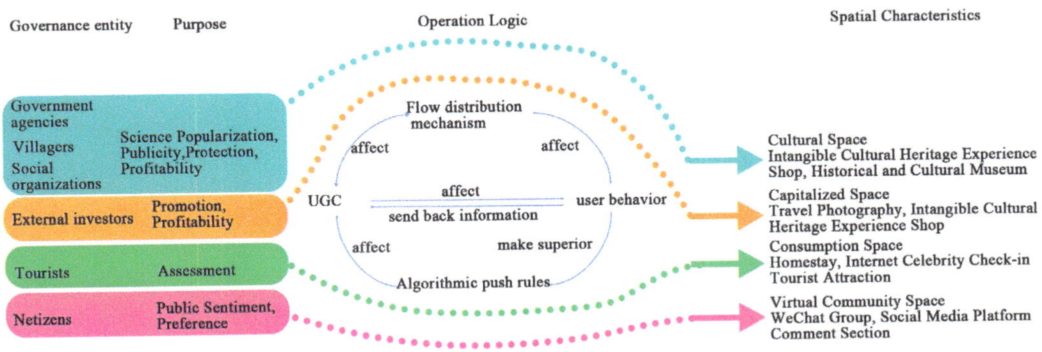

Figure 8. Governance patterns and spatial characteristics of fishing villages after multi-stakeholder participation.

4.4. The Meaning of Social Media Participation

XPV's transformation stems from the synergy between its unique ICH and contemporary trends. Fishing village spatial governance is a multi-stakeholder process involving traditional models centered on government, fishing communities, and clan authorities. The integration of social media has revitalized governance dynamics by introducing new participatory mechanisms.

In the governance system of fishing villages, the government consists of various types of administrative units at all levels. Their relationships are mainly manifested as leadership and subordination, guidance and being guided, supervision and being supervised, and collaboration and cooperation. These relationships exist not only between superiors and subordinates but also among peers [78]. The most significant aspect of social media's participation is that it breaks down the distortion, inefficiency, and inequality of information that occur during the information uploading and downloading processes in the traditional governance system. The state-led community participation model, without sufficient democratization and the inability to resolve structural contradictions, fails to achieve the expected community-building goals. Consequently, residents' cohesion and sense of belonging are limited.

During the process of community building and village regeneration, it is crucial to recognize that folklore is territorial, with each place having its unique folklore. The involvement of social media has provided a new platform for this local folklore in the era of rapid development. No matter how niche the folklore may be, it has the potential to thrive on the Internet, and building folklore into a local brand can therefore more effectively stimulate the local economy. The social Internet can amplify individual voices and demonstrate the strong vitality of collective power.

As direct beneficiaries and stakeholders in fishing village governance, villagers shape spatial dynamics through their daily production and livelihood activities. Social media platforms have empowered marginalized groups by providing channels for self-expression, transforming their cultural identities into assets that mitigate passive roles in traditional development narratives. The experiential and novel tourism demands amplified by social media align with villagers' unique cultural practices, which resist full commodification by capital. Despite economic transitions, residents maintain customary lifestyles, integrating traditional festivals and rituals into tourism offerings. This preserves authenticity while attracting visitors, creating a symbiotic relationship between heritage conservation and spatial transformation. Fishing village economies have transcended the reliance on fisheries, with tourism emerging as the catalyst for spatial reconfiguration. Tourism-driven development has shifted spatial functions from production to consumption, yet the industry adapts to modernity by intertwining ICH preservation with visitor experiences. This approach avoids abrupt ruptures with historical practices, ensuring continuity between tradition and modernity. Rather than polarizing locals and tourists, this model represents organic, community-led innovation that balances cultural integrity with economic viability. It exemplifies how grassroots agency, mediated through digital platforms, can redefine governance paradigms while safeguarding spatial and cultural authenticity.

However, it is worth noting that the process of digital engagement encounters various challenges such as the authenticity of platform data and the authenticity of information disseminated by users: 1. The authenticity of platform data: digital media platforms are profit-oriented, and their internal operation mechanism itself is a manifestation of power; for example, China's microblogging platform has established a heat list, which can be paid to move content to the list, thus increasing exposure [79]. 2. The authenticity of the information disseminated by the users: for example, some Netflix stars fabricate dramas and pose for photographs to attract traffic [80]. When someone consciously pushes or blocks comments or content (opinion control, platform traffic limitation, deletion of comments, platform intervention, etc.), the authenticity and effectiveness of digital engagement are therefore greatly reduced.

5. Conclusions

This study delves into the ways in which multiple stakeholders engage in the spatial planning and governance of fishing villages via social media and arrives at the following results: 1. Government agencies, social organizations, villagers, external investors, tourists and netizens jointly engage as multi-stakeholders. 2. Social media is the main platform for participation. 3. The representation of space has changed from concentrated zoning to decentralized dispersion. Under the joint governance of multiple stakeholders, there are further findings: 1. Social media addresses the deficiencies inherent in the traditional spatial governance model of fishing villages. E-participation disrupts the original structure of the discourse power system, offering effective means and methods for diverse stakeholders to partake in governance. 2. Social media shatters the monopoly of traditional fishing village governance, which was previously dominated by strong relationships. The advantages of weak relationships are significantly amplified, emerging as a crucial complement to the governance relationships within fishing villages. 3. Users can attain objectives such as increased visibility, follower growth, and business transformation through a series of mechanisms, including social media content dissemination, traffic competition, public opinion guidance, interaction, and data analysis. The integration of this operational mechanism with the spatial governance of fishing villages has become an essential supplement to the overall spatial governance framework of these villages.

However, the model of fishing village spatial governance with the involvement of multiple stakeholders under the influence of social media also has its drawbacks:

1. Dissolution and re-creation of ICH. Social media expedites the infiltration of foreign cultures. Traditional customs are simplified into performative forms, traditional costumes evolve with new aesthetic trends, and the traditional cultural space is transformed into a tourist symbol and a space for virtual consumption. For instance, the popular Zanhuawei from XPV circulating on the Internet often deviates from the most traditional styles. Even though local residents may not endorse this new aesthetic, they still recommend these new Zanhuawei styles to customers to meet their demands.
2. Community fragmentation and intergenerational conflicts. This is mainly manifested in the digital divide, which marginalizes the elderly from the decision-making process. While external forces such as fishermen, tourists, and non-governmental organizations (NGOs) now have more direct access to participate in governance, it also brings challenges such as information distortion and unequal participation opportunities. The weakening of traditional clan authority has exacerbated the imbalances within the community's power structure. Not all residents in XPV have embraced the tourism industry. There are still those who persist in their traditional livelihoods, such as prying oysters. They fail to comprehend the enthusiastic behavior of tourists and are unable to benefit from the booming tourism industry. Moreover, the occupation of fishing village spaces by foreign tourists and investors has undermined the spatial rights of some of the indigenous inhabitants.

The model of fishing village spatial governance through community media and the collaborative participation of all stakeholders has emerged as a paradigm that rectifies the shortcomings of previous fishing village regeneration cases. It showcases the outcomes of non-collective planning and restores the spatial development of fishing villages to the combined influence of human activities, the living environment, and economic development. This represents a form of collective wisdom of the people that is highly adaptive to the current situation and has emerged spontaneously. Finally, as a case study can be too one-sided and lack generalizability, we plan to use another larger study area or multiple communities as the subject of our future research to discuss the generalizability of the model in depth.

Supplementary Materials: The following supporting information can be downloaded at: https://www.mdpi.com/article/10.3390/w17111703/s1. Table S1. In-depth interview information.; Table S2. List of social platforms that have had XPV Zanhuawei posts.

Author Contributions: Conceptualization, S.-C.T.; data curation, J.W.; formal analysis, J.W.; funding acquisition, S.-C.T.; investigation, J.W., M.-M.H. and S.-C.T.; methodology, S.-C.T.; project administration, S.-C.T. and S.-H.L.; resources, J.W., S.-H.L. and M.-M.H.; software, J.W.; supervision, and S.-C.T.; validation, S.-C.T.; visualization, J.W.; writing—original draft, J.W. and S.-C.T.; writing—review and editing, S.-C.T. All authors have read and agreed to the published version of the manuscript.

Funding: This research was funded by the scientific research start-up fund of Jimei University, China, grant number Q202214.

Data Availability Statement: The data of this study are available from the corresponding author upon reasonable request.

Acknowledgments: This study benefited from all respondents, including Local villagers, workers, tourists, etc. At the same time, thanks to everyone who worked on this study. All deficiencies in the article are the responsibility of the authors.

Conflicts of Interest: The authors declare no conflicts of interest.

References

1. Vitruvius. *De Architectura*; Intellectual Property Publish: Beijing, China, 2001.
2. Li, X.; Li, B.; Jiang, W. State-led versus market-led: How institutional arrangements impact collaborative governance in participatory urban regeneration in China. *Habitat Int.* **2024**, *150*, 103134. [CrossRef]
3. Sanoff, H. *Community Participation Methods in Design and Planning*; John Wiley & Sons: Hoboken, NJ, USA, 1999.
4. Ji, J.; Heath, T. Community Participation of China's Urban Regeneration: A Case Study of Sanxue Historical District in Xi'an. In *Innovative Public Participation Practices for Sustainable Urban Regeneration*; Springer: Singapore, 2024; pp. 65–80.
5. Yang, J.; Yang, L.; Ma, H. Community Participation Strategy for Sustainable Urban Regeneration in Xiamen, China. *Land* **2022**, *11*, 600. [CrossRef]
6. Zhao, P.; Ali, Z.M.; Hashim, N.H.N.; Ahmad, Y.; Wang, H. Evaluating social sustainability of urban regeneration in historic urban areas in China: The case of Xi'an. *J. Environ. Manag.* **2024**, *370*, 122520. [CrossRef] [PubMed]
7. Li, Y.; Jia, L.; Wu, W.; Yan, J.; Liu, Y. Urbanization for rural sustainability—Rethinking China's urbanization strategy. *J. Clean. Prod.* **2018**, *178*, 580–586. [CrossRef]
8. Yang, X.; Li, W.; Zhang, P.; Chen, H.; Lai, M.; Zhao, S. The dynamics and driving mechanisms of rural revitalization in western China. *Agriculture* **2023**, *13*, 1448. [CrossRef]
9. Zhou, Y.; Shen, Y.; Yang, X.; Wang, Z.; Xu, L. Where to revitalize, and how? A rural typology zoning for China. *Land* **2021**, *10*, 1336. [CrossRef]
10. Tu, S.; Long, H.; Zhang, Y.; Ge, D.; Qu, Y. Rural restructuring at village level under rapid urbanization in metropolitan suburbs of China and its implications for innovations in land use policy. *Habitat Int.* **2018**, *77*, 143–152. [CrossRef]
11. Hu, Q.; Zhang, T.; Jiao, Z.; Duan, Y.; Dewancker, B.J.; Gao, W. The impact of fishery industrial transformation on rural revitalization at village level: A case study of a Chinese fishing village. *Ocean. Coast. Manag.* **2022**, *227*, 106277. [CrossRef]
12. Jiang, Y.; Mohabir, N.; Ma, R.; Wu, L.; Chen, M. Whose village? Stakeholder interests in the urban renewal of Hubei old village in Shenzhen. *Land Use Policy* **2020**, *91*, 104411. [CrossRef]
13. Yang, M. A brief discussion on the landscape regeneration design of abandoned traditional villages: Taking the Boao Forum for Asia theme park as an example. *Gard. Constr. Urban Plan.* **2022**, *4*, 112–115. [CrossRef]
14. Du, P.; Hou, X.; Xu, H. Dynamic Expansion of Urban Land in China's Coastal Zone since 2000. *Remote Sens.* **2022**, *14*, 916. [CrossRef]
15. Hou, X.; Wu, T.; Hou, W.; Chen, Q.; Wang, Y.; Yu, L. Characteristics of coastline changes in mainland China since the early 1940s. *Sci. China Earth Sci.* **2016**, *59*, 1791–1802. [CrossRef]

16. Rucińska, D.; Adinolfi, G.; Frigerio, I.; Gavinelli, D.; Zanolin, G.; Werner, W.; Rauscher, N.; Jaczewska, B.; Gręda, Ł. Case study in Poland: Understanding spatial diversity of social vulnerability to natural hazards based on local level assessments within the European Union. *Int. J. Disaster Risk Reduct.* **2023**, *96*, 103941. [CrossRef]
17. Zou, Z.; Zhang, Y.-Y.; Lee, S.-H.; Tsai, S.-C. The Transformation of Coastal Governance, from Human Ecology to Local State, in the Jimei Peninsula, Xiamen, China. *Water* **2023**, *15*, 2659. [CrossRef]
18. Zhang, Y.-Y.; Zou, Z.; Tsai, S.-C. From Fishing Village to Jimei School Village: Spatial Evolution of Human Ecology. *Int. J. Environ. Sustain. Prot.* **2022**, *2*, 33–43. [CrossRef]
19. Fabinyi, M. The role of land tenure in livelihood transitions from fishing to tourism. *Marit. Stud.* **2020**, *19*, 29–39. [CrossRef]
20. Lang, W.; Chen, T.; Li, X. A new style of urbanization in China: Transformation of urban rural communities. *Habitat Int.* **2016**, *55*, 1–9. [CrossRef]
21. Wang, X.-R.; Chu, T.-J.; Lin, T.-S. Key Sustainable Factors for Recreational Fishery Development Under Rural Revitalization Policy. *Int. J.* **2022**, *2*, 16–31. [CrossRef]
22. Liu, Y.; Sun, H. (Eds.) The Path Construction of the Evolution from Fishing Town to Leisure Island. In Proceedings of the 2023 2nd International Conference on Sport Science, Education and Social Development (SSESD 2023), Qingdao, China, 28–30 July 2023; Atlantis Press: Dordrecht, The Netherlands, 2023; pp. 266–274.
23. Shih, C.-H.; Wang, X.-R.; Lu, Y.-M.; Chu, T.-J. Assessing the Role of Policy in the Evolution of Recreational Fisheries in Chinese Fishing Villages: An Analytical Hierarchy Process (AHP) and Delphi Method Analysis. *Fishes* **2024**, *9*, 353. [CrossRef]
24. Tsai, S.-C.; Zhang, X.-F.; Lee, S.-H.; Wang, H. Urban Governance, Economic Transformation, and Land Use: A Case Study on the Jimei Peninsula, Xiamen, China, 1936–2023. *Water* **2024**, *16*, 913. [CrossRef]
25. Fagan, B. *Fishing: How the Sea Fed Civilization*; Gūsa Press; Walkers Cultural Enterprise Ltd.: New Taipei City, Taiwan, 2022.
26. Xia, D.; Mou, Q.; He, Y. The era textual research and cultural characteristics of Hainan fishermen's Genglubu. *J. South-Cent. Univ. Natl. (Humanit. Soc. Sci.)* **2016**, *36*, 54–59.
27. Huang, R.; Yue, Y.; Chen, Y.; Jin, Z.; Peng, Q. Mode selection of modernization transformation of traditional fishing villages from the perspective of complex adaptive system. *Oper. Manag.* **2020**, *1*, 14–19.
28. Yu, Y.; Han, X.-Y. Research on the relationship between fishermen's social interaction and social transformation of fishing villages. *Chin. Fish. Econ.* **2007**, *2*, 13–16.
29. Sundström, A. Corruption and regulatory compliance: Experimental findings from South African small-scale fisheries. *Mar. Policy* **2012**, *36*, 1255–1264. [CrossRef]
30. Min, L. How Community Co-management is Possible—A Case Study of Offshore Fisheries Resource Governance in Meicun. *Agric. Econ. Issues* **2023**, *12*, 85–99. [CrossRef]
31. Qing, W. New Research Progress on the Theory of Governing the Commons. *Econ. News* **2020**, *11*, 131–144.
32. Chen, J. National logic of modernizing the governance system. *Soc. Sci. China* **2019**, *5*, 23–39.
33. Cheng, Y.; Fei, X.; Luo, L.; Kong, X.; Zhang, J. Social network analysis of heterogeneous subjects driving spatial commercialization of traditional villages: A case study of Tanka Fishing Village in Lingshui Li Autonomous county, China. *Habitat Int.* **2025**, *155*, 103235. [CrossRef]
34. Näkki, P.; Bäck, A.; Ropponen, T.; Kronqvist, J.; Hintikka, K.A.; Harju, A.; Pöyhtäri, R.; Kola, P. *Social Media for Citizen Participation: Report on the Somus Project*; VTT Publications: Espoo, Finland, 2011; p. 755.
35. Pereira, G.C.; Rocha, M.C.F.; Poplin, A. (Eds.) E-Participation: Social media and the public space. In Proceedings of the Computational Science and Its Applications–ICCSA 2012: 12th International Conference, Salvador de Bahia, Brazil, 18–21 June 2012; Proceedings, Part I 12. Springer: Berlin/Heidelberg, Germnay, 2012; pp. 491–501.
36. He, M.-M.; Wang, J.; Lee, S.-H.; Tsai, S.-C. How Traditional Fishing Villages Move towards Sustainable Management: A Case Study of Industrial Transformation and Multi-Party Governance Models. *Sustainability* **2024**, *16*, 8532. [CrossRef]
37. Gibbons, J. "Placing" the relation of social media participation to neighborhood community connection. *J. Urban Aff.* **2020**, *42*, 1262–1277. [CrossRef]
38. Lin, Y.; Kant, S. Using social media for citizen participation: Contexts, empowerment, and inclusion. *Sustainability* **2021**, *13*, 6635. [CrossRef]
39. Effing, R.; Van Hillegersberg, J.; Huibers, T.W. (Eds.) Social media participation and local politics: A case study of the Enschede council in the Netherlands. In Proceedings of the Electronic Participation: 5th IFIP WG 8.5 International Conference, ePart 2013, Koblenz, Germany, 17–19 September 2013; Proceedings 5. Springer: Berlin/Heidelberg, Germnay, 2013; pp. 57–68.
40. Wang, D.; Zhu, L. "Be a Xunpu Woman for one day": How to Construct Collective Imagination by Shooting Short Video at an Internet Famous Site. *Media Watch* **2024**, *2*, 45–54. [CrossRef]
41. Ma, Y.; Zheng, W. Research on the Development of the East Coast Sea Tourism Complex in Quanzhou Central City. *J. Mudanjiang Univ.* **2013**, *22*, 106–108. [CrossRef]
42. UNESCO. Quanzhou: Emporium of the World in Song-Yuan China. Available online: https://whc.unesco.org/en/list/1561/ (accessed on 27 September 2024).

43. Quanzhou Municipal Natural Resources and Planning Bureau. *Quanzhou City's Quanpu Folk Culture Village Protection and Renova tion Plan*; Quanzhou Municipal Natural Resources and Planning Bureau: Quanzhou, China, 2013; Updated 2013/12/17. Available online: https://zyghj.quanzhou.gov.cn/xxgk/zdxxgk/ghcg/zxgh/201312/t20131217_2411074.htm (accessed on 17 November 2023).
44. The State Council of the PRC. Notice of the State Council on Promulgation of the Second Batch of National Intangible Cultural Heritage List and the First Batch of National Intangible Cultural Heritage Expansion Project List. Available online: https://www.gov.cn/zwgk/2008-06/14/content_1016331.htm (accessed on 4 May 2025).
45. Chen, Z.; Hong, W. Research on Cultural Inheritance of Women's Headdress in Xunpu. *Reg. Cult. Study* **2020**, *1*.
46. Wang, J.; He, M.-M.; Tsai, S.-C. Revitalization or Alienation: Reflections on Continuation of Traditional Culture. *Innov. Des. Cul* **2024**, *3*, 1–8. [CrossRef]
47. Xu, Y.; Lu, L. Probing the long-term evolution of traditional village tourism destinations from a glocalisation perspective: A case study of Wuzhen in Zhejiang province, China. *Habitat Int.* **2024**, *148*, 103073. [CrossRef]
48. Shen, H. Xunpu women's "head garden". *Friends Sci.* **2012**, *5*, 86–87.
49. *Custom of Xunpu Women*; China Intangible Cultural Heritage Network; China Intangible Cultural Heritage Digital Museum: Suzhou, China; Available online: https://www.ihchina.cn/project_details/15235/ (accessed on 2 June 2025).
50. *Ushering in the New Year "Off to a Good Start" the City's New Year's Day Holiday Main Tourism Economic Indicators Rise in Quantity and Quality*; Quanzhou Cultural Tourism Bureau Quanmedia: Quanzhou, China. Available online: https://www.quanzhou.gov.cn/zfb/xxgk/zfxxgkzl/qzdt/qzyw/202401/t20240103_2989069.htm (accessed on 2 June 2025).
51. *Provincial Government Circular Praise! One Place in Our City Was Selected*; Fengze District Government Office: Quanzhou, China, 2024.
52. Kapoor, K.K.; Tamilmani, K.; Rana, N.P.; Patil, P.; Dwivedi, Y.K.; Nerur, S. Advances in Social Media Research: Past, Present and Future. *Inf. Syst. Front.* **2018**, *20*, 531–558. [CrossRef]
53. Medaglia, R. Measuring the diffusion of eParticipation: A survey on Italian local government. *Inf. Polity* **2007**, *12*, 265–280. [CrossRef]
54. Shi, J.H.; Jing, M.J.; Zhu, Y. Social Media Interactive Circle Communication Patterns: Driving Force and Social Value—Analysis Based on Hot Social Events. *Journal. Enthus.* **2019**, *6*, 13–16.
55. Social Network Building Operations Guide 2025. Available online: https://max.book118.com/html/2025/0104/6135204142011021.shtm (accessed on 18 March 2025).
56. Li, W.; Zhang, P.; Zhao, K.; Chen, H.; Zhao, S. The Evolution Model of and Factors Influencing Digital Villages: Evidence from Guangxi, China. *Agriculture* **2023**, *13*, 659. [CrossRef]
57. He, Z.; Chen, M.; Gu, D. How digital village construction affects to the effectiveness of rural governance?—Research on the NCA and QCA methods. *Cities* **2025**, *156*, 105514. [CrossRef]
58. Ciolac, R.; Iancu, T.; Popescu, G.; Adamov, T.; Feher, A.; Stanciu, S. Smart Tourist Village—An Entrepreneurial Necessity for Maramures Rural Area. *Sustainability* **2022**, *14*, 8914. [CrossRef]
59. Visvizi, A.; Lytras, M.D. It's Not a Fad: Smart Cities and Smart Villages Research in European and Global Contexts. *Sustainability* **2018**, *10*, 2727. [CrossRef]
60. Roberts, E.; Anderson, B.A.; Skerratt, S.; Farrington, J. A review of the rural-digital policy agenda from a community resilience perspective. *J. Rural. Stud.* **2017**, *54*, 372–385. [CrossRef]
61. Rong, Z. From Attention to Circle: The Social Basis and Development Path of Traffic Production. *J. Shandong Univ. (Philos. Soc. Sci. Ed.)* **2025**, 53–63. [CrossRef]
62. Jia, W.L. Gansu Tianshui spicy hot pot "out of the circle" tips. *Cult. Ind.* **2024**, *31*, 97–100.
63. Yang, W.; Ruxin, W. The Integration of Traditional Costumes and Local Development in the Yellow River Basin—Taking Luoyang as an Example. *Chem. Fiber Text. Technol.* **2024**, *53*, 137–139.
64. Arnstein, S.R. A Ladder Of Citizen Participation. *J. Am. Inst. Plan.* **1969**, *35*, 216–224. [CrossRef]
65. Wang, Y.; Liu, H. Eu e-participation and its value review. *J. Xuehai* **2014**, *1*, 112–119. [CrossRef]
66. Liu, H.; Gao, X. Negative Public Opinion, Government Response and Discourse Reconstruction—An Analysis Based on 1711 Social Media Cases. *China Adm.* **2021**, *5*, 130–137. [CrossRef]
67. Granovetter, M.S. The Strength of Weak Ties. *Am. J. Sociol.* **1973**, *78*, 1360–1380. [CrossRef]
68. Fei, X.T. *Native China*; Shanghai People's Publishing House: Shanghai, China, 2006.
69. Li, W.; Liang, Y. Special Trust and Universal Trust: The structure and characteristics of Chinese trust. *Study Sociol.* **2002**, *17*, 11–22.
70. Huang, Z.H. Accurately grasping China's rural revitalization strategy. *China's Rural. Econ.* **2018**, *4*, 2–12.
71. Kong, J.; Jin, S.; He, C.; Qiu, H. *Community Economy*; China Machine Press: Beijing, China, 2015; p. 16.
72. Bond, R.M.; Fariss, C.J.; Jones, J.J.; Kramer, A.D.I.; Marlow, C.; Settle, J.E.; Fowler, J.H. A 61-million-person experiment in social influence and political mobilization. *Nature* **2012**, *489*, 295–298. [CrossRef]
73. Centola, D. An Experimental Study of Homophily in the Adoption of Health Behavior. *Science* **2011**, *334*, 1269–1272. [CrossRef]

74. Lin, R.; Utz, S. The emotional responses of browsing Facebook: Happiness, envy, and the role of tie strength. *Comput. Hum. Behav.* **2015**, *52*, 29–38. [CrossRef]
75. Wilcox, K.; Stephen, A.T. Are Close Friends the Enemy? Online Social Networks, Self-Esteem, and Self-Control. *J. Consum. Res.* **2013**, *40*, 90–103. [CrossRef]
76. Huang, B.Y.; Zhang, Y.X.; Lin, J.F. Beautiful! Xunpu Zanhuawei Hides the Most Tantalizing Spring Scenery in Quanzhou. *Quanzhou Evening Newspaper*. 2023. Available online: https://www.sohu.com/a/663540833_121124406 (accessed on 20 February 2025).
77. Yang, Y. Formation mechanism and development path of community economy: From the perspective of strong and weak relationship. *J. Shenzhen Univ. (Humanit. Soc. Sci. Ed.)* **2024**, *41*, 77–85.
78. Quanzhou Culture Radio; Television and Tourism Bureau. Quanzhou Culture, Radio, Television and Tourism Bureau Organization Functions Quanzhou Culture, Radio, Television and Tourism Bureau Official Website: cQuanzhou Culture, Radio, Television and Tourism Bureau. 2019. Available online: https://cbtb.quanzhou.gov.cn/zwgk/jgzn/ (accessed on 16 March 2025).
79. Qu, Z. Behind the microblogging hot search: Decoding the traffic business ecosystem. *Int. Brand Watch.* **2021**, 63–65.
80. Hu, X.H. "Qin Lang Lost His Homework" Farce Came to an End, in-Depth Reflection Can Not Stop Here. *Beijing Youth Daily*, bjyouth.ynet.com. 2024. Available online: https://www.xinhuanet.com/comments/20240415/5ef986e42291432a912da8723f42ae6c/c.html (accessed on 16 April 2025).

water

MDPI

Article

International Legal Systems in Tackling the Marine Plastic Pollution: A Critical Analysis of UNCLOS and MARPOL

Yen-Chiang Chang * and Muhammad Saqib *

School of Law, Dalian Maritime University, Dalian 116026, China
* Correspondence: ycchang@dlmu.edu.cn (Y.-C.C.); saqibayoub7890@gmail.com (M.S.)

Abstract: Marine plastic pollution (MPP) has become one of the most pressing environmental challenges, severely affecting marine ecosystems and human health. Even though international agreements such as UNCLOS and the International Convention for the Prevention of Pollution from Ships (MARPOL) of the International Maritime Organization (IMO) exist, the existing laws are often being introduced to question the inability of the present laws to do something about the escalating issue of plastic pollution. This study uses a doctrinal legal approach to examine how UNCLOS and MARPOL respond to marine plastic pollution (MPP) with a focus on their ability to handle land-based and ship-based MPP. Gaps in these frameworks are underlined, such as a lack of enforceable requirements under UNCLOS on reducing plastic rubbish from LBS and disparities in implementing MARPOL's regulations on plastic discharges from ships. This paper also explores activities of organizations such as the United Nations Environment Programme (UNEP), with a focus on the Regional Seas Programme and the Global Programme of Action. Although a lot has been achieved, much remains to be done to resolve the problem of marine plastic pollution. This paper concludes with a series of practical proposals aimed at refining international laws, strengthening enforcement, and encouraging collective action at the international level. The proposed measures are aimed at advancing a circular economy model, strengthening legal infrastructure, and fostering a single global response against marine plastic pollution.

Keywords: marine plastic pollution; UNCLOS; MARPOL; IMO; UN environment programme; best environmental practices; global programme of action

1. Introduction

Pollution is one of the most pressing issues faced by the contemporary world, and people have greatly contributed to pollution in different ways. It affects land and the atmosphere, but most prominently waters, which account for almost 71% of the earth's surface [1–3]. Water pollution is particularly alarming as it causes the spread of numerous viruses and diseases. Although land pollution is also the source of numerous diseases, oceanic water evaporates, mixes with the air, and returns as rain, bringing pollution back to the land. This cyclic process thus creates a dual burden on ecosystems [4,5]. Therefore, marine pollution does not just affect the marine ecosystem and marine life, but it also has a significant impact on human life [6,7].

The article of the United Nations Convention on the Law of the Sea (UNCLOS) defines pollution of the marine environment as "The introduction of materials or energy by humans, either directly or indirectly, into the marine environment, involving estuaries" [8]. This definition highlights three key components: elevated pollutant concentrations in

https://doi.org/10.3390/w17101547

various media, such as water, soil, and marine organisms, often termed contamination; the contaminants themselves; and the potential for significant ecological and human health impacts [9].

Marine plastic pollution (MPP) has emerged as one of the most urgent environmental issues of the 21st century. The rapid increase in plastic debris within oceans is having severe consequences for marine ecosystems, human health, and economic stability. Despite the widespread use and benefits, plastics have become a persistent pollutant due to their slow degradation, leading to extensive environmental damage over decades [10]. Studies have demonstrated that there is a substantial amount of plastic garbage in the world's oceans, which contributes to the loss of marine biodiversity and poses health hazards to humans through the food chain [11–14]. The production and consumption patterns of modern civilisation have exacerbated the problem, with approximately 80% of marine debris originating from land-based sources (LBS) [15]. This issue is further compounded by the increasing use of single-use plastics and the growing human population, which together contribute to the rising levels of marine litter [11].

MPP impacts are broad and severe. It poses ecological threats to marine life through entanglement, ingestion, and habitat disruption. Additionally, it facilitates the spread of invasive species and harmful substances [16]. The consequences on the economy are as severe, including industries like tourism, fishing, and maritime operations, with substantial expenses linked to cleanup activities and harm to marine sectors [17]. Furthermore, the contamination of human food sources with microplastics poses potential health risks, emphasising the urgent need for effective regulatory measures [18,19].

International legislation designed to tackle MPP has been instituted, although its efficacy is being questioned [20]. The United Nations Convention on the Law of the Sea (UNCLOS) and the International Maritime Organization's (IMO) International Convention for the Prevention of Pollution from Ships (MARPOL) are pivotal to these initiatives. UNCLOS provides a legal framework defining the rights and responsibilities of nations in using the oceans. Part XII focuses on protecting the marine environment and mandates states to prevent and control pollution from any source [8]. Similarly, MARPOL Annex V specifically prohibits the disposal of all plastics into the sea and regulates the disposal of other types of garbage from ships [14,21]. Notwithstanding these frameworks, the challenge of MPP persists, largely due to limitations and gaps in these regulations [22,23]. For example, while MARPOL has stringent rules for ship-based pollution, enforcement varies by country, and illegal dumping continues to be a problem [24,25]. Similarly, UNCLOS addresses pollution broadly but does not provide specific mechanisms for reducing plastic waste from LBS [13,14,24,26].

Furthermore, several challenges hinder the effectiveness of existing international regulations in addressing MPP. The fragmented framework of international law poses a significant challenge in addressing plastic pollution, as it overlaps with diverse fields like maritime regulations, environmental protection, and waste management, leading to inconsistencies and a lack of unified governance. The United Nations General Assembly (UNGA) has recognised this phenomenon of international law fragmentation, which complicates the implementation and enforcement of comprehensive solutions to complex environmental issues like MPP [27–30]. Another significant challenge is the focus of international maritime environmental law on downstream activities, often overlooking upstream sources of plastic pollution. While MARPOL addresses pollution from ships, it does not adequately tackle land-based sources, which contribute a substantial portion of marine plastic debris [31]. Effective management of plastic waste requires robust regulations that encompass the entire lifecycle of plastics, from production and consumption to disposal and recycling.

Internationally, the legal and governance frameworks addressing MPP have been fragmented and insufficient. While conventions such as UNCLOS and the 1974 Paris Convention have established some guidelines for managing marine pollution from LBS, these measures have often lacked binding obligations and comprehensive enforcement mechanisms [32]. The challenge of MPP requires a cohesive and multifaceted legal approach that integrates sustainable production, waste management, and marine environmental protection at global, regional, and national levels [33].

This paper utilises a doctrinal legal methodology to evaluate the efficacy of existing international legal frameworks in managing MPP, with a specific focus on UNCLOS and MARPOL. Through rigorous analysis, it identifies critical deficiencies and proposes enhancements aimed at bolstering regulatory effectiveness and safeguarding marine ecosystems from plastic contamination [34]. This issue's severity is highlighted by its significant effects on marine biodiversity, human health, and essential economic sectors, including fisheries and tourism. Plastic debris in marine environments degrades into microplastics, infiltrating the food chain and posing significant health risks [14,35]. Moreover, the economic ramifications encompass substantial costs associated with beach clean-ups, damage to fishing equipment, and reductions in tourism revenue [14]. Despite the existence of regulatory frameworks, such as UNCLOS and MARPOL, these instruments often suffer from inadequate enforcement mechanisms and lack binding obligations [32]. Effectively addressing MPP demands a unified global strategy that integrates sustainable production practices, robust waste management systems, and comprehensive marine environmental protection initiatives [33]. This research underscores the imperative for enhanced international collaboration and stringent regulatory measures to mitigate the pervasive threat posed by marine plastic pollution. Section 1 provides the study's background. Section 2 examines the impacts of MPP on marine ecosystems and human well-being, establishing a foundation for understanding its antecedents. Section 3 highlights the ecological and economic impacts of MPP. Section 4 analyses various international legal instruments and governance frameworks addressing MPP. Sections 5 and 6 evaluate UNCLOS and MARPOL frameworks, respectively. Section 7 offers a comprehensive discussion and conclusion.

2. From Source to Sea: Assessing the Ecological and Human Health Impacts of Micro-Plastics in Marine Ecosystems

Plastics are versatile materials that offer numerous benefits to society and individuals in their daily lives. However, the accumulation of plastics in the environment is a significant concern, and due to their slow degradation, this issue will persist for decades. The generation of litter is one of the primary consequences of the production and consumption models adopted by modern civilisation. Waste is a critical environmental problem that demands increased attention in the search for solutions, especially with regard to marine pollution [10]. Solid waste-related marine pollution is emerging as a global issue with consequences that extend across generations. There is evidence that this problem is persistent and worsening, despite decades of efforts in many countries to prevent and reduce marine litter. This trend can be attributed to the growing prevalence of single-use consumption practices and increasing human populations [11]. Approximately 80% of the billions of tonnes of rubbish that are dumped—intentionally or unintentionally—into the seas each year originate from LBS [15]. Due to the historical perception of the seas as places to dispose of trash, persistent pollutants have now become significant contributors to environmental issues.

Globally, plastic pollution is recognised as a significant problem caused by human activities, which has profound impacts on marine and coastal ecosystems [36]. Anthropogenic sources contribute to an unprecedented and continuous accumulation of plastic pollutants in various aquatic environments, resulting in the disruption of ecosystem structure, functioning, and, ultimately, the direct or indirect degradation of ecosystem services and values. The primary sources of these pollutants, which reach the ocean through various means, include both land-based and sea-based sources [37]. Although the complete effects of MPP on the environment and human health are still unknown, they have the potential for significant impacts, especially concerning "microplastics". These microplastics are increasingly being released into domestic wastewater streams and are formed as larger plastic debris degrades over time. These particles are also commonly found in skincare products and synthetic apparel [38]. Human exposure to and consumption of microplastics have been linked to an increasing number of disease disorders [6,7,39]. The magnitude of potential health effects on humans might be evaluated by the discovery made by the University of Ghent in 2014: an individual may invest up to 11,000 small pieces of plastic in their seafood annually [40].

The biological effects of MPP are numerous and diverse, including endangering wildlife through choking and starvation, facilitating the spread of invasive alien species (IAS) and other potentially harmful organisms to new areas, and transporting toxic chemicals and persistent organic pollutants (POPs), among others [16]. According to an investigation that collected samples of Arctic Sea ice from five different locations, an investigation that analyzed Arctic Sea ice samples from five distinct locations found concentrations exceeding 12,000 microplastic fragments per liter, indicating the widespread presence of plastic waste and microplastics throughout the world's oceans. are now widespread in the world's oceans [41]. Currently, an estimated 9–13 million tonnes of plastic waste enter the ocean every year. However, considering the reported plans of the widespread petrochemical industry to expand plastics production, partly as a precautionary measure due to the potential decrease in demand for their fuels in response to global warming, this figure could significantly increase. Plastic, along with microplastic debris, has a detrimental impact on society, the environment, and the economy. This includes the injury or death of a marine species and its entry into the food web, which raises health concerns [42].

The establishment of ideas and strategies for reducing marine microplastic contamination is in great demand [43]. The lack of expertise in this field is driving the expansion of research on the topic. Recent studies have revealed that certain types of polymers can undergo biodegradation by various bacteria, bacterial consortia, biofilm-forming bacteria, and fungi [44]. Biodegradation is influenced by various parameters, such as the type of microorganism, polymer form, their physicochemical properties, and environmental conditions. Plastic pollutants are dispersed throughout ecosystems in multiple forms and sizes, including megaplastics, macroplastics, mesoplastics, and microplastics.

Microplastics, including both primary and secondary particles, are extensively distributed in the water, sediments, and organisms of marine and coastal ecosystems [45]. Microplastic levels in coastal and marine ecosystems globally range from 0.001 to 140 particles/m^3 in water and 0.2 to 8766 particles/m^3 in sediments. The accumulation rate of microplastics in coastal and marine species varies from 0.1 to 15,033 counts. Consequently, plastic pollution has a broad range of negative consequences, including ecological and socioeconomic implications. Entanglement, toxicological consequences from plastic ingestion, asphyxia, hunger, organism dispersal, rafting, provision of novel habitats, and the emergence of invasive species are all significant ecological effects that pose increasing risks to biodiversity and trophic connections [46]. Degradation—alterations in ecosystem condition—and alterations to marine systems are related to the loss of ecosystem services and value. Consequently, this new pollutant has a detrimental impact on socioeconomic factors, including tourism, shipping, fisheries, and the well-being of people.

Presently, environmentally friendly alternatives to plastic, derived from recyclable materials, are entering the market [47]. Practical measures to combat plastic pollution include preventing the accumulation of plastic pollutants from various sources, promoting the 3Rs (reduce, recycle, reuse), raising awareness and enhancing capabilities, and enforcing manufacturer accountability [48]. Existing and implemented policies, laws, regulations, and efforts at the regional, global, and national levels are crucial in minimising plastic waste in marine and coastal areas.

3. Ecological and Economic Impacts of MPP

Although the majority of marine litter accumulates in coastal areas, plastic, including microplastics, is dispersed across the ocean, with higher deposition occurring in the convergence regions across all five subtropical ocean gyres [49]. Plastic pollution has long been recognised as a significant danger to the marine ecosystem. Specifically, the effects of MPP will be discussed in this section.

3.1. *Threats to Marine Life*

Marine organisms face significant threats from MPP, including entanglement and ingestion [50]. Entanglement in abandoned fishing gear, known as ghost nets, and other plastic debris, like packaging materials, can lead to severe physical harm or death for marine animals. Studies have shown that ghost nets continue to trap and kill marine life for extended periods, varying based on factors such as the type of gear and its location. This entanglement can result in injuries or death for various marine species, including whales, dolphins, seals, and seabirds, who often become trapped by their necks, flippers, or wings [39,51].

Ingestion of plastic debris also poses a critical threat to marine organisms. Animals may mistake plastic particles for food, leading to intestinal blockages or malnutrition due to the lack of nutrients, which can cause starvation [52]. Additionally, microplastics, tiny plastic particles resulting from the breakdown of larger debris, have been found to bioaccumulate across all levels of the food web. These microplastics can carry harmful pollutants, including persistent organic pollutants (POPs), heavy metals, and plastic additives, which can be toxic to marine life. The transfer of microplastics through trophic levels can amplify their harmful effects, impacting entire ecosystems [39,51].

3.2. *Contamination of Human Food Sources*

The pervasive presence of microplastics in foods consumed by humans, especially through wild-caught fish, raises significant concerns about the potential health implications. Recent studies revealing that microplastics were found in the digestive tracts of a substantial percentage of various fish species highlight the extent of MPP. Specifically, research showing that 66% of 498 studied fish species contained microplastics [53], alongside findings from New Zealand where three-quarters of commercially caught fish were contaminated [19], underscores a critical environmental and public health issue.

From a researcher's perspective, the ingestion of microplastics by marine organisms not only signifies a direct threat to marine life but also poses a significant risk to human health. The consumption of whole organisms such as sardines and shellfish, which are not gutted before consumption, exacerbates this risk by increasing exposure to both the physical presence of plastics and the toxic chemicals they may carry [18,19].

3.3. *Psychological and Emotional Effects*

There is substantial evidence indicating that marine litter can negatively impact the mental health benefits individuals derive from the ocean's aesthetic and therapeutic value. Therefore, a study highlights that the presence of marine litter can disrupt the mental health benefits associated with natural environments, as these spaces are often used for relaxation and recreation [17]. Likewise, plastic pollution and marine litter—global partnership facilitated by the UNEP—indicates that Marine Plastic Pollution (MPP) adversely affects not only the environment but also its cultural and spiritual activities. The existence of plastic litter in the sea can interfere with traditional practices and reduce cultural value that goes along with the coast, as has been reported by Trash Free Maryland [50]. However, the frustration of cultural rituals and failure in maintaining a spiritual relationship that many communities have with the ocean, which is vital for their cultural legacy, is caused by plastic pollution in the ocean.

3.4. Economic and Socio-Cultural Consequences

Marine litter also has considerable indirect impacts, particularly on small-scale fishing, tourism, and recreation industries. These indirect costs, though challenging to quantify, can disproportionately affect individuals whose livelihoods depend on coastal activities [54]. For example, small-scale fisheries face reduced catches and damage to vessels and gear, leading to devastating economic consequences. The European Union (EU) estimates that the fishing sector loses up to EUR 65 million annually due to vessel and gear damage, as well as decreased catches resulting from ghost fishing, where abandoned gear continues to trap marine species.

Tourism and recreation industries also experience significant losses due to marine litter. Polluted beaches can deter tourists, leading to a decline in local revenue. The costs associated with cleaning up beaches are substantial; the EU spends up to EUR 645 million per year on these efforts. In addition to these indirect impacts, marine litter incurs direct economic costs. Accidents caused by navigation hazards and fouling from marine debris can lead to substantial expenses for maritime operations. Ensuring the safety and cleanliness of coastal and marine environments is essential for maintaining the economic viability of activities such as fishing, tourism, and recreation [17].

In addition to these indirect impacts, marine litter poses direct economic costs. Accidents caused by navigation hazards and fouling from marine debris can incur significant expenses for maritime operations. Ensuring the safety and cleanliness of coastal and marine environments is crucial for maintaining the economic viability of activities such as fishing, tourism, and recreation [17,55]. New worldwide developments have greatly increased the global strategy of dealing with the problem of plastic pollution in various countries. The United Nations Environment Assembly (UNEA) has come through with many resolutions geared towards marine plastic pollution. Thus, it is evident that a global layered response is necessary. With the negotiations and the prospect of enacting the Global Plastics Treaty, which will establish enforceable rules for managing a state's plastic waste to occur across the world, it is evident that the international community is approaching a pivotal moment in addressing this urgent issue m. This current advancement reflects the transition to a larger regulatory regime with legally binding instruments supporting the United Nations Convention on the Law of the Sea (UNCLOS) and the International Convention for the Prevention of Pollution from Ships (MARPOL).

4. International Legal and Governance Frameworks for Addressing MPP

Sustainable production and consumption [56], the circular economy [33,57], waste management, freshwater resource management, biodiversity protection, and marine pollution are among the environmental issues that require a comprehensive blend of legislation to be effectively implemented at regional, global, and domestic levels of governance [33]. This is necessary to tackle the increasing challenge of plastic waste in the environment as a whole, and the multifaceted problem of MPP in particular [58]. Because of this, the MPP challenge is commonly linked to the extensively acknowledged phenomenon of international law [32] fragmentation, as well as related issues with legal coherence [27,59].

However, historically, the international legal approach has tended to focus on the framework provided by international maritime environmental law, often overlooking the need for control over the "upstream" activities that contribute to plastic waste. Nevertheless, it is increasingly evident that laws pertaining to freshwater resource environmental management will be crucial in combating MPP. The effective establishment and implementation of relevant international water law regulations will be crucial, especially in the case of vast transboundary watercourses that carry a significant portion of the plastic waste that ultimately reaches the seas.

4.1. Historical Legal Framework: Marine Pollution from LBS in International Law

MPP is primarily defined as a pollution issue arising from LBS within the framework of international marine environmental legislation, which has been subject to treaty protections for over 50 years. State parties were required by the Convention for the Prevention of Marine Pollution from Land-Based Sources, 1974 (Paris Convention 1974) [60,61] to take action to completely eradicate pollution of the maritime area concerning chemicals listed in Annex A, Parts I and III, as well as to severely restrict contamination of the marine area about substances included in Annex A, Part II. However, the Paris Convention 1974 was widely perceived as having limited legally binding obligations and a weak normative basis. Article 4(3) of the Paris Convention 1974 emphasised the importance of developing new guidelines and standards, including "specific regulations or standards governing the quality of the environment, discharges into the maritime area, such discharges into watercourses and emissions into the atmosphere as affect the maritime area, and the composition and use of substances and products" [60]. Therefore, while acknowledging the complexity of the challenge and the need for diverse legal approaches to tackle LBS, the Paris Convention 1974 explicitly recognised the substantial impact of pollutants transported through watercourses and the crucial role of freshwater pollution control measures. This fact remains applicable to MPP as well.

UNCLOS [62], which addresses the issue of LBS in Articles 207, 194, and 213 of its text, provides the overall worldwide legal framework that exists today. Nonetheless, UNCLOS is frequently viewed as legally ambiguous and devoid of explicit and mandatory obligations for the state parties, as it takes a similar approach to that outlined in the Paris Convention 1974. For example, Article 207 of UNCLOS only mandates that States "adopt laws and regulations to prevent, reduce, and control pollution of the marine environment from land-based sources, including rivers, estuaries, pipelines, and outfall structures, taking into account internationally agreed rules, standards, and recommended practices and procedures" [8]. Nevertheless, there have been limited globally recognised guidelines or norms that can guide national regulatory actions and be acknowledged under international law or state practice [63]. Commentators emphasise that, in contrast to this lack of normative clarity, UNCLOS does not impose any obligations to adhere to norms or standards, nor does it establish a specific timeframe for taking action. They also associate these shortcomings with an increased risk of legal fragmentation [64].

While the UNEP utilised the 1985 Montreal Guidelines [9] as "recommended practices and procedures" with the explicit aim of assisting states in developing national legislation and elaborating more specific international instruments, these guidelines were also perceived as overly general and lacking specificity. It seems that they have received relatively limited support in terms of state practice. The International Law Association (ILA) [65] first adopted model rules in 1972. Articles 207 and 213 of UNCLOS have faced criticism for attempting to "incorporate all possible contaminants originating from land under a single all-encompassing article", despite the fact that the various pollutants from LBS that harm the marine ecosystem have distinct origins, characteristics, and impacts [66]. As a consequence of this persistent issue, marine plastic has been classified under the broader category of "litter" and has not historically received prioritised attention, despite the continuous call to address it throughout the development of international standards on pollution from LBS [67].

Article 194(1) of UNCLOS says "to safeguard and sustain the marine environment", which amounts to a compelling due diligence requirement to "take, individually or jointly as appropriate, all measures consistent with this Convention that are necessary to prevent, reduce and control pollution of the marine environment from any source, using for this purpose the best practicable means at their disposal and in accordance with their capabilities" [8].

Article 194(2) of UNCLOS speech makes it abundantly evident that this expansive duty includes a general need for states to govern all pertinent, possibly harmful conduct under their control in a way that prevents contaminants of the marine environment within other maritime states' territorial waters in marine areas outside of their national jurisdiction: "States shall take all measures necessary to ensure that activities under their jurisdiction or control are so conducted as not to cause damage by pollution to other states and their environment and that pollution arising from incidents or activities under their jurisdiction or control does not spread beyond the areas where they exercise sovereign rights in accordance with this convention" [8].

It is evident that these fundamental marine pollution provisions of UNCLOS necessitate that states exercise due diligence concerning various inputs of plastic waste, with a specific emphasis on addressing pollution in significant drainage basins, including international watercourses. This focus is supported by established knowledge regarding global dispersal patterns, sources, and the extensive range of adverse impacts caused by MPP. Article 213 of UNCLOS explicitly states that when it comes to the prevention, mitigation, and control of pollution from LBS more generally, the responsibility of due diligence is both mandatory and crucial. "States shall enforce their laws and regulations adopted following Article 207 and shall adopt laws and regulations and take other measures necessary to implement applicable international rules and standards established through competent international organisations or diplomatic conference to prevent, reduce and control pollution of the marine environment from land-based sources" [8].

Current soft-law rules and standards are not completely underdeveloped regarding the management of pollution in international watercourses compilation of pertinent "internationally agreed rules, standards and recommended practices and procedures", which could help clarify the legal precautions required of states to meet the aforementioned diligence standards [8], which are generally lacking in detail. Especially, the Montreal Guidelines from UNEP in 1985 suggest in Guideline 5(3) that, "if discharges from a watercourse which flows through the territories of two or more States or forms a boundary between them are likely to cause pollution of the marine environment, the states concerned should cooperate in taking necessary measures to prevent, reduce and control such pollution" [68].

Furthermore, Guideline 5(2) urges non-coastal states to "cooperate in preventing, reducing, and regulating pollution of the marine ecosystem originating or predominantly originating from discharges within their territories into or through water basins or watercourses that flow into the marine environment" [68]. Chapter 17, Paragraph 24 of Agenda 21, which focuses on the protection of the seas and oceans, is valuable in that it recognises the need to adopt a comprehensive, multi-sectoral approach to safeguard the marine ecosystem as a whole. It emphasises the commitments that states have already made regarding marine biodiversity [69] and calls for additional action by the UNEP concerning the issue. Ultimately, this resulted in the adoption of the Global Programme of Action for the Preservation of the Marine Ecosystem from Land-Based Activities (GPA) in 1995 [70–72].

It has long been acknowledged that a more comprehensive global regulatory framework is needed for the distinct issue of LBS. In the long run, a global convention might be prepared "based on additional knowledge gained in the establishment and implementation [73] of regional, sub-regional, and bilateral treaties and taking into consideration principles or guidelines at the global scale developed within the framework of UNEP" [74], according to the 1982 Montevideo Programme for the expansion and periodic assessment of environmental law. The establishment of a comprehensive framework "for the long-term development of a worldwide treaty on pollution from terrestrial sources" is listed as one of the key objectives of the 1985 Montreal Guidelines. However, instead of adopting a legally binding global convention, the international community eventually embraced a non-binding instrument that included a formal declaration of standards and was accompanied by a detailed action plan [63]. This was a response to the UN Secretary–General's strong plea to the Committee on Preparation for the 1992 UNCED Conference in Rio to modify the global conventional framework by incorporating LBS and establishing "general principles for global application that would inspire, motivate, and guide national and regional measures" [69,75]. Similarly, the Table 1 summarizes key findings on marine plastic pollution (MPP) and its broad impacts. It highlights that 80% of ocean plastic originates from land-based sources, with serious economic consequences, such as EUR 65 million in annual losses to EU fisheries and EUR 645 million in beach cleanup costs. Additionally, it underscores the health risks of microplastic ingestion, threats to marine biodiversity, cultural disruptions, and the ineffectiveness of current legal frameworks like MARPOL and UNCLOS, due to weak enforcement and limited jurisdiction.

Table 1. Key findings on Marine Plastic Pollution (MPP) and economic impacts.

Aspect	Findings/Results	Sources
Marine Plastic Pollution (MPP)	80% of ocean plastic waste originates from land-based sources (LBS)	[10,15]
Economic Impact on Fisheries	EU fishing sector loses EUR 65 million annually due to damage from plastic waste	[17]
Economic Impact on Tourism	EU spends EUR 645 million annually on beach cleanups	[17]
Impact on Marine Biodiversity	Microplastics in marine organisms; 66% of 498 fish species studied contained microplastics	[19,53]
Health Impacts of Micro-plastics	Ingesting microplastics was linked to health issues such as endocrine disruption	[6,7]
Cultural Impact	Marine litter affects cultural practices and spiritual connections to oceans	[55]
International Legal Efforts	UNEA and Global Plastics Treaty negotiations aim to create enforceable global plastic waste management rules	[23,33]
MARPOL and UNCLOS Effectiveness	MARPOL bans ship-based plastic discharge, but enforcement varies by jurisdiction	[14,21]

Under the direction of UNEP [70,72,76,77], GPA was approved and is now in operation. The GPA is a soft-law agreement that reaffirms states' obligations under UNCLOS and other agreements. It is intended to serve as a "resource for national or regional authorities when developing and implementing long-term plans to prevent, reduce, regulate, or eliminate marine degradation caused by land-based activities" [70]. The goal of the GPA process is to foster and strengthen voluntary multi-stakeholder engagement in key sectoral areas that contribute to LBS [76]. To address this issue, a global multi-stakeholder collaboration on marine litter was formed in 2012 under the GPA's auspices. The Intergovernmental Review Meetings, which convene periodically to assess the implementation of the GPA [77], have endorsed numerous statements highlighting the importance of freshwater resource management in mitigating LBS pollution. For instance, the 2001 Montreal Declaration commits the 80 national government members to, among other things, "integrating and implementing coastal area and watershed management practices, and enhancing regional, global, and national governance processes", emphasising concerns about litter pollution and alterations in the quality of freshwater inflows [78].

An appeal was made to the UNEP Secretariat to conduct an analysis of alternatives and possibilities regarding the legal aspects of addressing LBS under the GPA. Additionally, a commitment to continue focusing on "enhancing the integrated approach to protecting coastlines and marine environments, particularly from environmental threats posed by excessive nutrients, sewage, marine litter, and microplastics" was included in the more recent 2018 Bali Declaration, which was ratified by the EU and 60 national governments. The 2030 Agenda for Sustainable Development Goals (SDGs), specifically SDG-14 (to conserve and sustainably use the oceans, seas, and marine resources), is connected to the problem of coastal and marine pollution caused by wastewater, marine litter, and microplastics [41] in agreements of the United Nations Environment Assembly (UNEA) that have been passed since the Bali Declaration. The fact that MPP, in particular, and land-based sources, in general, are specifically addressed in the 2030 Agenda speaks volumes about the current importance provided to these issues [34].

The due diligence norms anticipated by state actors are influenced by the extensive network of traditional methods addressing LBS in various regional seas despite persistent challenges in their effective implementation. Such instruments include:

- The 1976 Barcelona Convention for the Protection of the Mediterranean Sea Against Pollution to the 1980 Athens Protocol [61] and the 1996 Syracuse Protocol for the Preservation of the Mediterranean Sea [79] Against Pollution from LBS and Activities [80];
- The Quito Protocol of 1983 for the Safety of the South–East Pacific Against Contamination from LBS [81];
- The 1990 Protocol for the Preservation of the Marine Environment against Waste from LBS supplements the 1978 Kuwait Regional Convention for Cooperation on the Protection of the Marine Ecosystem from Pollution;
- The OSPAR Convention 1992 for the Preservation of the North-East Atlantic Marine Environment;
- The Helsinki Convention of 1992 for the Preservation of the Baltic Sea's Marine Environment;
- The Bucharest Convention 1992 on the Preservation of the Black Sea Against Pollution, which included the 1992 Protocol on Protection of the Black Sea Marine Environment Regarding Pollution from LBS;
- The 1983 Cartagena Convention for the Protection and Development of the Marine Environment of the Western Caribbean Region and the Aruba Protocol 1999 on LBS of Marine Pollution;
- The Jeddah Protocol 2005 for the Preservation of the Marine Environment from LBA in the Gulf of Aden and Red Sea;
- The 2010 Nairobi Protocol for Protecting the Marine and Maritime Environment of the Southwestern Indian Ocean from LBS; and
- The Abidjan Additional Protocol of 2012 to the Abidjan Convention Regarding Collaboration in the Southern African and Western Central Region's Protection and Reconstruction of the Marine and Coastal Ecosystem from Land-Based Activities.

Certain conventional arrangements have made significant progress in their practical implementation, although the outcomes achieved so far vary considerably. This discrepancy can be attributed primarily to differences among the involved state parties in terms of their financial, technical, and administrative capacities, as well as their normative and institutional development [82]. For instance, the LBS Protocol for the Mediterranean has developed comprehensive national and regional action plans with specific deadlines and action items. Additionally, the legally binding document, the Regional Strategies on Marine Litter Management in the Mediterranean Sea, sets forth precise measures and operational objectives to achieve good environmental status in the Mediterranean Sea. One of the targets is to reduce beach litter by 20% across the entire basin by 2024. These factors make the plan remarkable.

4.2. The GPA and Washington Declaration: International Efforts to Combat Land-Based Marine Pollution

Under Agenda 21, UN Environment received an invitation to organise an intergovernmental conference on safeguarding the marine environment from LBS. UN Environment accepted the offer and scheduled a meeting for 23 October to 3 November 1995, in Washington, D.C [69]. The GPA and the Washington Declaration were the two documents effectively endorsed by the EU and 109 other nations during the meeting [83].

Figure 1 illustrates a structured flowchart outlining the analytical framework for Marine Plastic Pollution (MPP), beginning with identifying sources and assessing ecological, economic, and cultural impacts. Similarly, it progresses through legal evaluations (UNCLOS, MARPOL), global initiatives (e.g., UNEA), and culminates in actionable policy recommendations. By making it easier for states to fulfil their UNCLOS [8], a mandated obligation to maintain and safeguard the marine environment, the GPA aims to halt the degradation of the marine environment resulting from LBS [84]. The GPA encompasses not only the prevention, mitigation, and management of marine pollution but also the remediation of pollution and measures that aid in the restoration of the marine ecosystem from the impacts of pollution [70]. Thus, the GPA surpasses the scope of relevant UNCLOS clauses. The objective of the Programme of Action is to serve as "a resource of conceptual and practical guidance" for the fulfilment of existing responsibilities and agreements, including those established after Agenda 21 and UNCLOS, as well as for the development of new initiatives. Three chapters of the GPA offer suggestions for regional, national, and international action [70].

Figure 2 illustrates the sources of Marine Plastic Pollution (MPP), showing that land-based sources contribute 80%, while ship-based sources account for 20%. This emphasizes the dominant role of land-based activities in driving oceanic plastic pollution. The GPA encourages states to develop national plans of action (NPAS) within the framework of integrated coastal area management at the national level. Similarly, provisions for identifying and evaluating issues, prioritising issues, choosing management approaches and measures, defining standards for judging the efficacy of strategies and programs, and guaranteeing program support elements (like funding, human resources, and lawful and enforcement mechanisms) should all be included in the NPAS [85]. Integrated coastal region management, watershed management, reducing poverty, assessment of environmental impacts, the preservation of important habitats and threatened animals, vertical policy integration, collaboration, caution, and equity between generations [70] are just a few of the principles and methods that the GPA calls on states to implement. States are encouraged to incorporate clean production practices, environmentally friendly and efficient technologies, product substitution, Best Available Techniques (BATS), and Best Environmental Practices (BEPS) into their policies and initiatives. However, potential actions encompass governmental measures, market-based tools that consider the polluter pays principle and cost internalisation, technical assistance and collaboration, education, and awareness-raising campaigns. However, waste reuse, recycling, and treatment are specifically mentioned, along with the significance of institutional frameworks, resource mobilisation [70], monitoring, and reporting. Some of the recommended criteria for evaluating the NPAs include flexibility, equity, cost-effectiveness, and environmental effectiveness.

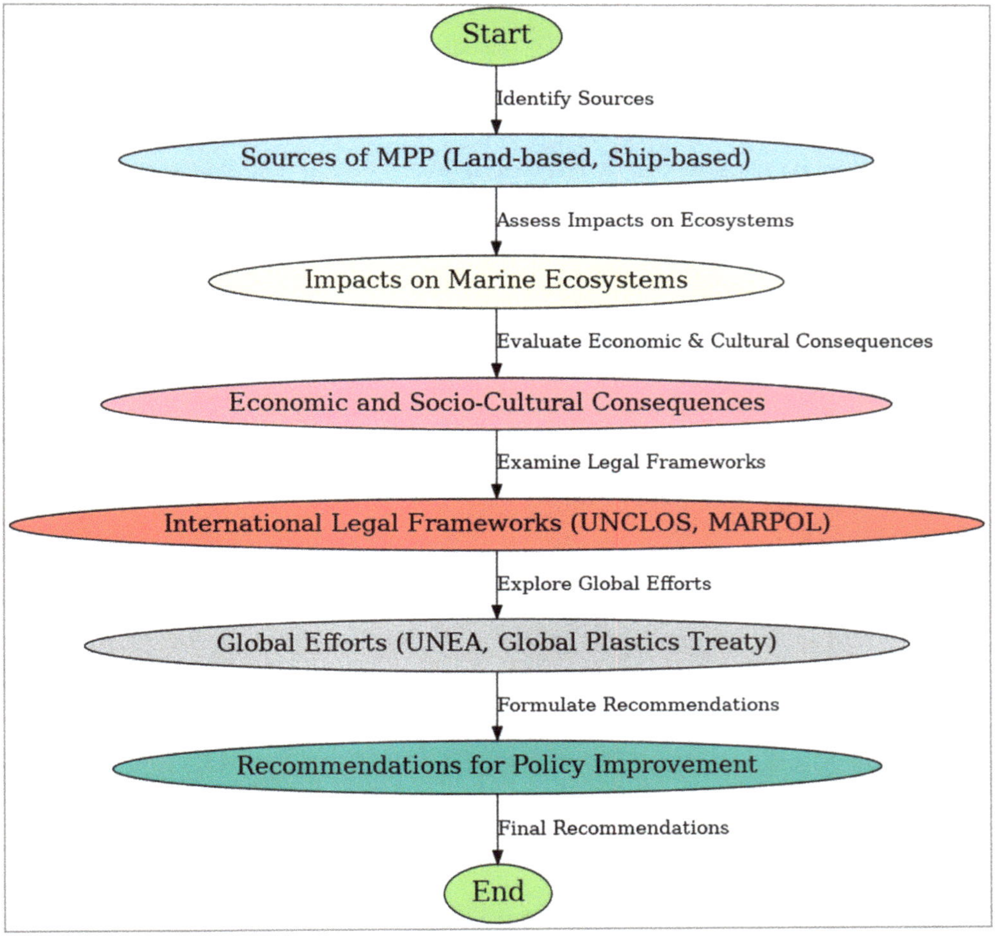

Figure 1. Flowchart for the Marine Plastic Pollution framework source (created by author).

Figure 3 highlights the levels of different marine species that have consumed microplastics. Among marine animals, fish take in the largest amount of microplastics (over 65%), with seals in second place, seabirds third, whales fourth and turtles fifth. Similarly, the GPA emphasises the need for enhanced regional collaboration to protect the marine environment from LBS. Thus, it encourages governments to strengthen existing regional agreements and initiatives, and, if required, establish new ones. Similarly, the GPA provides guidelines for developing regional action plans that align with the NPA approach. It highlights various considerations in this regard, including capacity-building plans and policy harmonisation. Thus, the GPA specifically encourages landlocked governments to actively engage in regional initiatives [70,86]. However, governments are advised to seek collaboration from multilateral finance organisations and other institutions in the planning and implementation of regional agreements, particularly in developing-country areas, regarding the institutional aspects of regional and sub-regional arrangements [70,72].

It further emphasises the critical importance of successful international collaboration in the implementation of the GPA, especially in areas such as financial support, technology transfer, and capacity building. One suggestion for mobilising expertise and skills is the establishment of a clearing-house system. Moreover, it is advised to regularly examine both the status of the marine environment and the GPA's implementation [70,72]. Under the GPA, the mobilisation of resources and the establishment of effective institutional frameworks are viewed as primary objectives of international cooperation. Similarly, the GPA also emphasised the need for the formulation of international legally binding instruments concerning the implementation of the prior informed consent (PIC) process for certain hazardous substances in international trade, as well as instruments addressing persistent organic pollutants (POPS), including a dedicated section on sewage and wastewater treatment. These tools were approved in 1998 and 2001, respectively [87].

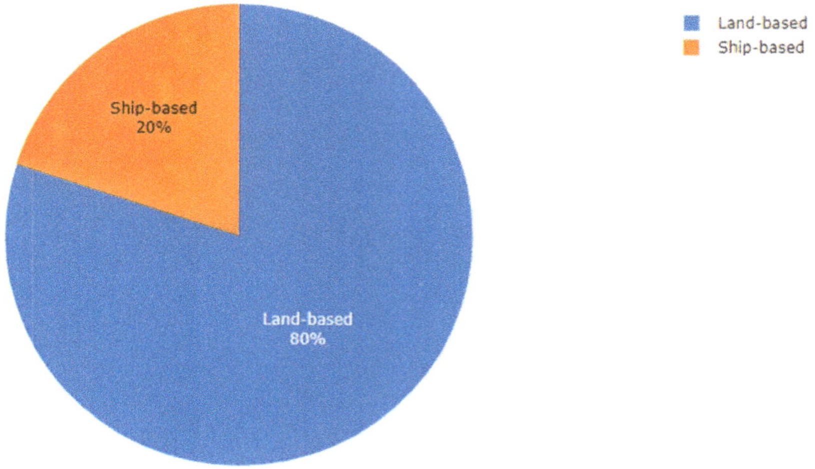

Figure 2. Plastic pollution sources.

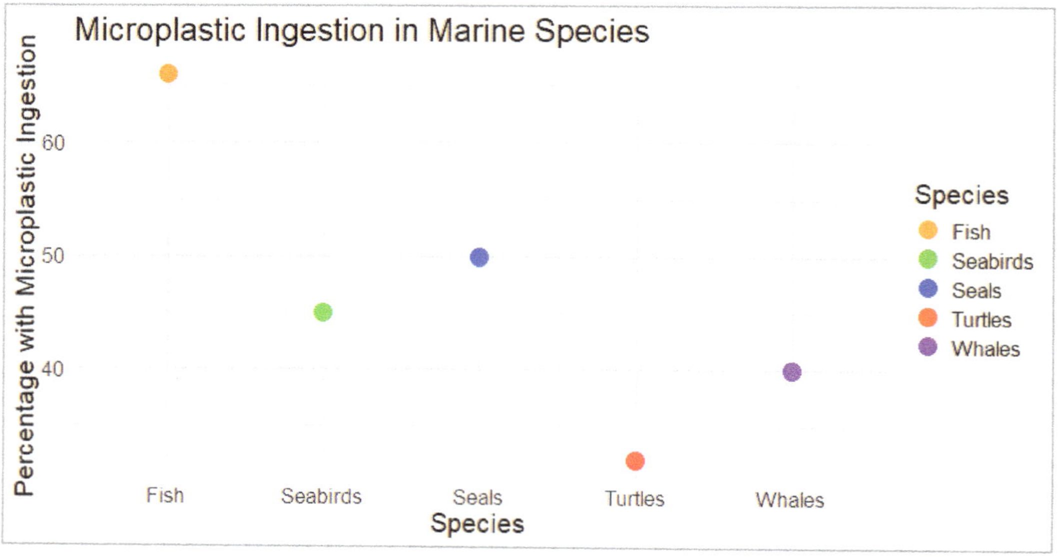

Figure 3. Marine biodiversity and microplastic ingestion.

4.3. *The Global Environment Facility's Strategic Focus Areas: Biodiversity Conservation and International Waterways*

The Global Environment Facility (GEF) plays a crucial role in implementing the GPA, particularly in the areas of international waters and biodiversity conservation. The GEF has been assigned a specific role in this context. It has been tasked with supporting the implementation of the GPA, focusing on priority areas such as biodiversity conservation and the protection of international waterways. The GEF's Operational Strategy, adopted in 1995, identifies the management of LBS of pollution as a top priority for action in its international waterways' focal region. The issue of MPP has also been addressed in GEF-sponsored publications, as well as during the GEF-6 and GEF-7 replenishment periods.

4.3.1. GEF Initiatives: International Water Law and Cross-Border Management

Projects addressing cross-border water management, safeguarding water supplies, and preserving important groundwater and surface water resources in Small Island Developing States (SIDS) are supported by the GEF's focus area on international waterways [88]. This aligns with the objectives of the GPA to prevent, minimise, and regulate marine debris originating from LBS, especially those that affect international waters. The International Law Commission (ILC) [8] worked on legislating the marine environment-related regulations of the 1994 draft articles [89] that would ultimately serve as the cornerstone of the 1997 UN Watercourses Convention, which was the initial globally applicable binding tool in the field that is also widely accepted as indicative of the context in customary international law [90]. The ILC was motivated by the state parties' tardiness in adopting meaningful measures for the implementation of what is currently known as Article 23 of the 1997 UN Watercourses Convention, which was initially enacted by the commission in 1991 and mandates that, "watercourse states shall, individually and, where appropriate, in cooperation with other states, take all measures with respect to an international watercourse that are necessary to protect and preserve the marine environment, including estuaries, taking into account generally accepted international rules and standards" [86].

Figure 4 presents the economic impact of marine plastic pollution, each year, losses in tourism and beach cleanups hit €600 million, while those in the fishing industry stay lower. It makes obvious that coastal areas face too much burden and costly maintenance for public buildings. As Article 23 is located in Part IV of the 1997 UN Watercourses Convention, which addresses the responsibilities of watercourse nations for ecosystem protection, it can be understood as an acknowledgment of the interconnections between freshwater and marine environments and the environmental issues associated with them. It essentially signifies the official recognition of "the increasingly significant problem of pollution transported into marine ecosystems through international watercourses" within the framework of customary international law pertaining to freshwater resources [89]. It is widely recognised that LBS, primarily rivers, contributes to approximately 85 percent of the pollution entering the marine environment. Therefore, this acknowledgment has been long overdue [91]. The importance of riverine pollution loading for LBS marine pollution is further made worse by the MPP problem. Article 23 of the UN Watercourses Convention also reaffirms duties that have long been recognised in international maritime environmental law despite their normative weaknesses and lack of specificity [8,60,92].

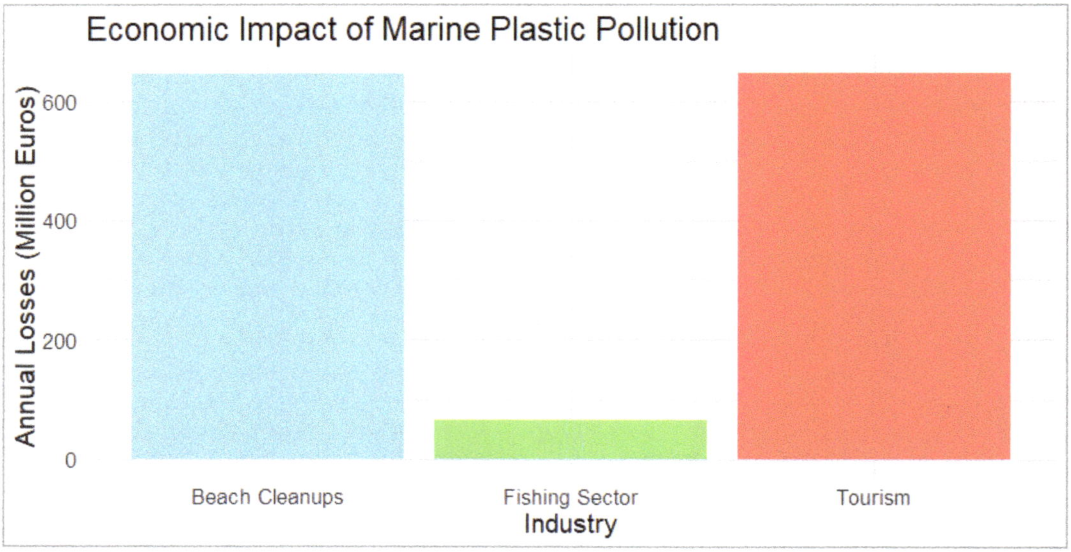

Figure 4. Economic impact.

Figure 5 shows a steady and significant rise in marine plastic pollution from 8 million in 2010 to over 21 million tons by 2020. Because of this trend, we must act quickly and implement tougher regulations on waste management everywhere. Article 23 of the UN Watercourses Convention, along with its other provisions, imposes a commitment of due diligence on watercourse states. It says that they must "take all of the steps required of which they are competent, financially and technologically" [86,89,93], either individually or jointly when appropriate. The precautionary principle is presumed to apply due to the interconnection with Articles 20–22 and the need that such action be done "on an equitable basis" [89,93]. Article 23 sets itself apart from other ecological obligations typically found in water resources conventions by encompassing the collective global interest of all states in safeguarding the broader marine environment, rather than solely focusing on preventing harm to adjacent watercourse states [82].

Conventions addressing marine pollution may directly place obligations on states regarding the management of transboundary rivers, in addition to international, regional, or basin-level water agreements. For instance, the 1992 OSPAR Convention specifically requires coastal state parties, such as Germany, France, and the Netherlands, as well as riparian countries along the Rhine, to restrict discharges into the river that eventually reach the North Sea. Although this provision, which can be found in Annex I of Article 2(1) of the Convention for the Protection of the Marine Environment of the North–East Atlantic, 1992 [84], does not place any new obligations on these states, it does clarify the normative status of their existing obligations regarding LBS pollution [94]. Similarly, discharges into the Danube River by Romania and Bulgaria would be subject to the requirements outlined in the 1992 Bucharest Convention on the Protection of the Black Sea from Pollution, as well as its Protocol on Pollution from LBS.

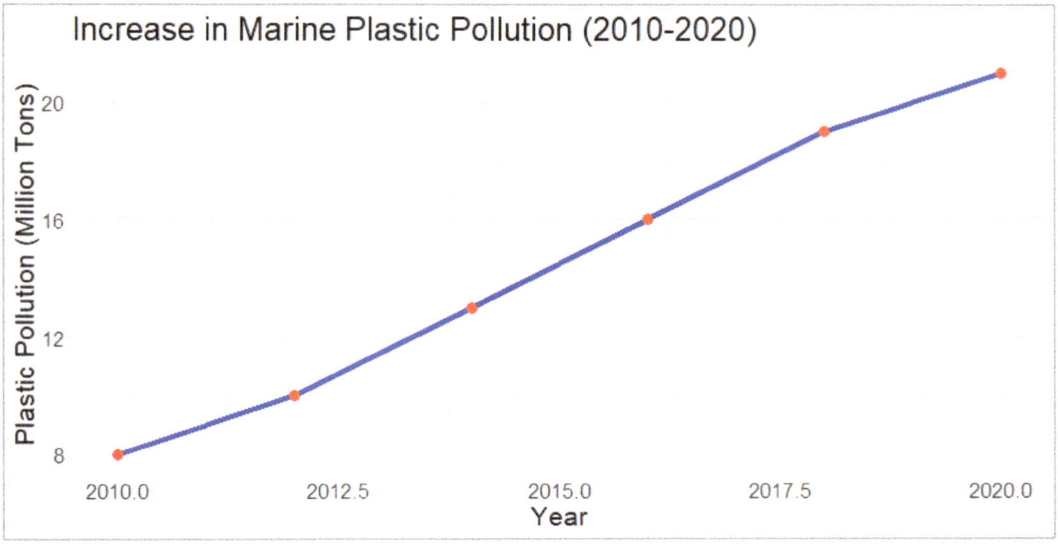

Figure 5. Marine Plastic Pollution over time.

4.3.2. Biodiversity Protection

The GEF is the primary multilateral fund dedicated to addressing this issue and has been a significant global source of funding for biodiversity conservation. Spanning from July 2022 to June 2026, the GEF's eighth programming term, known as GEF-8, is primarily focused on biodiversity protection. The success of achieving the post-2020 global biodiversity framework, including the implementation of the new framework at the UN Biodiversity Conference (COP-15) in Kunming, China, relies on this assistance. Rivers and streams play a crucial role in the transit of plastics from interior areas to the coast. According to research, approximately 34,000 out of the 100,888 river and stream outlets included in the study release macroplastic debris into the ocean. Consequently, an estimated 0.9 to 2.8 million tonnes of plastic waste entered the marine ecosystem in 2015. Eighty percent of the plastic debris released into the ocean by rivers worldwide comes from around 1700 rivers. Southeast Asia and West Africa's urban rivers, even the smaller ones, are the primary sources of plastic pollution [95]. Therefore, controlling and managing watercourses is essential to preventing marine plastic waste from LBS. Although environmental considerations had a significant role in the creation of this body of legislation, its regulatory effects on the reduction of plastic pollution are still very negligible.

Although UNCLOS provides for detailed legal framework for the protection of marine environments, its guidance on how to address land-based sources of plastic pollution is vague and lacking in deterrence [81]. Because there are no clear-cut enforcement policies and definite targets, UNCLOS has a hard time making inroads against MPP. What are the promises of influence and policy across this research? Therefore, it is essential to clarify and evaluate which regulatory policies are most suitable, based on the findings presented in this study. It is recommended that this information be discussed as a separate sub-section before the conclusion.

4.3.3. Collaborative Action

In order to conserve biodiversity, the GEF supports feasible measures that promote the ethical utilisation of resources, enhance public awareness of their significance, and foster collaboration among important organisations and individuals to drive action. This aligns with the objective of the GPA to encourage cooperation between nations and stakeholders in combating marine pollution and preserving biodiversity.

4.4. Policy and Practical Implications

This study highlights several critical policy implications for international and national marine governance. It promotes the value of legally binding agreements that are specific to combat marine plastic pollution (MPP). Critical recommendations include enhancing the surveillance of already established maritime conventions, including UNCLOS and MARPOL, ensuring that such standards have amended enforceable provisions that target plastic emission reduction from land- and sea-based activities [82]. Adopting a comprehensive and legally enforceable Global Plastics Treaty based on common global benchmarking should be high on the agenda for international plastic waste management issues. The treaty should provide proactive measures to avoid pollution and effective ways of removal while creating measurable tasks for every providing nation. Thus, the report advocates for improved integrated freshwater–marine pollution governance, while understanding the pivotal role freshwater streams play in facilitating plastics into the oceans. Further, the strategy needs to promote the use of global Best Available Techniques (BATs) and Best Environmental Practices (BEPs) so as to significantly reduce plastic waste at its source [58]. Developing countries need to build their capacities to enable meaningful cooperation and successful implementation in any part of the world. This study will also align with the goals that are established in SDG-14, which aims at protecting and wisely managing our seas. Progression of marine sustainability and pollution management on a worldwide scale can be positively impacted, in large measure, through the reform of economics and by promoting a circular economy.

5. UNCLOS: International Legal Framework for MPP

This section will examine the relevant provisions of UNCLOS, which serve as the primary international legal framework for protecting the marine environment from plastic pollution originating from LBS. UNCLOS is an international legal framework that serves as a comprehensive organisation for the preservation and sustainable use of the oceans and their resources [28,69,96,97]. It also establishes various guidelines for the preservation and protection of the marine realm. The convention was adopted on 10 December 1982, during the third UNCLOS (1973–1982) and came into effect on 16 November 1994 There are 170 parties in it right now, with the EU among them [8]. Although the agreement "establishes the legal framework within which all activities in the oceans and seas must be conducted", it also encompasses LBSs that may have an impact on the marine environment within its scope. Due to its extensive coverage and comprehensive nature, UNCLOS is often referred to as the "constitution that governs the oceans" [98,99].

Figure 6 displays a histogram illustrating the distribution of microplastic particle counts in marine samples. It appears from the data that the most common particles counted in multiple samples were between 35 and 50, meaning there is considerable to high contamination of microplastics in many areas. UNCLOS adoption has played a crucial role in the development and codification of the law of the sea. Numerous substantive provisions of the agreement are acknowledged as reflecting customary international law, either as a result of their adoption or during the negotiations, or because they are a formalisation of prior customary rules [100,101]. The convention, referred to as the zonal management strategy, aims to harmonise and balance the interests of individual nations across its various sections. In order to protect the common interests of the global community, it also provides a framework for international cooperation in maritime affairs (integrated management approach). This dual strategy is yet another unique feature of UNCLOS [102,103].

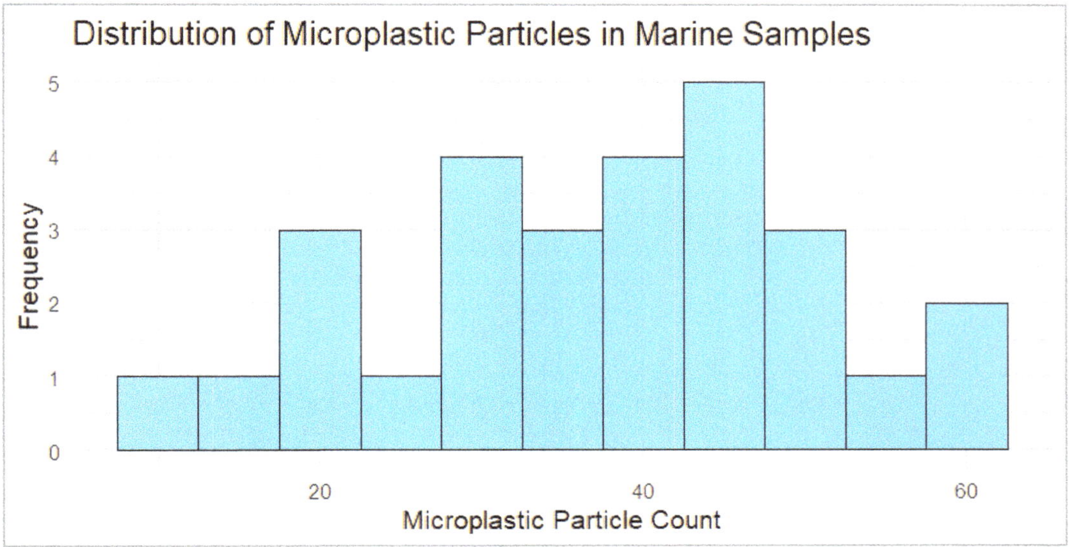

Figure 6. Histogram of microplastic ingestion counts.

The convention has several rules that address environmental protection. The subject is primarily covered in UNCLOS Part XII, which offers a unified framework for the preservation of maritime environments [58]. UNCLOS is the first international agreement in the world that explicitly declares the general obligation of states to protect and preserve the marine environment by addressing all forms of pollution. UNCLOS introduced these wide-ranging duties, leading to a substantial shift in promoting marine sustainability from a state of considerable freedom to pollute the oceans to a global framework of stringent pollution control. Under UNCLOS, a healthy marine environment is recognised as a matter of shared concern that transcends the interests of individual nations [101–104].

5.1. Normative Framework for MPP Mitigation: Analysing UNCLOS Part XII

UNCLOS Part XII is divided into 11 sections and comprises 46 articles. The initial section includes several legal principles that govern the Part XII system. The wording and spirit of Principle 7 of the Stockholm Declaration 1972 [105] and other contemporaneous publications were followed in the drafting of the principles. Article 192, which constitutes the opening sentence of Part XII, establishes the general obligation of governments to protect and preserve the marine environment. The global legal framework for mitigating MPP revolves around this obligation. The following subsection focuses on the specific requirements outlined in UNCLOS Part XII (3) and elucidates its normative content in Part XII (2).

5.2. Institutions Associated with UNCLOS

While UNCLOS must be interpreted as a whole, it is composed of various sections, some of which are considered constitutive in nature, while others amend and codify existing laws [102,106–108]. These elements subsequently establish new institutions, enact new laws, and introduce new concepts. In this regard, UNCLOS has adopted numerous resolutions that highlight the importance of operational protocols and early investments in pioneering research on polymetallic nodules [102,106–108]. Furthermore, UNCLOS Part XII addresses the protection and preservation of the marine environment, including the need for effective measures to combat marine plastic pollution. This issue is increasingly critical as plastic pollution threatens marine biodiversity, disrupts ecosystems, and impacts human health, underscoring the urgency for robust legal frameworks and international cooperation [109].

5.2.1. Division for Ocean Affairs and the Law of the Sea (DOALOS)

The Division for Ocean Affairs and the Law of the Sea (DOALOS), operating under the Office of Legal Affairs, has consistently received recognition for its role in promoting the wider acceptance and consistent implementation of UNCLOS. It also plays a critical role in addressing MPP through its support of the convention. Following the convention's adoption, DOALOS was entrusted with the responsibilities assigned to the Secretary–General and the tasks related to its entry into force during the 52nd UNGA in 1998. Specifically, the division monitors advancements in all relevant fields to provide an annual report to the UNGA on matters concerning ocean affairs and the legal status of the sea.

Moreover, the division formulates recommendations to the UNGA and other intergovernmental forums to enhance comprehension of the convention. This ensures that the division is capable of addressing requests for advice and assistance from states in implementing the convention. Additionally, the division acts as the secretariat of UNCLOS, offering information, advice, and support to states to foster a better understanding of the convention, related agreements, their broader acceptance, uniform and consistent application, and effective implementation. This includes addressing pollution from land-based sources, such as plastics, which significantly impact marine ecosystems. In addition, by providing guidance, facilitating international cooperation, and promoting the effective enforcement of these legal frameworks, DOALOS helps member states meet their obligations to control and reduce MPP [110].

Additionally, it creates suggestions for the UNGA and other intergovernmental forums to enhance public knowledge of the convention. This guarantees that the division can react to requests from states for guidance and support in putting the convention into practice [111]. Furthermore, the division functions as the UNCLOS secretariat, offering states guidance, support, and information to enhance public knowledge of the convention and its associated agreements, as well as to ensure their widespread ratification, uniform and consistent application, and efficient execution.

Furthermore, DOALOS contributes to mitigating marine plastic pollution through various activities. It assists countries in developing and enforcing national regulations aligned with UNCLOS provisions related to marine pollution. This includes providing technical support and capacity-building to improve national and regional approaches to plastic waste management. DOALOS also supports international initiatives, such as the Global Partnership on Marine Litter (GPML), which focuses on collaborative efforts to tackle marine litter, including plastics. By facilitating these global partnerships and promoting best practices, DOALOS enhances the effectiveness of international and national strategies to address plastic pollution in marine environments [109,112].

5.2.2. International Tribunal for the Law of the Sea (ITLOS)

Established in 1982 by UNCLOS, the International Tribunal for the Law of the Sea (ITLOS) is an unbiased court that adjudicates cases related to the interpretation and application of the convention. The tribunal consists of 21 independent members who are selected based on their high standing in the field of maritime law, ensuring their impartiality, integrity, and expertise [113]. UNCLOS grants eligibility to the state parties to utilise the tribunal, and it may also be accessed by other organisations or bodies under specific circumstances. The tribunal possesses jurisdiction over any matter brought before it in accordance with the convention. Furthermore, the tribunal's jurisdiction extends to subjects explicitly mentioned in any other agreement that confers authority upon the tribunal. In certain situations, the tribunal may also offer recommendations under international agreements regarding the objectives of the convention.

Furthermore, ITLOS's jurisdiction under UNCLOS extends to disputes concerning the protection and preservation of the marine environment, a category that encompasses marine plastic pollution. Part XII of UNCLOS, which addresses marine environmental protection, mandates states take measures to prevent, reduce, and control pollution from various sources, including plastics [114]. Although ITLOS has not yet specifically adjudicated a case solely focused on plastic pollution, its role in interpreting Part XII provides a legal framework that could address such issues.

The tribunal's potential involvement in marine plastic pollution disputes could involve determining whether states have fulfilled their obligations under UNCLOS to manage plastic waste effectively and to cooperate in the prevention and control of marine pollution [115]. ITLOS's decisions and advisory opinions on these matters could significantly influence state practices and international standards related to marine plastic pollution. For instance, the tribunal's approach to general pollution cases and its interpretation of UNCLOS principles could set precedents for addressing the specific challenges posed by marine plastic pollution, including the legal responsibilities of states to mitigate and manage plastic waste [116]. Such precedents could enhance global efforts to combat plastic pollution and strengthen the implementation of international environmental agreements.

6. IMO's MARPOL Framework: Addressing Marine Pollution from Sea-Based Sources

The IMO is the principal entity responsible for the continuous evolution of laws and policies related to ship pollution. This specialised UN agency plays a pivotal role in overseeing several key treaties that address civil responsibility, marine pollution control from ships, and dumping at sea. Among these, the MARPOL [117] and the London Convention on the Prevention of Marine Pollution by Dumping of Wastes and Other Matter [118], along with its 1996 Protocol [119], are particularly significant in combating plastic pollution from maritime sources [117,120]. MARPOL, initially adopted in 1973 and modified by the Protocol of 1978, constitutes the primary international framework for preventing pollution from ships. Its Annex V specifically targets the prevention of pollution by garbage from ships, including plastics, which are a major component of marine litter [121]. The London Convention, established in 1972, and its 1996 protocol provide comprehensive guidelines for the control of marine pollution by prohibiting the deliberate disposal of wastes at sea. These instruments collectively form the backbone of international efforts to mitigate marine pollution, particularly from plastics.

The implications of UNCLOS for the IMO and the instruments agreed upon under its auspices have been extensively analysed by the IMO Secretariat. According to a 1987 evaluation by the IMO Secretariat, UNCLOS "depends on them for the effective implementation of its general principles", but rather assumes the existence of specific regulations and anticipates future ratifications by the IMO [122]. This indicates that UNCLOS provides a broad legal framework, while the detailed regulatory measures are left to the IMO's conventions and protocols. In 2018, the IMO's Marine Environment Protection Committee (MEPC) established an action plan specifically aimed at preventing and minimising marine litter from ship-based sources, supporting SDG 14, which focuses on conserving and sustainably using the oceans, seas, and marine resources [123]. This action plan includes several measures, such as enhancing the enforcement of existing regulations, promoting the development of port reception facilities, and encouraging research and innovation to address the issue of marine litter from ships [21,124].

The relationship between UNCLOS and the IMO conventions underscores a complementary framework where UNCLOS sets out broad principles, and the IMO develops specific regulatory measures to implement these principles. For instance, Article 211 of UNCLOS specifically mandates that states shall establish international rules and standards to prevent, reduce, and control pollution of the marine environment from vessels, taking into account IMO standards [123]. This article effectively integrates the IMO's regulatory mechanisms into the overarching legal framework provided by UNCLOS. Moreover, the IMO's continuous efforts in updating MARPOL and other related conventions reflect an adaptive approach to emerging environmental challenges. The adoption of amendments to MARPOL Annex V, which prohibits the discharge of all plastics from ships, exemplifies the dynamic nature of maritime environmental law in response to the growing problem of plastic pollution [54,121].

The 2018 MEPC action plan is a testament to the IMO's proactive stance in addressing marine litter. It aligns with the precautionary principle, a key tenet of international environmental law, which advocates for preventive measures in the face of potential environmental harm [52]. By fostering international cooperation and encouraging member states to adopt stringent measures against marine litter, the IMO's initiatives contribute significantly to global marine environmental protection efforts.

7. Discussion

MPP poses a significant environmental and legal challenge that extends beyond national borders, requiring a multifaceted approach. Although international legal frameworks such as UNCLOS and the MARPOL exist, they are currently insufficient in addressing the scope of the problem. This analysis evaluates these frameworks, highlights their shortcomings, and suggests necessary improvements. Historically, the recognition of marine pollution from LBSs has evolved significantly. The UN Environment's establishment of the Regional Seas Programme and the GPA were pivotal in addressing this issue. The GPA remains a crucial platform, offering guidance for action at all governance levels. Awareness of MPP expanded rapidly in subsequent years, with the international community emphasising the issue at forums such as the UNCSD in 2012, the UNEA, the UNGA, and Ocean Conferences. States established procedures under UNEA to explore potential solutions, recognising the need for action and setting objectives and targets. Numerous nations and stakeholders now advocate for an international accord to address this global crisis.

Biodiversity conventions and chemicals and waste conventions also address MPP from unique perspectives. The urgency of addressing MPP is underscored by its pervasive impact on marine ecosystems, public health, and economic stability. To effectively mitigate this crisis, the international community must adopt a more integrated and proactive approach, emphasising precise, enforceable regulations and innovative solutions. UNCLOS provides a comprehensive legal framework for marine environmental protection, outlining states' responsibilities to prevent and control pollution. However, its provisions are broadly defined and lack the specificity required to address the unique challenges posed by plastic pollution. Due to the scale and impact of LBS on marine environments, this oversight is significant. Moreover, UNCLOS relies heavily on state implementation, resulting in varied enforcement and effectiveness. States often prioritise economic interests, leading to insufficient measures against MPP. Therefore, UNCLOS should include clear provisions for definitions, responsibilities, and enforcement to address plastic pollution effectively.

MARPOL, particularly Annex V, focuses on ship-generated waste and explicitly prohibits the disposal of plastics at sea. However, enforcement of these provisions is inconsistent across jurisdictions. Many states lack the infrastructure or political will to enforce MARPOL's regulations effectively, leading to continued illegal dumping and poor waste management practices. Additionally, MARPOL's narrow focus on maritime sources neglects the significant contributions of land-based activities to MPP. Strengthening MARPOL Annex V requires stringent compliance measures, improved monitoring, and expanded coverage of land-based sources. Furthermore, the ecological impacts of MPP extend beyond visible pollution, with microplastics presenting an insidious threat to marine life and ecosystems. Current international legal instruments do not adequately address the pervasive nature of microplastics, which bioaccumulate and biomagnify through the food web, ultimately impacting human health. This deficiency calls for a comprehensive legal approach, including stringent regulations on the production, use, and disposal of plastics, and measures to mitigate microplastic pollution.

Similarly, MPP poses significant health risks due to the presence of microplastics in marine organisms consumed by humans. Studies have shown widespread contamination of seafood with microplastics, carrying toxic chemicals and potentially harmful organisms. The ingestion of microplastics through seafood has been linked to various health disorders, including endocrine disruption, carcinogenic effects, and other chronic health issues. Current international frameworks do not adequately address these human health risks, underscoring the need for more stringent and comprehensive regulations on plastic production and disposal to safeguard public health. Addressing health risks requires research on microplastics, integrating findings into policies, and educating the public on responsible plastic use. Additionally, MPP disrupts socio-cultural practices and the cultural significance of marine environments, negatively affecting mental health and diminishing the aesthetic and therapeutic value of coastal areas. These socio-economic impacts highlight the urgent need for robust regulatory measures and effective waste management systems.

In addition, the fragmented nature of international governance exacerbates the challenge of addressing MPP. The lack of coordination among various international agreements and organizations leads to overlapping jurisdictions and regulatory gaps. A cohesive global strategy is essential, integrating the principles of the circular economy and emphasising the 3Rs (reduce, reuse, recycle). This strategy should involve binding international agreements with clear, enforceable obligations and a unified approach harmonizing national regulations with international standards. However, it is crucial that effective management of MPP requires coordinated efforts among nations, international bodies, and stakeholders. Enhanced international collaboration is crucial, fostering partnerships for scientific research, sharing best practices, and developing innovative solutions to reduce plastic waste. Legal frameworks must facilitate such collaboration, creating platforms for dialogue, cooperation, and mutual assistance. Monitoring systems must track pollution accurately and ensure transparency. Furthermore, a novel integrated and comprehensive legal instrument is needed to reduce plastic production, enhance waste management, and promote a circular economy, with mechanisms for regular review. Policies should adopt the 3Rs (reduce, reuse, recycle), incentivise sustainable production, and invest in research for biodegradable solutions. National laws must align with international standards, with better enforcement through infrastructure, funding, and training.

Protection of marine plastics from pollution (MPP) is regarded as one of the most important global measures to be implemented to resolve the increasing plastic pollution issue. In answer to the urgent need to cooperate, the United Nations Environment Programme (UNEP) has reacted by supporting the treaty, with emphasis on establishing law-enforceable rules for the reduction and management of plastic waste abroad. Now, with the continued devastation wrought to marine ecosystems by MPP, the treaty offers an essential opportunity to establish strong international rules for reducing plastic waste between states. Through the creation of standard global management, recycling, and the reduction of plastic waste, the treaty is the secret to harmonizing policies within arrangements like UNCLOS and MARPOL. Such strategies would allow for countries to follow standard rules in the sense that regulatory standards will be effectively followed across borders. Beyond dealing with land-based plastics, the treaty aims to curb lesser-known plastic-causing sources of debris in the ocean, arising from ships and other maritime activities. Moreover, the Global Plastics Treaty may promote the large-scale adoption of Best Available Techniques (BATs) and Best Environmental Practices (BEPs) by industries, encouraging the circular economy transition. Under these standards, the treaty would promote the innovation of recycling and reduce reliance on disposable plastics and sustainably manufactured industrial products.

The Global Plastics Treaty, which learns from the successful action taken by countries such as Kenya and Norway on plastic waste, can heighten this momentum by establishing a global template for concerted action. This treaty is a critical step towards SDG 14—the conservation of marine ecosystems in a sustainable way.

8. Conclusions

This study critically examined the effectiveness of international legal frameworks, particularly UNCLOS and MARPOL, in addressing the growing challenge of marine plastic pollution (MPP). The primary objective was a determination of the sufficiency of the existing laws with regard to the collection of plastic waste from both shoreline and vessel waste dumps. An examination of legal doctrines revealed significant problems with enforcement, narrow jurisdictional reach, and the absence of meaningful responsibilities to control marine plastic pollution. The study findings revealed that although the UNCLOS and the MARPOL provide basic structures, they lack the required mechanisms to address the current plastic waste challenges decisively. This study highlighted the socio-economic and cultural impact of MPP, especially its impact on fisheries, tourism, and people living along the coast. Moreover, such efforts as UNEA resolutions and the proposed Global Plastics Treaty suggest great progress is being made toward more effective governance. It is obvious that current international laws should be changed so as to be more effective in addressing the magnitude and weight of the plastic crisis. This involves taking enforceable legal instructions, promoting the use of circular economy practices, and promoting international cooperation among others. Improving institutional coordination and involving all involved stakeholders at all levels will be imperative in making effective and sustainable countermeasures against marine plastic pollution.

Author Contributions: Writing—original draft, M.S.; Supervision, Y.-C.C. All authors have read and agreed to the published version of the manuscript.

Funding: This research was funded by Shenzhen Philosophy and Social Science Planning 2024 Annual Project, China. "Study on Coordination of Rule of Law in Guangdong-Hong Kong-Macao Greater Bay Area" (Approval number: SZ2024B024).

Data Availability Statement: The original contributions presented in this study are included in the article. Further inquiries can be directed to the corresponding author.

Conflicts of Interest: The authors declare no conflicts of interest..

References

1. World Economic Forum, Air Pollution Is Killing Millions—It's Time to Hold Ourselves Accountable for the Harm It Causes. 2022. Available online: https://www.weforum.org/agenda/2022/05/air-pollution-killing-millions-how-can-we-tackle/ (accessed on 4 July 2024).
2. Chang, Y.-C.; Khan, M. The Maritime Labour Convention 2006 in human rights context: An appraisal. *Mar. Policy* **2023**, *154*, 105688. [CrossRef]

3. Sanchez-Triana, E. Fighting Air Pollution: A Deadly Killer and a Core Development Challenge. World Bank Blogs. 2023. Available online: https://blogs.worldbank.org/en/health/fighting-air-pollution-deadly-killer-and-core-development-challenge (accessed on 4 July 2024).
4. World Economic Forum. Why Ocean Pollution Is a Clear Danger to Human Health. 2021. Available online: https://www.weforum.org/agenda/2021/02/ocean-pollution-human-health-coasts-microplastics/ (accessed on 4 July 2024).
5. National Geographic. Pollution. (n.d.). Available online: https://education.nationalgeographic.org/resource/pollution (accessed on 4 July 2024).
6. Dewailly, E.; Furgal, C.; Knap, A.; Galvin, J.; Baden, D.; Bowen, B.; Depledge, M.; Duguay, L.; Fleming, L.; Ford, T.; et al. Unluata, Indicators of Ocean and Human Health. *Can. J. Public Health* **2002**, *93*, S34–S38. [CrossRef]
7. Landrigan, P.J.; Stegeman, J.J.; Fleming, L.E.; Allemand, D.; Anderson, D.M.; Backer, L.C.; Brucker-Davis, F.; Chevalier, N.; Corra, L.; Czerucka, D.; et al. Rampal, Human Health and Ocean Pollution. *Ann. Glob. Health* **2020**, *86*, 151. [CrossRef]
8. United Nations Convention on the Law of the Sea. 1982. Available online: http://treaties.un.org/doc/Publication/UNTS/Volume%201833/volume-1833-A-31363-English.pdf (accessed on 2 June 2024).
9. Wong, M.; Lanzoni, N. Land-based sources of marine pollution and dumping at sea. In *Research Handbook on Ocean Governance Law*; Edward Elgar Publishing: Northampton, MA, USA, 2023; pp. 109–127. Available online: https://www.elgaronline.com/edcollchap/book/9781839107696/book-part-9781839107696-20.xml (accessed on 4 July 2024).
10. Wysocki, I.T.-V.; Le Billon, P. Plastics at sea: Treaty design for a global solution to marine plastic pollution. *Environ. Sci. Policy* **2019**, *100*, 94–104. [CrossRef]
11. Lestari, P.; Trihadiningrum, Y. The impact of improper solid waste management to plastic pollution in Indonesian coast and marine environment. *Mar. Pollut. Bull.* **2019**, *149*, 110505. [CrossRef]
12. Fava, M.F. Plastic Pollution in the Ocean: Data, Facts, Consequences, Ocean Lit. Portal. 2022. Available online: https://oceanliteracy.unesco.org/plastic-pollution-ocean/ (accessed on 4 July 2024).
13. IUCN, The Plastic Pollution Crisis, IUCN. 2022. Available online: https://www.iucn.org/story/202207/plastic-pollution-crisis (accessed on 4 July 2024).
14. Kumar, V. Ocean Plastic Pollution an Overview: Data and Statistics, Infin. Lab. 2023. Available online: https://infinitalab.com/plastics/ocean-plastic-pollution-an-overview-data-and-statistics/ (accessed on 4 July 2024).
15. Liu, Z.; Adams, M.; Walker, T.R. Are exports of recyclables from developed to developing countries waste pollution transfer or part of the global circular economy? *Resour. Conserv. Recycl.* **2018**, *136*, 22–23. [CrossRef]
16. Barnes, D.K.A.; Galgani, F.; Thompson, R.C.; Barlaz, M. Accumulation and fragmentation of plastic debris in global environments, Philos. *Trans. R. Soc. B Biol. Sci.* **2009**, *364*, 1985–1998. [CrossRef]
17. Nash, K.L.; van Putten, I.; Alexander, K.A.; Bettiol, S.; Cvitanovic, C.; Farmery, A.K.; Flies, E.J.; Ison, S.; Kelly, R.; Mackay, M.; et al. Oceans and society: Feedbacks between ocean and human health. *Rev. Fish Biol. Fish.* **2022**, *32*, 161–187. [CrossRef]
18. Frontiers. High Levels of Microplastics Found in Northwest Atlantic Fish, PHYS.ORG. 2018. Available online: https://phys.org/news/2018-02-high-microplastics-northwest-atlantic-fish.html (accessed on 4 July 2024).
19. RNZ. Three Quarters of Fish in Southern NZ Contain Microplastics—Study, Radio N. Z. 2022. Available online: https://www.rnz.co.nz/news/national/475298/three-quarters-of-fish-in-southern-nz-contain-microplastics-study (accessed on 4 July 2024).
20. Mukherjee, P.K.; Mejia, M.Q.; Xu, J. (Eds.) *Maritime Law in Motion*; Springer International Publishing: Cham, Switzerland, 2020. [CrossRef]
21. International Maritime Organisation. Action Plan to Address Marine Plastic Litter from Ships. 2018. Available online: https://wwwcdn.imo.org/localresources/en/MediaCentre/HotTopics/Documents/IMO%20marine%20litter%20action%20plan%20MEPC%2073-19-Add-1.pdf (accessed on 1 January 2024).
22. Xanthos, D.; Walker, T.R. International policies to reduce plastic marine pollution from single-use plastics (plastic bags and microbeads): A review. *Mar. Pollut. Bull.* **2017**, *118*, 17–26. [CrossRef]
23. Thompson, T. Plastic Pollution: Three Problems that a Global Treaty Could Solve, Nature. 2022. Available online: https://www.nature.com/articles/d41586-022-03835-w (accessed on 4 July 2024).
24. Galgani, F.; Hanke, G.; Maes, T. Global Distribution, Composition and Abundance of Marine Litter. In *Marine Anthropogenic Litter*; Bergmann, M., Gutow, L., Klages, M., Eds.; Springer International Publishing: London, UK, 2015; pp. 29–56. [CrossRef]
25. van Sebille, E.; Wilcox, C.; Lebreton, L.; Maximenko, N.; Hardesty, B.D.; van Franeker, J.A.; Eriksen, M.; Siegel, D.; Galgani, F.; Law, K.L. A global inventory of small floating plastic debris. *Environ. Res. Lett.* **2015**, *10*, 124006. [CrossRef]
26. Jambeck, J.R.; Geyer, R.; Wilcox, C.; Siegler, T.R.; Perryman, M.; Andrady, A.; Narayan, R.; Law, K.L. Plastic waste inputs from land into the ocean. *Science* **2015**, *347*, 768–771. [CrossRef] [PubMed]
27. United Nations General Assembly. Fragmentation of International Law: Difficulties Arising from the Diversification and Expansion of International Law. Report of the Study Group of the International Law Commission, Finalized by Martti Koskenniemi. 2006. Available online: https://legal.un.org/ilc/documenta (accessed on 22 June 2024).

28. United Nations. The Future We Want. 2012. Available online: https://www.un.org/en/development/desa/population/migration/generalassembly/docs/globalcompact/A_RES_66_288.pdf (accessed on 1 June 2024).
29. United Nations General Assembly. Transforming Our World: The 2030 Agenda for Sustainable Development. 2015. Available online: https://www.un.org/en/development/desa/population/migration/generalassembly/docs/globalcompact/A_RES_70_1_E.pdf (accessed on 18 May 2024).
30. Mendenhall, E. Building a regime complex for marine plastic pollution. *Camb. Prisms Plast.* **2023**, *1*, e12. [CrossRef]
31. Khan, M.; Butt, M.J.; Chang, Y.-C. Maritime Law in Motion: Book Review. *China Oceans Law Rev.* **2023**, *19*, 142–144.
32. Young, M.A. Fragmentation and International Environmental Law. 2019. Available online: https://papers.ssrn.com/abstract=3441535 (accessed on 4 July 2024).
33. Shahzabeen, A.; Ghosh, A.; Pandey, B.; Shekhar, S. Circular Economy and Sustainable Production and Consumption. In *Green Circular Economy: A New Paradigm for Sustainable Development*, 1st ed.; Singh, P., Yadav, A., Chowdhury, I., Singh, R.P., Eds.; Springer International Publishing: Cham, Switzerland, 2023; pp. 43–65. [CrossRef]
34. UN Environment. *Global Partnership on Plastic Pollution and Marine Litter*; U.N. Environment Programme: Athens, Greece, 2017. Available online: http://www.unep.org/explore-topics/oceans-seas/global-partnership-plastic-pollution-and-marine-litter (accessed on 4 July 2024).
35. Rukikaire, K. Comprehensive Assessment on Marine Litter and Plastic Pollution Confirms Need for Urgent Global Action, UN Environ. 2021. Available online: http://www.unep.org/news-and-stories/press-release/comprehensive-assessment-marine-litter-and-plastic-pollution (accessed on 4 July 2024).
36. Soares, J.; Miguel, I.; Venâncio, C.; Lopes, I.; Oliveira, M. Perspectives on Micro(Nano)Plastics in the Marine Environment: Biological and Societal Considerations. *Water* **2020**, *12*, 3208. [CrossRef]
37. Azfaralariff, A.; Lazim, A.M.; Amran, N.H.; Mukhtar, N.H.; Bakri, N.D.; Azrihan, N.N.; Mohamad, M. Mini review of microplastic pollutions and its impact on the environment and human health. *Waste Manag. Res.* **2023**, *41*, 1219–1226. [CrossRef]
38. Carrington, D. Microplastics May Be Linked to Inflammatory Bowel Disease, Study Finds, The Guardian. 2021. Available online: https://www.theguardian.com/society/2021/dec/22/microplastics-may-be-linked-to-inflammatory-bowel-disease-study-finds (accessed on 4 July 2024).
39. Kühn, S.; Rebolledo, E.L.B.; van Franeker, J.A. Deleterious Effects of Litter on Marine Life. In *Marine Anthropogenic Litter*; Bergmann, M., Gutow, L., Klages, M., Eds.; Springer International Publishing: Cham, Switzerland, 2015; pp. 75–116. [CrossRef]
40. Francisco, B.; Gutiérrez-Bouzán, C.; Álvarez-Sánchez, A.; Vilaseca, M. Textile microfibers reaching aquatic environments: A new estimation approach. *Environ. Pollut.* **2020**, *265*, 114889. [CrossRef]
41. McIntyre, O. International Water Law's Role in Addressing the Problem of Marine Plastic Pollution: A Vital Piece in a Complex Puzzle! Chin. *J. Environ. Law* **2022**, *6*, 218–252. [CrossRef]
42. Gardiner, B. The Plastics Pipeline: A Surge of New Production Is on the Way, Yale Environ. 2019. Available online: https://e360.yale.edu/features/the-plastics-pipeline-a-surge-of-new-production-is-on-the-way (accessed on 4 July 2024).
43. Calero, M.; Godoy, V.; Quesada, L.; Martín-Lara, M.Á. Green strategies for microplastics reduction. *Curr. Opin. Green Sustain. Chem.* **2021**, *28*, 100442. [CrossRef]
44. Yuan, J.; Ma, J.; Sun, Y.; Zhou, T.; Zhao, Y.; Yu, Y. Microbial degradation and other environmental aspects of microplastics/plastics. *Sci. Total Environ.* **2020**, *715*, 136968. [CrossRef]
45. Díaz-Mendoza, C.; Mouthon-Bello, J.; Pérez-Herrera, N.L.; Escobar-Díaz, S.M. Plastics and microplastics, effects on marine coastal areas: A review. *Environ. Sci. Pollut. Res.* **2020**, *27*, 39913–39922. [CrossRef]
46. Iroegbu, A.O.C.; Ray, S.S.; Mbarane, V.; Bordado, J.C.; Sardinha, J.P. Plastic Pollution: A Perspective on Matters Arising: Challenges and Opportunities. *ACS Omega* **2021**, *6*, 19343–19355. [CrossRef] [PubMed]
47. Moshood, T.D.; Nawanir, G.; Mahmud, F.; Mohamad, F.; Ahmad, M.H.; Airin, A. Biodegradable plastic applications towards sustainability: A recent innovations in the green product. *Clean. Eng. Technol.* **2022**, *6*, 100404. [CrossRef]
48. Sonu; Rani, G.M.; Pathania, D.; Abhimanyu; Umapathi, R.; Rustagi, S.; Huh, Y.S.; Gupta, V.K.; Kaushik, A.; Chaudhary, V. Agro-waste to sustainable energy: A green strategy of converting agricultural waste to nano-enabled energy applications. *Sci. Total Environ.* **2023**, *875*, 162667. [CrossRef]
49. UN Environment. Global Environment Outlook 6, U.N. Environment Programme. 2019. Available online: http://www.unep.org/resources/global-environment-outlook-6 (accessed on 4 July 2024).
50. Sheriff, S.S.; Yusuf, A.A.; Akiyode, O.O.; Hallie, E.F.; Odoma, S.; Yambasu, R.A.; Thompson-Williams, K.; Asumana, C.; Gono, S.Z.; Kamara, M. A comprehensive review on exposure to toxins and health risks from plastic waste: Challenges, mitigation measures, and policy interventions. *Waste Manag. Bull.* **2025**, *3*, 100204. [CrossRef]
51. Benson, N.U.; Agboola, O.D.; Fred-Ahmadu, O.H.; De-la-Torre, G.E.; Oluwalana, A.; Williams, A.B. Micro(nano)plastics Prevalence, Food Web Interactions, and Toxicity Assessment in Aquatic Organisms: A Review. *Front. Mar. Sci.* **2022**, *9*, 851281. [CrossRef]

52. Convention on Biological Diversity. Precautionary Approach. 2006. Available online: https://www.cbd.int/marine/precautionary.shtml (accessed on 8 July 2024).
53. Wootton, N.; Reis-Santos, P.; Gillanders, B.M. Microplastic in fish—A global synthesis. *Rev. Fish Biol. Fish.* **2021**, *31*, 753–771. [CrossRef]
54. International Maritime Organisation. Prevention of Pollution by Garbage from Ships. 2018. Available online: https://www.imo.org/en/OurWork/Environment/Pages/Garbage-Default.aspx (accessed on 8 July 2024).
55. Trash Free Maryland. Can Litter Affect Mental Health? 2019. Available online: https://www.trashfreemaryland.org/blog/2019/05/23/can-litter-affect-mental-health-what-were-starting-to-learn-about-a-connection-between-clean-communities-and-wellness (accessed on 4 July 2024).
56. Andersen, I. *Marine Litter and the Challenge of Sustainable Consumption and Production*; United Nations Environment Programme: Athens, Greece, 2020. Available online: http://www.unep.org/news-and-stories/speech/marine-litter-and-challenge-sustainable-consumption-and-production (accessed on 4 July 2024).
57. European Parliament. Circular Economy: Definition, Importance and Benefits. 2023. Available online: https://www.europarl.europa.eu/topics/en/article/20151201STO05603/circular-economy-definition-importance-and-benefits (accessed on 5 July 2024).
58. Nordquist, M.; Rosenne, S.; Yankov, A. (Eds.) *United Nations Convention on the Law of the Sea 1982: A Commentary*; Brill: Leiden, The Netherlands, 2011. Available online: https://brill.com/edcollbook/title/11294 (accessed on 7 July 2024).
59. Frisso, G.M.; Kirk, E.A. Changing role of law-making in responding to planetary boundaries? In *Research Handbook on Law, Governance and Planetary Boundaries*; Edward Elgar Publishing: Cheltenham, UK, 2021; pp. 147–166. Available online: https://www.elgaronline.com/edcollchap/edcoll/9781789902730/9781789902730.00016.xml (accessed on 5 July 2024).
60. Convention for the Prevention of Marine Pollution from Land-Based Sources. 1974. Available online: https://treaties.un.org/doc/Publication/UNTS/Volume%201546/volume-1546-I-26842-English.pdf (accessed on 1 January 2024).
61. Maletto, F. The application of the Principle of Prevention to land-based sources of marine pollution: An International Law approach. *Opol. Stud. Adm.-Prawne* **2023**, *21*, 125–148. Available online: https://www.ceeol.com/search/article-detail?id=1165715 (accessed on 5 July 2024). [CrossRef]
62. Sakhuja, D.V.; Francis, A.M. *India and Australia: Strengthening International Cooperation Through the Indo Pacific Oceans Initiative*; Centre for Public Policy Research and Monash University: Clayton, Australia, 2023. Available online: https://books.google.com.hk/books?hl=en&lr=&id=Z6C5EAAAQBAJ&oi=fnd&pg=PA7&dq=Ocean+and+Climate+as.+India+&+Australia:+Strengthening+International+Cooperation+Through+the+Indo+Pacific+Oceans+Initiative&ots=tY-6Wwt6b_&sig=Wd7JmboRteVoap3WvK3VnXQ9tLA&redir_esc=y#v=onepage&q=Ocean%20and%20Climate%20as.%20India%20&%20Australia:%20Strengthening%20International%20Cooperation%20Through%20the%20Indo%20Pacific%20Oceans%20Initiative&f=false (accessed on 5 January 2024).
63. McIntyre, O. Addressing Marine Plastic Pollution as a 'Wicked' Problem of Transnational Environmental Governance. 2020. Available online: https://papers.ssrn.com/abstract=3637482 (accessed on 5 July 2024).
64. Kirk, E.A.; Popattanachai, N. Marine plastics: Fragmentation, effectiveness and legitimacy in international lawmaking. *Rev. Eur. Comp. Int. Environ. Law* **2018**, *27*, 222–233. [CrossRef]
65. Bognanovic, S. *International Law of Water Resources: Contribution of the International Law Association (1954–2000)*; BRILL: The Hague, The Netherlands, 2021.
66. Finska, L.; Howden, J.G. Troubled waters—Where is the bridge? Confronting marine plastic pollution from international watercourses. *Rev. Eur. Comp. Int. Environ. Law* **2018**, *27*, 245–253. [CrossRef]
67. Wang, S. International law-making process of combating plastic pollution: Status Quo, debates and prospects. *Mar. Policy* **2023**, *147*, 105376. [CrossRef]
68. United Nations Environment Programme. *Montreal Guidelines for the Protection of the Marine Environment Against Pollution from Land-Based Sources*; UNEP: Nairobi, Kenya; The University of California: Oakland, CA, USA, 1985.
69. United Nations Conference on Environment and Development. *Agenda 21: Programme of Action for Sustainable Development*; United Nations: Rio de Janeiro, Brazil, 1992. Available online: https://sustainabledevelopment.un.org/outcomedocuments/agenda21 (accessed on 4 July 2024).
70. United Nations Environment Programme. *Global Programme of Action for the Protection of the Marine Environment from Land-Based Activities*; United Nations Environment Programme: Washington, DC, USA, 1995. Available online: https://www.unep.org/resources/toolkits-manuals-and-guides/global-programme-action-protection-marine-environment-land (accessed on 19 June 2024).
71. Gosovic, B. *The Quest for World Environmental Cooperation: The Case of the UN Global Environment Monitoring System*, 1st ed.; Routledge: London, UK, 2019. [CrossRef]

72. United Nations Environment Programme. *Intergovernmental Conference to Adopt a Global Programme of Action for the Protection of the Marine Environment from Land-Based Activities*; United Nations Environment Programme: Nairobi, Kenya, 1995. Available online: https://digitallibrary.un.org/record/198932/ (accessed on 10 March 2024).
73. United Nations Environment Programme. *Montevideo Programme for the Development and Periodic Review of Environmental Law*; United Nations Environment Programme: Montevideo, Uruguay, 1982. Available online: https://wedocs.unep.org/bitstream/handle/20.500.11822/20587/MontevideoProgrammeI.pdf?sequence=1&isAllowed=y (accessed on 12 May 2024).
74. Nollkaemper, A. Marine pollution from land-based sources: Towards a global approach. *Mar. Pollut. Bull.* **1992**, *24*, 8–12. [CrossRef]
75. United Nations General Assembly. *Report of the Preparatory Committee for the United Nations Conference on Environment and Development*; United Nations: Rio de Janeiro, Brazil, 1992. Available online: https://www.un.org/esa/dsd/agenda21/Agenda%2021.pdf (accessed on 4 March 2024).
76. VanderZwaag, D.L.; Wells, P.G.; Karau, J. Environment—The Global Programme of Action for the Protection of the Marine Environment from Land-Based Activities: A Myriad of Sounds, Will the World Listen? *Ocean Yearb. Online* **1998**, *13*, 183–210. [CrossRef]
77. United Nations Environment Programme. *Third Intergovernmental Review Meeting on the Implementation of the Global Programme of Action for the Protection of the Marine Environment from Land-Based Activities*; United Nations: Manila, Brazil, 2012. Available online: https://stg-wedocs.unep.org/handle/20.500.11822/10697 (accessed on 5 July 2024).
78. United Nations Environment Programme. *Montreal Declaration for the Development of Environmental Norms in International Law*; International Environmental Law Research Centre: Montreal, QC, Canada, 2001.
79. United Nations Environment Programme. *Convention for the Protection of the Marine Environment and the Coastal Region of the Mediterranean and Its Protocols*; United Nations Environment Programme: Athens, Greece, 2019. Available online: https://wedocs.unep.org/bitstream/handle/20.500.11822/31970/bcp2019_web_eng.pdf (accessed on 18 June 2023).
80. United Nations Environment Programme. *Protocol for the Protection of the Mediterranean Sea Against Pollution from Land-Based Sources and Activities*; United Nations Environment Programme: Athens: Greece, 1980. Available online: https://resolutions.unep.org/uploads/1980-protection_of_mediterranean_sea_against_pollution.pdf (accessed on 12 April 2024).
81. Protocol for the Protection of South-East Pacific Against Pollution from Land-Based Sources. 1983. Available online: https://www.ecolex.org/details/treaty/protocol-for-the-protection-of-south-east-pacific-against-pollution-from-land-based-sources-tre-000768/ (accessed on 8 June 2024).
82. Tanzi, A.; Arcari, M. *The United Nations Convention on the Law of International Watercourses: A Framework for Sharing*; BRILL: The Hague, The Netherlands, 2021.
83. Recognition the Principle of State Responsibility in Land based Marine Pollution as a Response to State Sovereignty. International Research Conference Articles (KDU IRC). 2023. Available online: http://ir.kdu.ac.lk/handle/345/7317 (accessed on 5 July 2024).
84. Convention for the Protection of the Marine Environment of the North-East Atlantic. 1992. Available online: https://www.ospar.org/convention/text (accessed on 4 July 2024).
85. Guggisberg, S. Finding equitable solutions to the land-based sources of marine plastic pollution: Sovereignty as a double-edged sword. *Mar. Policy* **2024**, *159*, 105960. [CrossRef]
86. United Nations. Convention on the Law of the Non-Navigational Uses of International Watercourse. 1997. Available online: https://legal.un.org/ilc/texts/instruments/english/conventions/8_3_1997.pdf (accessed on 21 August 2024).
87. Hough, P. Rotterdam Convention on the Prior Informed Consent Procedure in Trade. In *Essential Concepts of Global Environmental Governance*, 2nd ed.; Routledge: London, NY, USA, 2020; p. 221.
88. Akinrinade, O.E.; Agunbiade, F.O.; Alani, R.; Ayejuyo, O.O. Implementation of the Stockholm Convention on persistent organic pollutants (POPs) in Africa—Progress, challenges, and recommendations after 20 years. *Environ. Sci. Adv.* **2024**, *3*, 623–634. [CrossRef]
89. International Law Commission. Non-Navigational Uses of International Watercourses, Report of the International Law Commission on the Work. 1994. Available online: https://legal.un.org/ilc/texts/instruments/english/commentaries/8_3_1994.pdf (accessed on 8 March 2024).
90. McIntyre, O. *Environmental Protection of International Watercourses under International Law*; Routledge: London, UK, 2016. [CrossRef]
91. Ramcharan, B.G. The International Law Commission and international environmental law. *Ocean Manag.* **1975**, *2*, 315–322. [CrossRef]
92. International Law Association. *Articles on Marine Pollution of Continental Origin*; International Law Association: New York, NY, USA, 1972. Available online: https://www.internationalwaterlaw.org/documents/intldocs/ILA/ILA-Articles_on_Marine_Pollution_of_Continental_Origin-New_York1972.pdf (accessed on 14 June 2024).
93. Convention on the Law of the Non-Navigational Uses of International Watercourses. 1997. Available online: https://treaties.un.org/doc/Publication/UNTS/Volume%202999/Part/volume-2999-I-52106.pdf (accessed on 2 July 2024).

94. Nollkaemper, A. Legal Protection of the Marine Environment from Pollution of International Watercourses: Recent Developments. *Mar. Pollut. Bull.* **1993**, *26*, 298–301. [CrossRef]
95. Meijer, L.J.J.; van Emmerik, T.; van der Ent, R.; Schmidt, C.; Lebreton, L. More than 1000 rivers account for 80% of global riverine plastic emissions into the ocean. *Sci. Adv.* **2021**, *7*, eaaz5803. [CrossRef]
96. United Nations. UN Doc A/67/79, Report of the UN Secretary-General, Oceans and the Law of the Sea. 2012. Available online: https://docs.un.org/en/A/67/79 (accessed on 22 June 2024).
97. United Nations. Oceans and the Law of the Sea. 2015. Available online: https://www.un.org/en/global-issues/oceans-and-the-law-of-the-sea (accessed on 1 July 2024).
98. Chandra, V. The United Nations Convention on the Law of the Sea (UNCLOS) and Maritime Boundary Disputes in Areas of Hydrocarbon Potential. Doctoral Thesis, Deakin University, Geelong, Australia, 2021.
99. Yu, C. An Analysis of the Approach to Interpreting Pertinent UNCLOS Provisions. In *Marine Scientific Research and the Regulation of Modern Ocean Data Collection Activities Under UNCLOS*; Brill: Leiden, The Netherlands, 2022; pp. 13–39. [CrossRef]
100. Boyle, A.E. Land-based sources of marine pollution: Current legal regime. *Mar. Policy* **1992**, *16*, 20–35. [CrossRef]
101. Boyle, A.E.; Redgwell, C. *Birnie, Boyle, and Redgwell's International Law and the Environment*, 4th ed.; Oxford University Press: Oxford, UK, 2021.
102. Rothwell, D.R.; Stephens, T. The International Law of the Sea. 2011. Available online: https://papers.ssrn.com/abstract=1739387 (accessed on 7 July 2024).
103. Tanaka, Y. The International Law of the Sea. *Chin. J. Int. Law* **2011**, *10*, 173–175. [CrossRef]
104. Boyle, A.E. Marine Pollution under the Law of the Sea Convention. *Am. J. Int. Law* **1985**, *79*, 347–372. [CrossRef]
105. United Nations General Assembly. United Nations Conference on the Human Environment. 1972. Available online: https://treaties.un.org/doc/Publication/UNTS/Volume%20824/volume-824-I-11817-English.pdf (accessed on 8 August 2024).
106. Boyle, A.; Freestone, D. *International Law and Sustainable Development: Past Achievements and Future Challenges*; Oxford University Press: Oxford, UK, 1999. [CrossRef]
107. Churchill, R.R.; Lowe, A.V. *The Law of the Sea*, 3rd ed.; Manchester University Press: Manchester, UK, 1999.
108. Uriz, G. International Law and Sustainable Development. Past Achievements and Future Challenges, edited by Alan Boyle and David Freestone. *Leiden J. Int. Law* **2001**, *14*, 939–946. [CrossRef]
109. Almroth, B.C.; Eggert, H. Marine Plastic Pollution: Sources, Impacts, and Policy Issues. *Rev. Environ. Econ. Policy* **2019**, *13*, 317–326. [CrossRef]
110. Young, M.A. Implementing international law: Capacity-building, coordination and control. *Camb. Int. Law J.* **2023**, *12*, 4–23. [CrossRef]
111. United Nations Oceans and Law of the Sea. The Division for Ocean Affairs and the Law of the Sea, Its Functions and Activities. 2013. Available online: https://www.un.org/Depts/los/doalos_activities/about_doalos.htm (accessed on 8 July 2024).
112. Young, M.A. Strengthening Capacity in Ocean Governance. *Asia-Pac. J. Ocean. Law Policy* **2023**, *8*, 5–24. [CrossRef]
113. Jaenicke, G. The Interpretation of the Law of the Sea Convention in the Jurisprudence of the International Tribunal for the Law of the Sea. In *Liber Amicorum Judge Shigeru Oda*; Brill: The Hague, The Netherlands, 2022; pp. 683–695. [CrossRef]
114. Hao, Y. Breakthrough or Slowdown? The International Legal Development of the. *China Oceans Law Rev.* **2023**, *47*, 100–122. Available online: https://www.airitilibrary.com/Article/Detail/P20201016006-N202404020001-00006 (accessed on 20 August 2024).
115. Klein, N. *Maritime Security and the Law of the Sea*; Oxford University Press: New York, NY, USA, 2011.
116. Rao, P.C.; Gautier, P. *Preliminary Material*; Brill: Leiden, The Netherlands, 2006. Available online: https://brill.com/display/book/edcoll/9789047410201/Bej.9789004152403.i-521_001.xml (accessed on 20 August 2024).
117. International Convention for the Prevention of Pollution from Ships. 1973. Available online: https://treaties.un.org/Pages/showDetails.aspx?objid=0800000280291139 (accessed on 8 July 2024).
118. Convention on the Prevention of Marine Pollution by Dumping of Wastes and Other Matter. 1972. Available online: https://treaties.un.org/Pages/showDetails.aspx?objid=08000002800c34d0 (accessed on 9 March 2024).
119. Protocol to the Convention on the Prevention of Marine Pollution by Dumping of Wastes and Other Matter. 1996. Available online: https://www.imo.org/en/OurWork/Environment/Pages/London-Convention-Protocol.aspx (accessed on 8 July 2024).
120. International Maritime Organisation. Marine Environment. 2011. Available online: https://www.imo.org/en/OurWork/Environment/Pages/Default.aspx (accessed on 8 July 2024).
121. SAFETY4SEA. Revised MARPOL Annex V "Regulations for the Prevention of Pollution by Garbage from Ships". 2012. Available online: https://safety4sea.com/revised-marpol-annex-v-regulations-for-the-prevention-of-pollution-by-garbage-fr/ (accessed on 8 July 2024).
122. Joseph, A.; Dalaklis, D. The international convention for the safety of life at sea: Highlighting interrelations of measures towards effective risk mitigation. *J. Int. Marit. Saf. Environ. Aff. Shipp.* **2021**, *5*, 1–11. [CrossRef]

123. Christodoulou, A.; Fernández, J.E. Maritime Governance and International Maritime Organization Instruments Focused on Sustainability in the Light of United Nations' Sustainable Development Goals. In *Sustainability in the Maritime Domain: Towards Ocean Governance and Beyond*; Carpenter, A., Johansson, T.M., Skinner, J.A., Eds.; Springer International Publishing: Cham, Switzerland, 2021; pp. 415–461. [CrossRef]
124. International Maritime Organisation. Addressing Marine Plastic Litter from Ships—Action Plan Adopted, "IMO Pledges to Further Address the Significant Problem Posed by Plastics to the Marine Environment Through a Number of Measures Identified in an Action Plan". 2018. Available online: https://www.imo.org/en/MediaCentre/PressBriefings/Pages/20-marinelitteractionmecp73.aspx (accessed on 8 July 2024).

water

MDPI

Article

Efficient Naval Surveillance: Addressing Label Noise with Rockafellian Risk Minimization for Water Security

Gabriel Custódio Rangel [1,2], Victor Benicio Ardilha da Silva Alves [1,2], Igor Pinheiro de Araújo Costa [1,3,*], Miguel Ângelo Lellis Moreira [1,3], Arthur Pinheiro de Araújo Costa [4], Marcos dos Santos [3,4] and Eric Charles Eckstrand [2]

1 Operational Research Department, Naval Systems Analysis Center, Rio de Janeiro 20091-000, RJ, Brazil; gabriel.custodio@marinha.mil.br (G.C.R.); benicio@marinha.mil.br (V.B.A.d.S.A.); miguellellis@hotmail.com (M.Â.L.M.)
2 Operational Research Department, Naval Postgraduate School, Monterey, CA 93943, USA; eric.eckstrand@nps.edu
3 Production Department, Fluminense Federal University, Niterói 24210-346, RJ, Brazil; marcosdossantos@ime.eb.br
4 Systems and Computing Department, Military Institute of Engineering, Rio de Janeiro 22290-270, RJ, Brazil; arthurcosta.araujo@ime.eb.br
* Correspondence: costa_igor@id.uff.br

Abstract: This study proposes developing a resilient machine learning algorithm based on neural networks to classify naval images used in surveillance, search, and detection operations in vast coastal and marine environments. Coastal areas critical for water resource management often face challenges such as illegal fishing, trafficking, piracy, and other illicit activities that require robust monitoring systems powered by computer vision. However, real-world datasets in such environments can be compromised by label noise due to random inaccuracies or deliberate adversarial attacks, leading to decreased accuracy in machine learning models. Our innovative approach employs Rockafellian Risk Minimization (RRM) to mitigate the impact of label noise contamination, crucial to maintaining data integrity in water-related security and governance operations. Unlike existing methodologies that rely on extensively cleaned datasets, our two-step process adjusts neural network weights and manipulates nominal probabilities of data points to isolate potential data corruption effectively. This technique reduces dependence on meticulous data cleaning, thereby increasing data processing efficiency in water resources and coastal management. To validate the effectiveness and reliability of the proposed model, we apply RRM in various parameter settings to datasets specific to naval environments and evaluate its classification accuracy against traditional methods. By leveraging the proposed model, we aim to reinforce the robustness of ship detection models, ultimately contributing to developing more reliable automated maritime surveillance systems. Such systems are essential for strengthening governance, security, and water management and curbing illegal activities at sea.

Keywords: computer vision; neural networks; machine learning; Rockafellian risk minimization

1. Introduction

The field of machine learning (ML) has been expanding rapidly, transforming how machines learn from data to make data-driven predictions and decisions. Within ML, computer vision (CV) has gained prominence as a specialized area that enables machines to interpret visual data at an unprecedented level of complexity. This capability has profound implications across sectors, creating applications in areas like autonomous vehicles, medical diagnostics, security, and military operations. Using sophisticated mathematical models, CV allows for machines to "see" and understand the visual world, creating new pathways for technological advancement and real-world impact.

In the military, CV applications play a critical role in enhancing decision-making processes by improving situational awareness and supporting autonomous systems in dynamic environments. For Brazil, a country with an extensive coastline of approximately 3.6 million square kilometers that spans 200 nautical miles within the Brazilian Exclusive Economic Zone (EEZ) [1], these advancements are essential. The Brazilian Navy (BN) is responsible for safeguarding Brazil's maritime borders, where challenges such as piracy, smuggling, and illegal pollution persist. The BN launched the Blue Amazon Management System (SisGAAz) to address these threats, leveraging a sophisticated network of satellites, radars, and multi-platform systems for continuous monitoring of Brazil's jurisdictional waters [2].

Automatic ship detection, an essential component of maritime surveillance, typically depends on meticulously curated datasets. However, gathering and processing high-quality data can be both labor-intensive and time-consuming. This study introduces an innovative approach to improving the robustness of image detection models in naval applications, particularly for binary classification tasks. Building on the Rockafellian relaxation methodology, this approach combines classical convolutional neural networks (CNNs) with stochastic gradient descent (SGD) training, allowing for refined adjustments of nominal probabilities for each image. The Rockafellian method strengthens CNN models by addressing label corruption issues common in ML datasets, which may arise from random inaccuracies or targeted adversarial attacks. This model aims to reinforce data integrity in maritime security operations by reducing dependency on pre-cleaned datasets and improving efficiency in data processing [3,4].

The remainder of this study is organized as follows: Section 2 provides a literature review to support the objectives of this research, offering a detailed summary of relevant concepts, conditions, and practices related to the research aim. Also, Section 2 presents the mathematical formulation and methodologies, discussing the foundational mathematics of the ML algorithms and emphasizing the distinctions between models. Section 3 showcases the results of our three unique methodologies, focusing on model testing accuracy, AUC curve, and computational runtime, with a brief analysis of the outcomes. Section 4 concludes with a summary of the findings and suggestions for future research directions.

2. Materials and Methods

2.1. Background

2.1.1. The Blue Amazon

With its 7500 km Atlantic coastline, Brazil has a vested interest in developments concerning the Atlantic Ocean. In 2004, the Brazilian Navy (BN) introduced the term "Blue Amazon", a vast oceanic area extending from Brazil's coastline to the edge of its continental shelf, covering the surface, waters, seabed, and subsoil. Registered as a trademark in 2010, the Blue Amazon is a significant region rich in resources like marine biodiversity, minerals, oil, and natural gas [1,2]. Figure 1 compares the dimensions of both Amazons.

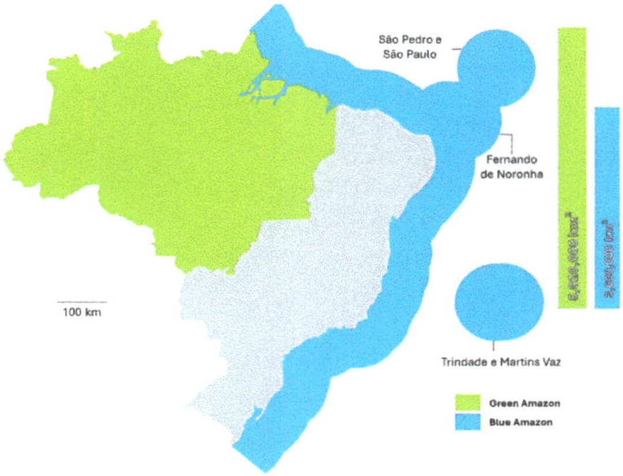

Figure 1. Amazons dimensions. Adapted from [3].

2.1.2. Surveillance System

Protecting the Blue Amazon poses challenges, as insufficient security could lead to criminal activities, including piracy, smuggling, and illegal pollution. To counteract these threats, the BN established the Blue Amazon Management System (SisGAAz), a strategic initiative focused on safeguarding this extensive region [4]. SisGAAz employs satellite technology, multi-platform systems, radars, and sensors for continuous monitoring and management of Brazil's jurisdictional waters and Search and Rescue (SaR) region, integrating data and decision-making for effective oversight [5,6]. Figure 2 illustrates the SisGAAz system.

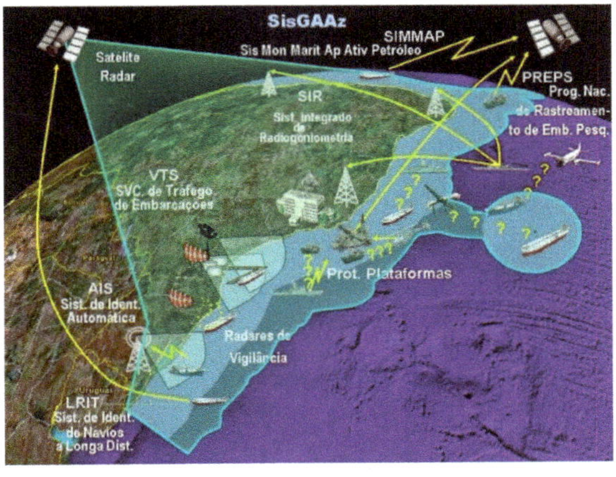

Figure 2. SisGAAz system. Adapted from [2].

2.2. Concepts

2.2.1. Classification

Classification is a key task focused on predicting categorical outcomes or labels by learning boundaries that differentiate classes within the data. Unlike regression, which estimates continuous values, classification assigns each data point to a distinct category, aiding in tasks such as image recognition and spam detection [7–9].

Classification tasks are divided into binary and multiclass types based on the number of target categories. This article focuses on binary classification, where data are assigned to one of two categories, typically labeled as 1 and -1 (e.g., predicting purchase likelihood or detecting spam) [10]. In contrast, multiclass classification categorizes data into more than two classes.

2.2.2. Learning and Optimization

Learning and optimization are essential in ML, with learning improving optimization by leveraging past experiences and data patterns. While learning algorithms assume data reflects the real-world problem, optimization algorithms focus on finding the best solution without making such assumptions [11]. The primary goal in ML is to minimize expected generalization error, or risk, which measures performance differences between training and unseen test data [12].

To address this, expected loss minimization is commonly used, calculating average loss across inputs based on the true data distribution. However, with large datasets, this may be computationally impractical. In such cases, empirical risk minimization (ERM) approximates the true distribution using the training data's empirical distribution, assigning equal probabilities to observed data points [13].

2.2.3. Neural Networks

Neural networks (NNs), a popular type of ML model, are widely used in fields such as image recognition, natural language processing, and pattern recognition. This article primarily focuses on convolutional neural networks (CNNs), a specialized form of NN. Neural networks, also known as artificial neural networks (ANNs), are inspired by the human brain's structure and consist of three main layers: input, hidden, and output. Figure 3 illustrates this structure.

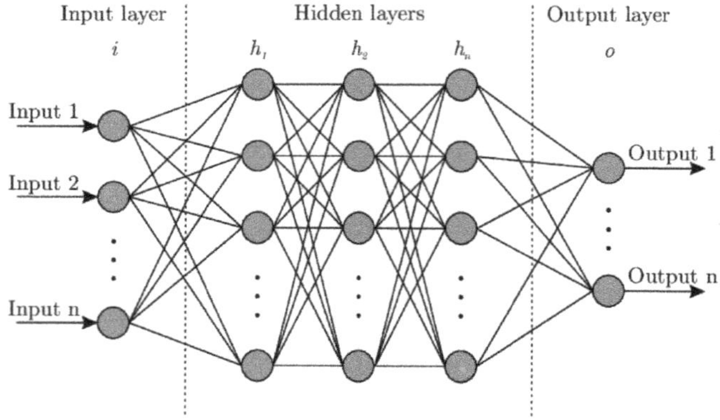

Figure 3. Basic ANN and its layers. Source: [14].

Neural network layers consist of interconnected neurons that process and learn from data through backpropagation, which calculates error gradients, compares predictions to the target, and adjusts network weights. This iterative process continues until the network minimizes error, achieving accurate predictions. A perceptron neuron, shown in Figure 4, combines weighted inputs, adds a bias, and applies an activation function to generate the final output [14].

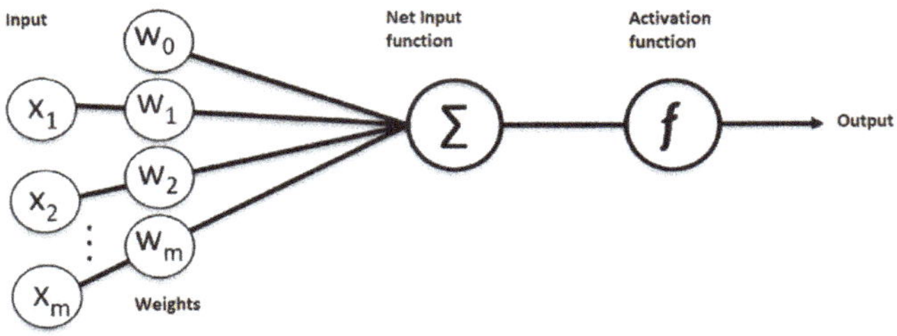

Figure 4. Perceptron model adapted from [15].

Activation functions are essential in training neural networks, as they adjust gradients and introduce nonlinearity, enabling deep networks to learn complex functions [16]. Two widely used activation functions are Rectified Linear Units (ReLU) and Sigmoid.

The ReLU activation function is defined as

$$\mathrm{f(x)} = \max(0, \mathrm{x}). \tag{1}$$

The sigmoid activation function is defined as

$$f(x) = \frac{1}{(1 + \exp(-x))}. \tag{2}$$

Figure 5 shows graphic representations of these activation functions.

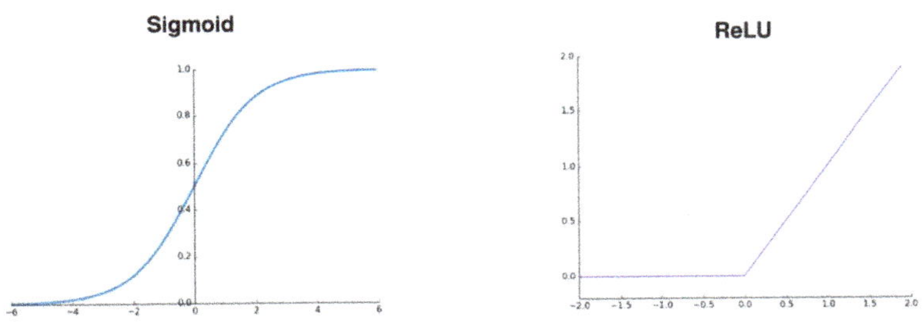

Figure 5. Activation functions. Adapted from [17].

Typically, in multilabel classification networks, the final layer utilizes another activation function called softmax. This function converts the real-valued activations into probabilities for different classes [18]; the softmax function is defined as

$$f(x) = \frac{\exp\left(x_i\right)}{\sum_{j=1}^{K} \exp\left(x_j\right)}, \tag{3}$$

where x_i represents the i-th element of the input vector and $\sum_{j=1}^{K} \exp\left(x_j\right)$ denotes the sum of exponential values over all elements in x. Each probability in the result is in the range 0 to 1, and the sum of the probabilities is 1, representing a valid probability distribution over the K classes.

2.2.4. Computer Vision and Convolutional Neural Networks

Computer vision (CV) is a field that develops algorithms to analyze and interpret visual data, enabling machines to perceive and understand the visual world much like human perception [19]. CV applications focus on extracting and categorizing information from images and videos. In this research, we concentrate on images, which are collections of pixels arranged in two dimensions, typically containing color channels such as red, green, and blue (RGB). Each color channel holds specific color intensity values that can be processed to gain insights and make decisions in CV tasks [20].

Convolutional neural networks (CNNs) are the most effective tools in CV for extracting and learning key features from images [21]. CNNs, like traditional neural networks, use weights, biases, and nonlinear functions to produce outputs. However, CNNs uniquely perform convolutions in place of matrix multiplications, using filters or kernels to scan the input image and generate feature maps. Each kernel's size and stride determine how it navigates the image, influencing output resolution and downsampling when strides greater than one are applied [22].

Pooling layers, typically placed between convolution layers, further downsample feature maps by selecting maximum or average values within a window, reducing spatial dimensions and computational load [23]. After convolutional and pooling operations, the multi-dimensional output is flattened into a one-dimensional array that passes into fully connected (FC) layers, responsible for final predictions. These dense layers function similarly to hidden layers in traditional neural networks, with each neuron fully connected to the previous layer's neurons [24].

After processing the fully connected layer, the output is passed through a softmax activation function (1.3). This function converts the output values into probabilities for each class. The input is then classified based on the class with the highest probability [25].

2.3. *Literature Review*

2.3.1. Classification Problem

Early research began with Maron's Naïve Bayes classifier in 1961, applying Bayes' theorem with feature independence for classification [26]. In 1967, Cover et al. introduced the k-nearest neighbors (k-NN) algorithm, classifying new data by majority vote among nearby points [27].

In 1986, Quinlan's Iterative Dichotomiser 3 (ID3) algorithm introduced decision trees to ML, leveraging information gain to recursively split datasets for improved classification [28]. Support Vector Machines (SVMs), introduced by Cortes and Vapnik in 1995, created hyperplanes that separate classes in a way that maximizes margins, proving robust across a variety of data conditions [29]. Ensemble methods like Boosting (Freund and Schapire, 1997) and Random Forests (Breiman, 2001) further improved classification accuracy by combining multiple base models [30,31].

The rise of deep learning transformed classification through neural networks (NNs), with Hinton et al.'s 2006 work on deep belief networks marking a milestone. These deep models excel at identifying complex patterns in high-dimensional data, providing powerful tools for modern classification tasks [32].

2.3.2. Neural Networks Structures

The concept of computational neurons began with McCulloch et al. in 1943, who modeled binary neurons inspired by brain activity [33]. In 1958, Rosenblatt's perceptron model expanded on this, calculating weighted sums of inputs and forming the basis of artificial neural networks (ANNs) when combined [34].

Despite initial progress, NN development slowed until 1989, when Rumelhart et al. introduced backpropagation, a method for adjusting weights to minimize errors, enabling more complex applications [35,36]. Convolutional neural networks (CNNs) later emerged to tackle image recognition by extracting features through convolutional layers. LeNet, an early CNN by Yann LeCun in 1998, achieved high accuracy in digit recognition, used by the U.S. Postal Service on the MNIST dataset [37], as presents Figure 6.

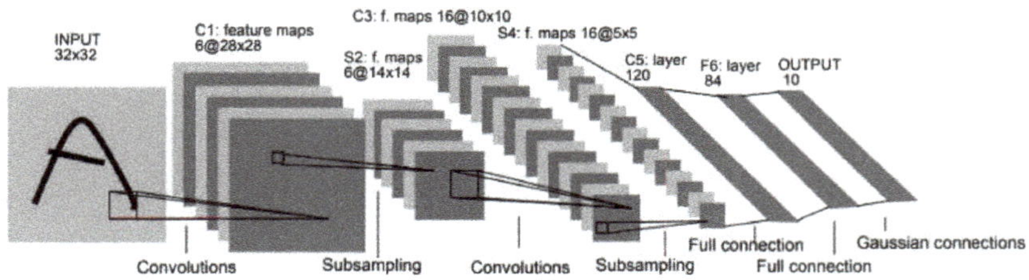

Figure 6. LeNet CNN architecture. Source: [37].

The development of CNNs, supported by academia and industry, has included deeper architectures, residual connections, and attention mechanisms, enhanced by GPUs and large datasets like ImageNet. Deep networks faced issues like the vanishing gradient problem, where gradient values become too small to update weights effectively, as highlighted by Glorot et al. in 2010 [38].

Major advancements came from the ImageNet competition, with AlexNet (2012) showing deeper layers improve performance when overfitting is controlled [39], and GoogLeNet (2014) enhancing results by increasing channels per layer [40]. Figure 7 shows the AlexNet architecture.

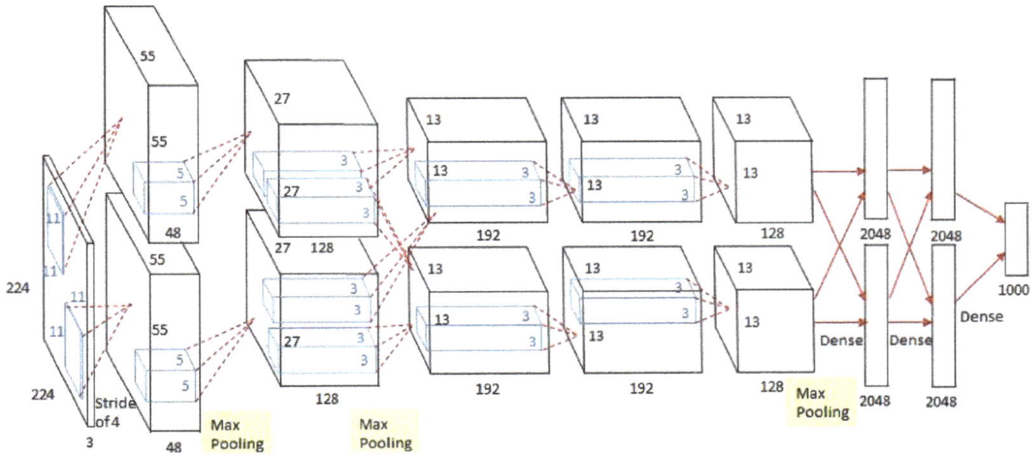

Figure 7. AlexNet CNN architecture. Adapted from [39].

A key similarity between AlexNet and GoogLeNet is their use of the ReLU activation function (1.1), introduced by Krizhevsky et al. [39], which accelerates training and helps mitigate the vanishing gradient problem. The main innovation in GoogLeNet is its inception module, which captures features at multiple scales using various filter sizes and includes 1×1 convolutions to reduce channels, enhancing computational efficiency.

The VGG architecture, developed by Simonyan et al. at the University of Oxford, gained significant recognition in the 2014 ILSVRC for its impressive performance, despite placing second to GoogLeNet [41]. Known for its straightforward design, VGG has had a lasting impact on the deep learning community, with two main variants, VGG-16 and VGG-19, indicating the number of layers in each model. Figures 8 and 9 illustrate the structure of VGG-16 and VGG-19, respectively.

Figure 8. VGG-16 structure. Adapted from [42].

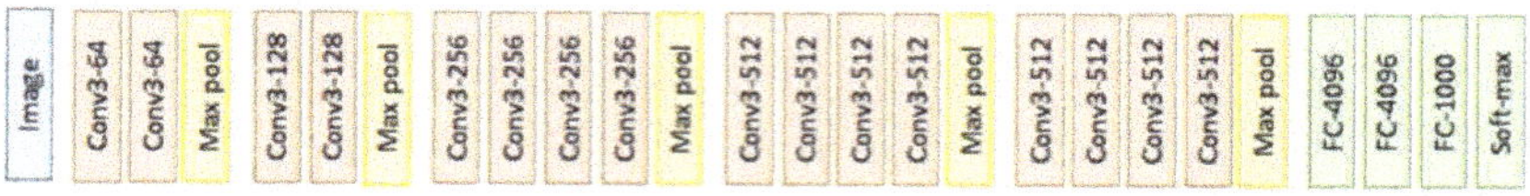

Figure 9. VGG-19 structure. Adapted from [42].

In 2015, He et al. introduced ResNet, a deep learning architecture featuring residual learning through shortcut (or skip) connections [43]. These connections enable gradients to bypass certain layers, making it feasible to train much deeper networks than before. ResNet's innovative design, with residual blocks to capture complex features, quickly gained popularity and won the 2015 ILSVRC classification task [44].

VGG and ResNet share a notable similarity in their depth, with VGG reaching up to 19 layers and ResNet extending this concept to 152 layers. However, they differ in their approaches to addressing the vanishing gradient problem. While VGG suffers from this issue due to its uniform structure, ResNet mitigates it with residual learning via skip connections, allowing for gradients to bypass certain layers [45,46].

Additionally, VGG has a straightforward architecture of 3×3 convolutions and 2×2 pooling, while ResNet's design is more complex due to its use of residual blocks. Together, these architectures represent advances achieved through both theoretical and empirical exploration in CNN development.

2.3.3. Maritime Computer Vision

Computer vision (CV) aims to replicate and enhance human visual interpretation capabilities, bringing significant advancements to maritime operations. Traditional navigation tools—such as vessel traffic services (VTSs), radar, and automatic identification systems (AISs)—have been foundational in maritime monitoring but face limitations. Research by Yassir et al. highlighted that camera surveillance systems can bridge these gaps, enhancing threat prediction and mitigating illegal activities [47–50].

The integration of CNN-based CV has significantly advanced maritime image interpretation, as shown in studies using specialized datasets for target classification. One study achieved a 94% success rate in classifying marine targets using a CNN with data augmentation [51]. Another, by Gallego et al., utilized the MASATI dataset and transfer learning, boosting accuracy to 99.76% with pre-trained models like VGG and ResNet [52]. Fang et al. further enhanced detection of small targets using CNNs with infrared imaging, demonstrating improvements in robustness and accuracy [53].

Recent developments in UAVs and sensor technology, paired with increased computational power, have propelled CV applications in maritime surveillance. UAVs equipped with CV have shown success in object detection, utilizing models like YOLO for real-time performance [54]. Lo et al. and Lygouras et al. demonstrated YOLO's efficiency in UAV-based surveillance, showing the potential of deep learning for dynamic object tracking and search-and-rescue (SAR) tasks [55–58].

Reflecting this growing importance, the Maritime Computer Vision (MaCVi) workshop debuted in 2023, emphasizing the rising role of CV in maritime applications [59].

2.3.4. Label Noise

To develop high-performance deep learning models, experts employ various techniques, including preprocessing, data augmentation, and transfer learning from established architectures. However, a common issue in real-world data is label noise, which can significantly impact model outputs. Label noise, where labels are incorrectly assigned due to errors or deliberate alterations, affects real-world datasets with an estimated 8–38.5% corruption rate [60,61]. Errors arise from random noise during data collection or labeling, and intentional data poisoning, where attackers manipulate data to degrade model performance [62,63].

Researchers have proposed various methods to mitigate label noise's effects without major structural changes. For instance, Ren et al. introduced a meta-learning algorithm that weights training samples by gradient orientation to improve performance under noisy labels [64]. Thulasidasan's deep abstaining classifier (DAC) selectively ignores confusing samples during training, enhancing robustness [65]. Chen et al. implemented a hierarchical structure to adapt loss based on noise ratios, bolstering model resilience [66]. Narasimhan et al. combined learning-to-reject (L2R) with out-of-distribution (OOD) detection to handle outliers and complex samples, allowing for the model to abstain from such instances effectively [67–70].

This study builds on Royset et al.'s Rockafellian relaxation [25], which optimizes models by allowing flexibility in decision-making under label noise, enhancing stability. Rockafellian relaxation analyzes the sensitivity of assumptions and parameters, supporting robust optimization [71]. We apply this framework to improve the reliability of computer vision (CV) in maritime contexts, focusing on label accuracy and model stability, aiming to contribute to resilient deep learning models in noisy environments.

2.4. Rockafellian Risk Minimization

This section develops Rockafellian relaxation in the context of ML, leading to Rockafellian Risk Minimization (RRM) as a means for training in settings with corrupt data. We contrast the approach with ERM and discuss computational aspects.

2.4.1. Formulation

In a labeled set $\{x_j, y_j, j = 1, \ldots, n\}$ of images, x_j specifies the attributes of each pixel in the j-th image and y_j is the corresponding label we seek to determine a parameter vector w in a neural network (NN) so that the prediction corresponds to the label. Considering two labels, -1 and 1, and $g_w(x_j)$ as the NN prediction for the j-th image's label, the binary cross-entropy loss function (Equation (4)) is used to determine the vector w that minimizes the loss:

$$f_j(w) = \begin{cases} -\ln g_w(x_j) \; if \; y_j = 1 \\ -\ln(1 - g_w(x_j)) \; if \; y_j = -1. \end{cases} \tag{4}$$

Classical ERM then amounts to solving the optimization problem:

$$\underset{w \in \mathbb{R}^d}{minimize} \sum_{j=1}^{n} p_j \, f_j(w), \tag{5}$$

where $p_j = 1/n$ typically but the data points can also be weighted differently.

In many applications, data corruption can occur due to various reasons such as data collection and labelling entry errors, or even adversarial attacks intended to poison the training data and disrupt the system. As pointed out by Royset et al. [25], and elaborated in section D of chapter II, an NN that has been trained using ERM may not perform well if the training data are corrupted. This observation provides a strong motivation for RRM, an adaptive method to identify and remove corrupted data points, hence improving the overall performance and reliability of the model [72].

RRM leverages auxiliary decision variables $u_1, \ldots, u_n$, to adjust the probability associated with each data point. This leads to the formulation

$$\underset{w\in\mathbb{R}^d, u\in U}{\text{minimize}} \sum_{j=1}^{n} \left((p_j + u_j) f_j(w) + \theta |u_j| \right), \tag{6}$$

where θ represents a penalty parameter, u is a perturbation vector with n dimensions that alters the nominal probability vector p, and the set

$$U = \left\{ u \in \mathbb{R}^n \middle| u_j \geq -p_j, \sum_{j=1}^{n} u_j = 0 \right\}. \tag{7}$$

We adjust the probability associated with each data point by adding u_j to p_j. To identify corrupted data points, we optimize u based on their calculated loss with the goal of reducing the probability that affects $f_j(w)$ to zero. This effectively eliminates the data point from being considered in the training process.

2.4.2. Training Algorithm

The RRM method consists of the usage of an Alternating Direction Heuristic based on Linear Programming (ADH-LP), which alternates between optimizing different sets of variables while keeping the others fixed, hence the name "alternating direction". This strategy begins by optimizing w, which adjusts the neural network weights. Subsequently, we optimize u to modify each data point's probability, isolating and discarding potential data corruption. The model is refined progressively in each cycle until it reaches the number of iterations defined by the user. Linear programming optimizes u through an objective function with linear constraints, and the parameter is the step size $\mu \in (0,1]$ that updates u at each iteration i. We note that components of u^i sum to zero and are always greater than $-p_j$. The details of the algorithm are given next.

Alternating Direction Heuristic (ADH-LP)

Data. Number of epochs κ, number of iterations τ, initial weights w^0, and stepsize μ.

Step 0. Set iteration counter $i = 1,\ p^1 = p,\ u^1 = 0$.

Step 1. Starting from w^{i-1}, apply SGD-type algorithm for κ epochs to the problem

$$\underset{w\in\mathbb{R}^d}{\text{minimize}} \sum_{j=1}^{n} p_j^i f_j(w). \tag{8}$$

Let w^i be resulting solution.

Step 2. Select u^{i+1} for application in the LP based on w^i and u^i.

Step 3. Solve the linear optimization problem

$$\underset{u\in U, v\in \mathbb{R}^n}{\text{minimize}} \sum_{j=1}^{n} \left(u_j f_j\left(w^i\right) + \theta v_j \right)\ s.t. u_j \leq v_j,\ -u_j \leq v_j,\ j = 1, \ldots, n.$$

Let (u^*, v^*) be a minimizer.

Step 4. Set $u^{i+1} = \mu u^* + (1-\mu) u^i$. Go to Step 3.

Step 5. If $i < \tau$, set $p^{i+1} = p + u^{i+1}$, replace i by $i + 1$, and back to Step 1.

Else, stop.

3. Results

In this section, we delve into the model's results, beginning with a discussion on datasets, data preprocessing, optimizers, and network structures. We then present the results achieved by our ERM and RRM models both without label contamination and under varying levels of label contamination in training data. Following this, we analyze the behavior of our perturbation vector values in specific cases. Finally, we evaluate the adaptability and robustness of RRM compared to ERM across different contamination levels.

We utilize two datasets: the Airbus Ship Detection (AIRBUS) [73] and Maritime Satellite Imagery (MASATI) [52] datasets.

Each dataset is examined in its own subsection, detailing unique characteristics and specific challenges. In both, we consider an environment without label contamination and another where label contamination in the training set is gradually increased, starting from 10%, 20%, 30%, and 40% by randomly swapping training example labels. Higher contamination levels of 50% and 60% were tested, but the CNNs struggled to learn patterns at these levels.

The CNN topology used in this study was selected through a combination of empirical experimentation and systematic hyperparameter tuning to balance performance and computational efficiency. The primary focus of this research is not the CNN architecture itself but the application of the RRM methodology to mitigate label noise in maritime datasets. However, to ensure that the CNN was capable of effectively extracting relevant features, the spatial resolution of the input images was carefully considered. Both datasets used in this study—MASATI and Airbus Ship Detection—contain RGB images resized to the same dimensions, ensuring consistency in preprocessing and compatibility with a single CNN topology. This standardization eliminated the need to design separate architectures for each dataset.

The choice of resizing was guided by the need to maintain visual clarity of small vessels, which are critical targets for detection. The resolution was adjusted to a level where ships could still be visually distinguishable to the human eye, preserving key spatial features while reducing computational demands. Although the datasets originally differ in resolution—768×768 pixels for AIRBUS and 512×512 pixels for MASATI—resizing them to a common input dimension ensured that the CNN could process both datasets effectively without requiring topology modifications. This approach provided a consistent framework for evaluating the RRM methodology across datasets with varying characteristics.

The relationship between spatial resolution and CNN topology is particularly relevant for detecting ships of varying sizes. High-resolution images allow for the identification of smaller ships, while maintaining a balance between resolution and computational efficiency ensures practical applicability in real-world scenarios. By adopting a standardized preprocessing approach and ensuring that the resized resolution was sufficient to retain critical features, the CNN topology was made robust and generalizable. This design decision highlights the emphasis on evaluating the RRM's effectiveness, while ensuring the CNN's ability to process maritime data with diverse spatial resolutions.

Our experiments are conducted using Tensorflow version 2.11.0. For w-optimization in the first phase of the ADH process, we employ two NN optimizers: Adam [74] and SGD [75]. Adam is set with a learning rate of $1.0{\cdot}10^{-3}$, while SGD uses a learning rate of 0.1 for the AIRBUS dataset and 0.02 for the MASATI dataset, with a momentum of 0.9 for both datasets. All experiments run on a single Nvidia Tesla V100 GPU with 32 GB of memory.

For u-optimization in the second phase of the ADH process, we use the ADH-LP algorithm, implemented with Pyomo version 6.4.0 and the CPLEX solver.

We investigate various network architectures, learning rates, and other hyperparameters, though our goal is not to enhance performance by altering network architecture. Instead, we aim to show that an RRM approach can offer benefits over an ERM approach regardless of network configuration. For comparison, we select two network configurations. Both CNNs take resized $128 \times 128 \times 3$ images as input, with ReLU activation in hidden layers and two softmax-activated output units for binary classification. The Adam-optimized

CNN has 2,228,002 trainable parameters, while the SGD-optimized CNN has 8,409,026. Tables 1 and 2 detail the CNN structures used with each optimizer. Both networks use binary cross-entropy as the loss function.

Table 1. Description of the CNN used for Adam.

#	Layer	Filters	Kernel Size	Output Size	# Parameters
1	Convolution	16	3×3	$128 \times 128 \times 16$	448
	Max-Pooling		2×2	$64 \times 64 \times 16$	
2	Convolution	32	3×3	$64 \times 64 \times 32$	4640
	Max-Pooling		2×2	$32 \times 32 \times 32$	
3	Convolution	64	3×3	$32 \times 32 \times 64$	18,496
	Max-Pooling		2×2	$16 \times 16 \times 64$	
4	Convolution	128	3×3	$16 \times 16 \times 128$	73,856
	Max-Pooling		2×2	$8 \times 8 \times 128$	
5	Fully connected	256		1×256	2,097,408
6	Fully connected	128		1×128	32,896
7	Softmax (Output)	2		1×2	258

Table 2. Description of the CNN used for SGD.

#	Layer	Filters	Kernel Size	Output Size	# Parameters
1	Convolution	32	3×3	$128 \times 128 \times 32$	896
2	Batch Normalization Activation				128
3	Max-Pooling		2×2	$64 \times 64 \times 32$	-
4	Convolution Activation	64	3×3	$64 \times 64 \times 64$	18,496
	Max-Pooling		2×2	$32 \times 32 \times 64$	
5	Fully-connected	128		1×128	8,388,736
6	Batch Normalization Activation				512
7	Softmax (Output)	2		1×2	258

After establishing a baseline model with ERM, we apply the RRM algorithm ADH-LP using the same network configurations.

For ERM, we set the number of epochs $\kappa = 500$. For RRM algorithm, we establish $\kappa = 10$ and iterations $\tau = 50$, which guarantees that all heuristic runs will include 500 epochs. The ADH-LP uses the default step size parameter $\mu = 0.5$. Concerning the penalty parameter θ in both ERM and RRM, we show 5 different levels of θ to evaluate their performance in the u-optimization across the values of $\theta \in \{0.15, 0.20, 0.25, 0.30, 0.35\}$. Table 3 summarizes the parameters described.

Table 3. Parameters for ERM and RRM.

	Parameters			
Algorithm	**Epochs (κ)**	**Iterations (τ)**	**Stepsize (μ)**	**Penalty (θ)**
ERM	500	1	-	-
RRM(ADH-LP)	10	50	0.5	0.15, 0.20, 0.25, 0.30, 0.35

To compare results, we ensure that the random seed is the same at two specific times for all runs: when we divide the data into training and testing sets and when we execute closed-set contamination through label swapping.

Table 4 summarizes the algorithm runtimes over 500 epochs, separating the time used for the w-optimization and for u-optimization processes in both datasets.

Table 4. Computational time of algorithms.

	Dataset							
	MASATI				**AIRBUS**			
	CNN Optimizer				**CNN Optimizer**			
	Adam		**SGD**		**Adam**		**SGD**	
Optimization phase	w	u	w	u	w	u	w	u
Algorithm	Total runtime in seconds over optimization phases							
ERM	500	-	1400	-	22,500	-	68,500	-
RRM (ADH-LP)	500	600	1400	600	22,500	1000	68,500	1000

There are noticeable differences between the MASATI and AIRBUS datasets and the Adam and SGD optimizers. When using the Adam optimizer on the MASATI dataset, the ADH-LP method takes 1100 s. However, if we switch to the SGD optimizer, ADH-LP takes 2000 s. Looking at the AIRBUS dataset, with the Adam optimizer, RRM with ADH-LP takes 23,500 s to complete. Under the SGD optimizer, ADH-LP requires 69,500 s. It is worth noting that the ADH-LP method takes longer in the u-optimization process, especially in the larger AIRBUS dataset.

3.1. *MASATI Dataset*

The MASATI dataset [52] provides a rich collection of maritime scenes captured through optical aerial imagery in the visible spectrum. Designed to evaluate ship detection and classification methods, MASATI stands out for its representation of dynamic marine environments. Each image in the dataset is a color (RGB) image with a fixed resolution of 512×512 pixels, stored in the PNG format, which ensures lossless image quality and preserves visual details critical for machine learning applications. The dataset incorporates significant variability in weather, lighting, and the presence of multiple targets, making it a challenging yet valuable resource for robust model evaluation.

The full MASATI dataset consists of 7389 labeled images distributed across seven distinct classes: coast, land, ship, sea, coast–ship, multi, and detail. These classes are designed to represent the diverse maritime conditions and scenes captured within the dataset, ranging from open water to coastlines and complex multi-target scenarios. For this study, however, the dataset was adapted to a binary classification problem. Specifically, two classes were selected: "ship", comprising 1027 images, and "sea", containing 1022 images.

This reduced the dataset to a total of 2049 images, enabling a focused investigation into the presence or absence of ships in maritime environments.

Despite its relatively small size, this binary subset of the MASATI dataset offers a unique opportunity to assess the robustness and adaptability of ship detection algorithms. The limited number of images emphasizes the need for efficient model training and testing, ensuring that the results are not overly reliant on large-scale data. This makes MASATI particularly valuable for evaluating the performance of algorithms in scenarios where data availability is constrained, a common challenge in real-world maritime applications.

Furthermore, the inclusion of challenging environmental factors, such as varying weather conditions, illumination changes, and the presence of multiple targets within a single image, highlights the dataset's complexity. These characteristics ensure that any proposed method must be versatile and robust to deliver accurate results. Figure 10 of the study illustrates some example images from the dataset, showcasing the diversity and dynamic nature of the scenes captured.

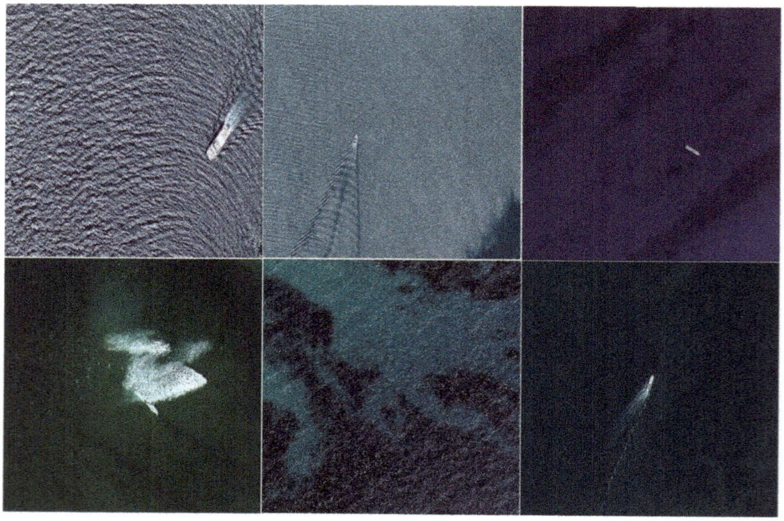

Figure 10. MASATI dataset images. Source: [52].

3.1.1. Accuracy Results with MASATI

In this section, we show our results in the MASATI dataset, explicitly focusing on the accuracy score, which indicates how effectively our models can classify images within the test set. Along with presenting the results, we also provide a thorough analysis to interpret the observed performance.

In MASATI, the train/test split is achieved using a 90/10 proportion, which results in 1844 images in the training set and 205 in the test set.

The final test accuracy results for ERM and RRM with ADH-LP using Adam optimizer are shown in Table 5 at various levels of corruption.

Table 5. Final test accuracy in MASATI for ERM and RRM (ADH-LP/Adam).

	Corrupted Training Data Percentage				
Method	40%	30%	20%	10%	0%
ERM	0.624	0.653	0.785	0.858	0.941
RRM ($\mu = 0.5$)					
$\theta = 0.15$	0.624	0.668	0.800	0.863	0.951
$\theta = 0.20$	0.609	0.726	0.848	0.878	0.931
$\theta = 0.25$	0.668	0.682	0.814	0.863	0.941
$\theta = 0.30$	0.604	0.702	0.804	0.843	0.926
$\theta = 0.35$	0.614	0.692	0.756	0.834	0.921

Note: The values highlighted in gray represent the cases where RRM outperform or match ERM.

The RRM method exhibits performance at par or even superior to ERM at various levels of data corruption, especially for specific penalty values. Remarkably, when the values of 0.15, 0.20, and 0.25 are chosen for θ, they yield the best overall performance. For example, θ = 0.15 performs exceptionally well at lower corruption levels, θ = 0.20 outperforms all other θ values at a 20% corruption level, and θ = 0.25 performs best at a 30% corruption level.

The performance difference is relatively minimal in scenarios where the RRM method does not surpass ERM. This finding implies that even with non-optimal selections of θ, the RRM method can deliver performance nearly equivalent to ERM. It accentuates the versatility of the RRM method as a promising alternative to ERM, especially in environments fraught with corrupted data.

The topology of the CNN used in this study was primarily selected based on empirical results obtained during the experimentation phase and through a systematic hyperparameter search, ensuring a balance between performance and computational efficiency. Although the primary focus of this research is the mathematical model for mitigating errors in noisy datasets rather than the CNN architecture itself, the network design aimed to effectively capture image patterns, including small vessels. To achieve this, the spatial resolution of the images from both the MASATI and AIRBUS datasets was uniformly resized to a level where small vessels could still be visually distinguishable to the human eye. This resizing criterion ensured that critical visual features were preserved during preprocessing.

Furthermore, as both datasets were resized to the same dimensions before being fed into the CNN, there was no need to adapt the network topology to account for differences in the original resolutions. This standardized approach simplified implementation and allowed for a focused evaluation of the RRM methodology under varying levels of label noise. The selected CNN topology and preprocessing pipeline proved sufficient to support the primary objective of this study: assessing the robustness of the proposed RRM framework in scenarios with corrupted datasets.

The results of the experimentation phase suggest that a set of $\theta \in \{0.15, 0.20, 0.25\}$ provides an optimal balance in modulating the penalty for model complexity versus the fit to training data. This balance contributes to superior generalization of the test data, particularly at intermediate corruption levels.

Now, turning to the SGD optimizer, Table 6 displays the final test accuracy results for ERM and RRM in ADH-LP at various levels of corruption.

Table 6. Final test accuracy in MASATI for ERM and RRM (ADH-LP/SGD).

	Corrupted Training Data Percentage				
Method	40%	30%	20%	10%	0%
ERM	0.556	0.546	0.581	0.663	0.648
RRM (μ = 0.5)					
θ = 0.15	0.604	0.648	0.648	0.639	0.634
θ = 0.20	0.663	0.692	0.648	0.648	0.609
θ = 0.25	0.639	0.687	0.658	0.629	0.614
θ = 0.30	0.668	0.585	0.600	0.643	0.634
θ = 0.35	0.624	0.556	0.639	0.648	0.653

Note: The values highlighted in gray represent the cases where RRM outperforms or matches ERM.

From Table 6, the RRM method, when compared with ERM, delivers better or comparable performance at different levels of data corruption.

Examining the selection of the θ ∈ {0.20, 0.25, 0.30}, we observe that these selections yield the highest accuracies across the range of data corruption levels.

At a 40% corruption level, the RRM outperforms the ERM method. For θ = 0.20, the accuracy improvement is a substantial 10.7%. Similarly, for θ = 0.30, the performance is enhanced by 11.2% compared to ERM. When faced with 30%, the accuracy improvement for θ = 0.20 over ERM is an impressive 14.6%, the highest observed in this dataset. In the 20% corruption level, the RRM with a θ = 0.25 exhibits a promising 7.7% improvement in accuracy over the ERM.

Interestingly, at lower corruption levels of 10% and 0%, the performance of RRM remains comparable to ERM. This reveals the versatility of the RRM method, maintaining robust performance even as the corruption level diminishes.

The superior performance of RRM at higher corruption levels (20%, 30%, and 40%) across all θ values, combined with its comparable performance at lower corruption levels, indicates its resilience.

3.1.2. U-Optimization Analysis with MASATI

In the following section, we aim to examine the u-vector behavior in some MASATI cases. We show the dynamic evolution of the perturbation vector values across the iterative steps of the RRM algorithm and how it influences performance and resilience, thereby illustrating its crucial role within this methodology.

The u-vector helps improve the model's performance by adjusting the probability of each data point. It assigns lower values to mislabeled examples excluding them from the NN training process. The penalty parameter θ regulates the u-values and determines the amount of non-zero u-values. It is important to note that increasing θ value results in a higher penalty for non-zero u-values. However, setting θ to excessively low values can cause too many examples to be assigned low u-values, resulting in a small training set for the NN that cannot effectively learn patterns. Conversely, high θ values may incentivize the model to select zero for all u-values, making the RRM model work as the traditional ERM. Therefore, one must search for a θ value that can outperform ERM.

Table 7 illustrates the relationship between θ and u-values; we take one experiment using RRM (ADH-LP) in various contamination levels and compare θ with mislabeled images excluded from the NN training. An image is considered excluded from training if its associated u-value achieves a value of $-1/N$, where N represents the total number of training observations and $1/N$ is the nominal probability.

Table 7. Penalty parameter and perturbation vector relationship in MASATI.

ADH-LP/SGD RRM (μ = 0.5)	Contamination Levels							
	40% (737 Mislabeled Images)		30% (553 Mislabeled Images)		20% (368 Mislabeled Images)		10% (184 Mislabeled Images)	
Value of Penalty θ	Model Accuracy	# of Mislabeled Images Excluded	Model Accuracy	# of Mislabeled Images Excluded	Model Accuracy	# of Mislabeled Images Excluded	Model Accuracy	# of Mislabeled Images Excluded
0.15	0.604	327	0.648	286	0.648	211	0.639	99
0.20	0.663	338	0.692	290	0.648	197	0.648	91
0.25	0.639	318	0.687	263	0.658	163	0.629	94
0.30	0.668	295	0.585	230	0.600	159	0.643	87
0.35	0.624	253	0.556	213	0.639	140	0.648	71

Looking at the 30% label contamination column in Table 7, we take a closer look at the cases where the best (θ = 0.20) and the worst (θ = 0.35) test accuracies were achieved. Starting with the worst case, where θ = 0.35, Figure 11 displays the accuracy of training (red curve) and test (blue curve) of ERM (left) and RRM (right) over 500 epochs. The test accuracies are very comparable in both approaches, where RRM achieves 0.556 against 0.546 of ERM.

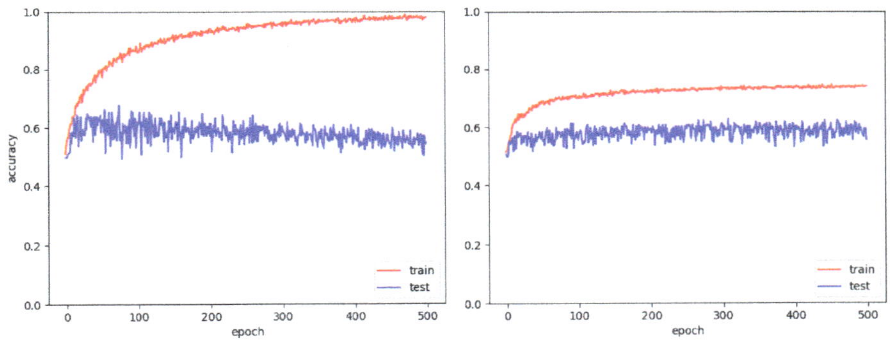

Figure 11. Training and test accuracy for ERM (**left**) and RRM ADH-LP/θ = 0.35 (**right**) on MASATI with 30% of contamination.

Table 8 reports the evolution of the u-vector across the initial updates and its final update. The columns, labeled by the i-counter value, display the distribution of u_i-values after each u-optimization, covering 553 mislabeled and 1291 correctly labeled images.

Table 8. Evolution of u-vector across ADH-LP/SGD in MASATI (30% contamination; θ = 0.35).

Nominal Probability (1/N) $5.4 \cdot 10^{-4}$	Iteration Number					
	i = 1		i = 2		i = 49	
u_i-Values	Mislabeled Images	Correct Labeled Images	Mislabeled Images	Correct Labeled Images	Mislabeled Images	Correct Labeled Images
>>0	0	1	0	1	1	1
≈0	304	813	266	782	338	1005
$-1.5 \cdot 10^{-4}$	0	0	37	92	1	5
$-2.7 \cdot 10^{-4}$	249	477	38	41	0	1
$-4.0 \cdot 10^{-4}$	0	0	212	375	0	1
$-5.4 \cdot 10^{-4}$	0	0	0	0	213	278
Total of images	553	1291	553	1291	553	1291

During the initial iteration, some images from both label categories receive a u_i-value of $-2.7 \cdot 10^{-4}$. A larger number from both groups retains their initial u_i-value of zero, indicating no changes. With the second update, there is a discernible shift within the range of u-values as the algorithm seeks to adjust probabilities in its u-optimization process.

Upon the final update, the algorithm identifies 213 out of 553 incorrectly labeled images. These images are assigned a u_i-value of $-5.4 \cdot 10^{-4}$, effectively excluding them from further NN training. This exclusion stems from the fact that this u_i-value nullifies the nominal probability assigned to the images at the inception.

While a minor proportion of the correctly labeled images receive this lowest u_i-value, the majority (1005 out of 1291) retain their initial probability or a value close to it and continue contributing to the NN training process.

The scenario is different when we shift to the best accuracy of RRM where θ = 0.20. Figure 12 displays the accuracy of training (red curve) and test (blue curve) of ERM (right) and RRM (left) over 500 epochs. The test accuracies are different, and we clearly note the improvement of RRM against ERM. In that case, RRM achieves 0.692 against 0.546 of ERM. The accuracy plots show much less noise and more stability in the RRM test accuracy curve compared to the ERM and the previous RRM test accuracy curve.

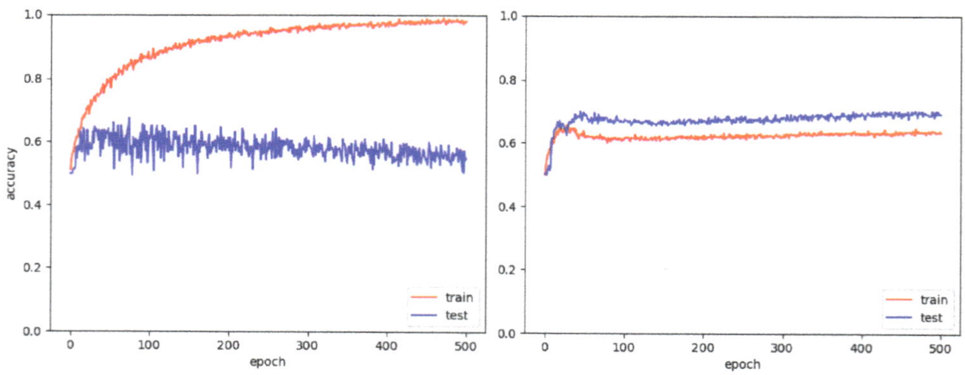

Figure 12. Training and test accuracy for ERM (**left**) and RRM ADH-LP/θ = 0.20 (**right**) on MASATI with 30% of contamination.

Table 9 displays the u-vector evolution of the RRM best case scenario in the two early updates and how it finished in its last update.

Table 9. Evolution of u-vector across ADH-LP/SGD in MASATI (30% contamination; θ = 0.20).

Nominal Probability (1/N) $5.4 \cdot 10^{-4}$	Iteration Number					
	$i = 1$		$i = 2$		$i = 49$	
u_i-Values	Mislabeled Images	Correct Labeled Images	Mislabeled Images	Correct Labeled Images	Mislabeled Images	Correct Labeled Images
>>0	0	1	0	1	0	1
≈0	133	453	93	370	260	863
$-1.5 \cdot 10^{-4}$	0	0	39	105	3	1
$-2.7 \cdot 10^{-4}$	420	837	40	83	0	0
$-4.0 \cdot 10^{-4}$	0	0	381	732	0	0
$-5.4 \cdot 10^{-4}$	0	0	0	0	290	426
Total of labels	553	1291	553	1291	553	1291

During the initial iteration, we have more images from both label categories receiving the u_i-value of $-2.7 \cdot 10^{-4}$, and the minority from both groups retaining their initial u_i-value of zero. In the second update just 93 of the 553 incorrectly labeled images remained with the initial u_i-value of zero. Observing RRM test accuracy in Figure 12, we notice that it stabilizes around epoch 75 and even slightly improves with additional epochs. This suggests that RRM may benefit from additional training past 500 epochs. Moreover, the stabilization of the test accuracy corresponds to the stabilization of u-value assignments.

Upon the final update, the algorithm identifies 290 out of 553 incorrectly labeled images, 77 more than the worst case. These images are assigned a u_i-value of $-5.4 \cdot 10^{-4}$, effectively excluding them from further NN training.

While a considerable proportion of the correctly labeled images receive this lowest u_i-value, the majority (863 out of 1291) retain their initial probability or a value close to it and continue contributing to the NN training process.

3.2. AIRBUS Dataset

The Airbus Ship Detection dataset [73], initially introduced as part of a Kaggle competition in 2018, is a large-scale collection of satellite imagery provided by Airbus (Blagnac, France). It serves as a benchmark for testing and improving machine learning models for ship detection in maritime environments. The dataset includes 18,392 satellite images, all formatted as RGB files with a fixed resolution of 768 × 768 pixels. These images are stored in JPG format, which, while efficient for storage, introduces compression artifacts that can pose challenges for precise image analysis. Each image is accompanied by annotations in the form of segmentation masks that highlight the exact location of ships when present.

For the purpose of this study, a balanced subset of the dataset was extracted to ensure equal representation of both categories: "ship" and "no ship". From the original dataset, a total of 10,428 images were selected, evenly divided between the two classes, resulting in 5214 images for each. This random sampling approach not only maintains a balance in class distribution but also ensures that the evaluation metrics reflect an unbiased performance across both detection and non-detection scenarios. Such a balanced setup is critical for reducing model bias and improving generalizability to real-world data where ship presence might vary significantly.

The dataset also captures a wide variety of maritime conditions, including different lighting, weather, and environmental scenarios. This variability introduces significant challenges for computer vision algorithms, as models must contend with factors such as shadows, partial occlusions, and atmospheric distortions. Additionally, smaller vessels are particularly difficult to detect due to their reduced size relative to the overall image dimensions. The inclusion of these complexities makes the Airbus Ship Detection dataset an excellent choice for evaluating the robustness and performance of advanced machine learning approaches, such as the RRM method employed in this study. Figure 13 illustrates some images of AIRBUS.

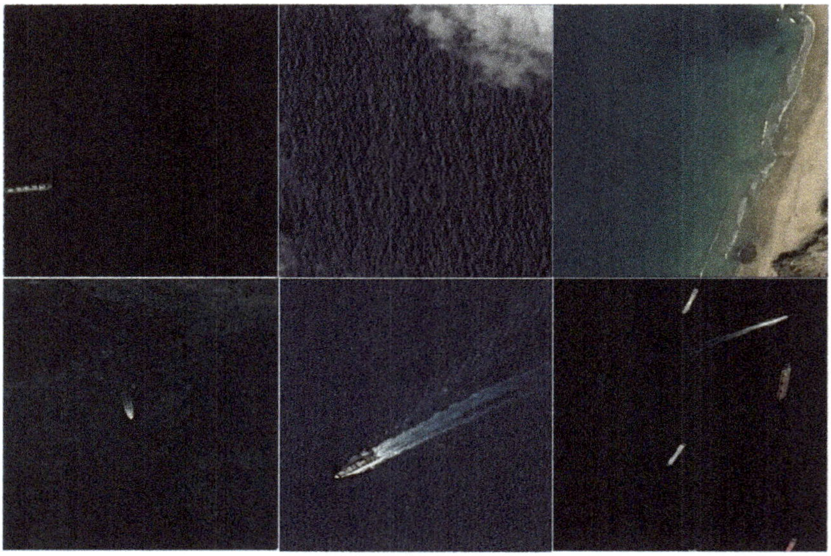

Figure 13. AIRBUS dataset images. Source: [73].

The MASATI dataset contains 2049 images. In contrast, the AIRBUS dataset, containing nearly five times more images, allows for us to evaluate our algorithm in a more data-rich environment. Both data-limited and data-rich situations are likely in practical scenarios. Therefore, analyzing our algorithms in a more data-rich environment adds an important dimension.

3.2.1. Accuracy Results with AIRBUS

In this section, we present the results of the AIRBUS dataset, focusing on the test accuracy score, which measures how effectively our models can classify images in the test set. We also provide a comprehensive analysis to interpret the observed performance.

In this dataset, the train/test split was achieved using an 80/20 proportion, which resulted in 8340 images in the training set and 2086 in the test set.

The final test accuracy results for ERM and RRM with ADH-LP using Adam optimizer are shown in Table 10 at various levels of corruption.

Table 10. Final test accuracy in AIRBUS for ERM and RRM (ADH-LP/Adam).

	Corrupted Training Data Percentage				
Method	40%	30%	20%	10%	0%
ERM	0.588	0.657	0.729	0.797	0.867
RRM (μ = 0.5)					
θ = 0.15	0.602	0.692	0.758	0.831	0.868
θ = 0.20	0.607	0.668	0.751	0.819	0.875
θ = 0.25	0.616	0.690	0.763	0.823	0.867
θ = 0.30	0.619	0.686	0.783	0.824	0.872
θ = 0.35	0.602	0.695	0.767	0.819	0.872

Note: The values highlighted in gray represent the cases where RRM outperform or match ERM.

Despite the data corruption level, RRM consistently delivers better accuracy than ERM. This performance advantage holds across all tested θ values.

In all corruption levels, the superior performance of RRM is evident, even in a no-corruption setting. This highlights the effectiveness of the RRM methodology not just in handling corrupted data but also in solving complex hidden data patterns.

At the 20% corruption level, RRM with a θ = 0.30 demonstrates a substantial increase in accuracy, as high as 5.4% over ERM, which showcases the potential effectiveness of RRM when dealing with moderate data corruption. Similarly, the most remarkable improvement for the 30% corruption scenario was 3.8% with a θ = 0.35.

Looking at the overall performance across different corruption levels, the θ = 0.25 setting emerges as a consistently strong choice. This configuration delivers higher or equivalent accuracy compared to ERM across all contamination levels.

In summary, these findings underscore RRM's potential as a resilient and practical approach to handling various levels of data corruption, consistently matching or even slightly outperforming the ERM. The analysis supports the robustness of RRM and its broader applicability in tasks that involve varying degrees of data corruption.

Now, turning to the SGD optimizer, Table 11 displays the final test accuracy results for ERM and RRM in ADH-LP at various levels of corruption, respectively.

Table 11. Final test accuracy in AIRBUS for ERM and RRM (ADH-LP/SGD).

	Corrupted Training Data Percentage				
Method	40%	30%	20%	10%	0%
ERM	0.560	0.629	0.681	0.735	0.769
RRM ($\mu = 0.5$)					
$\theta = 0.15$	0.603	0.739	0.745	0.774	0.764
$\theta = 0.20$	0.671	0.755	0.753	0.775	0.769
$\theta = 0.25$	0.687	0.733	0.757	0.769	0.767
$\theta = 0.30$	0.684	0.744	0.764	0.765	0.774
$\theta = 0.35$	0.661	0.747	0.764	0.770	0.779

Note: The values highlighted in gray represent the cases where RRM outperforms or matches ERM.

Similar to the previous example the table shows that the RRM consistently delivers better accuracy than ERM. This performance advantage holds across all tested θ values.

At the 20% corruption level, the RRM method exhibits superior performance, particularly when configured with a θ value of 0.30. It shows a remarkable 8.3% improvement in accuracy over ERM. As the corruption level rises to 30% and θ = 0.20, RRM demonstrates an impressive accuracy improvement of 12.6% compared to ERM. Finally, at the 40% corruption level, the robustness of RRM becomes even more evident. In the set θ ∈ {0.20, 0.25, 0.30}, the RRM method delivers substantial accuracy improvements of 11.1%, 12.7%, and 12.4% over ERM, respectively.

In the lower corruption levels of 10% and 0%, the performance of RRM remains comparable or slightly superior to ERM. This underlines the flexibility of the RRM method, as it maintains strong performance even when the level of corruption decreases.

The most consistently effective choice of θ across the range of corruption levels appears to be θ = 0.30. It is the best choice for managing different levels of corruption in the AIRBUS dataset when using ADH-LP with SGD, as it consistently performs well, even in high corruption levels, and maintains similar results to other values in scenarios with 10% or no corruption.

3.2.2. U-Optimization Analysis with AIRBUS

In the following section, we aim to examine the u-vector behavior in some AIRBUS cases. Table 12 illustrates the relationship between θ and excluded images by the perturbation vector using RRM (ADH-LP) in different levels of label contamination.

Table 12. Penalty parameter and perturbation vector relationship in AIRBUS.

ADH-LP/SGD	Contamination Levels in AIRBUS							
	40% (3336 Mislabeled Images)		30% (2502 Mislabeled Images)		20% (1668 Mislabeled Images)		10% (834 Mislabeled Images)	
Value of Penalty θ	**Model Accuracy**	**# of Mislabeled Images Excluded**	**Model Accuracy**	**# of Mislabeled Images Excluded**	**Model Accuracy**	**# of Mislabeled Images Excluded**	**Model Accuracy**	**# of Mislabeled Images Excluded**
0.15	0.603	1985	0.739	1812	0.745	1210	0.774	621
0.20	0.671	2108	0.755	1767	0.753	1200	0.775	587
0.25	0.687	2127	0.733	1701	0.757	1158	0.769	573
0.30	0.684	2019	0.744	1636	0.764	1154	0.765	564
0.35	0.661	1737	0.747	1635	0.764	1109	0.770	538

Looking at the 40% label contamination column in Table 12, when $\theta = 0.25$, we have the highest number of mislabeled images excluded (2127 images) and the best test accuracy achieved (0.687) at this contamination level. The worst accuracy (0.603) is achieved when $\theta = 0.15$. Figure 14 displays the accuracy of training (red curve) and test (blue curve) of ERM (left) and RRM (right) over 500 epochs. The test accuracies differ in 4.3%, where RRM achieves 0.603 against 0.560 of ERM.

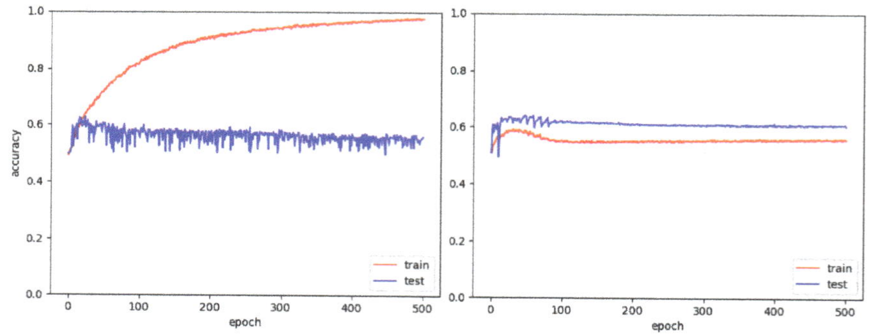

Figure 14. Training and test accuracy for ERM (**left**) and RRM ADH-LP/θ = 0.15 (**right**) on AIRBUS with 40% contamination.

Table 13 reports the u-vector evolution of u-optimization across 3336 mislabeled images and 5004 correctly labeled images, when $\theta = 0.15$.

Table 13. Evolution of u-vector across ADH-LP/SGD in AIRBUS (40% contamination; $\theta = 0.15$).

Nominal Probability (1/N) $12.0 \cdot 10^{-5}$	Iteration Number					
	$i = 1$		$i = 2$		$i = 49$	
u_i-Values	**Mislabeled Images**	**Correct Labeled Images**	**Mislabeled Images**	**Correct Labeled Images**	**Mislabeled Images**	**Correct Labeled Images**
>>0	0	1	0	1	0	2
≈0	131	264	64	194	1346	3135
$-3.0 \cdot 10^{-5}$	0	0	24	63	2	1
$-6.0 \cdot 10^{-5}$	3205	4739	67	153	3	6
$-9.0 \cdot 10^{-5}$	0	0	3181	4593	0	0
$-12.0 \cdot 10^{-5}$	0	0	0	0	1985	1860
Total of images	3336	5004	3336	5004	3336	5004

During the initial iteration, as the penalty is very low (θ = 0.15), the vast majority of images from both label categories receives the u_i-value of $-6.0 \cdot 10^{-5}$, and the rest retain their initial u_i-value of zero. In the second update, just 64 of the 3336 incorrectly labeled images remained with the initial u_i-value of zero. Eventually, over iterations, the u-value distribution stabilizes, along with test accuracy.

At the final update, the algorithm identifies 1985 out of 3336 incorrectly labeled images. These images are assigned a u_i-value of $-12 \cdot 10^{-5}$, effectively excluding them from further NN training, as this value represents the nominal probability in AIRBUS dataset. In the correct labeled images column, we notice a very similar number of images (1860) receives the lowest u_i-value and the majority (3135 out of 5004) retain their initial probability or a value close to it and contributes to the NN training process.

We also analyze the case where θ = 0.25, and it shows how crucial is the choice of our penalty parameter θ. Using RRM, this experiment reaches the best test accuracy of 0.687, an improvement of 12.7% in accuracy compared to the 0.560 of ERM. Figure 15 shows the accuracy of training (red curve) and test (blue curve) of ERM (left) and RRM (right) over 500 epochs.

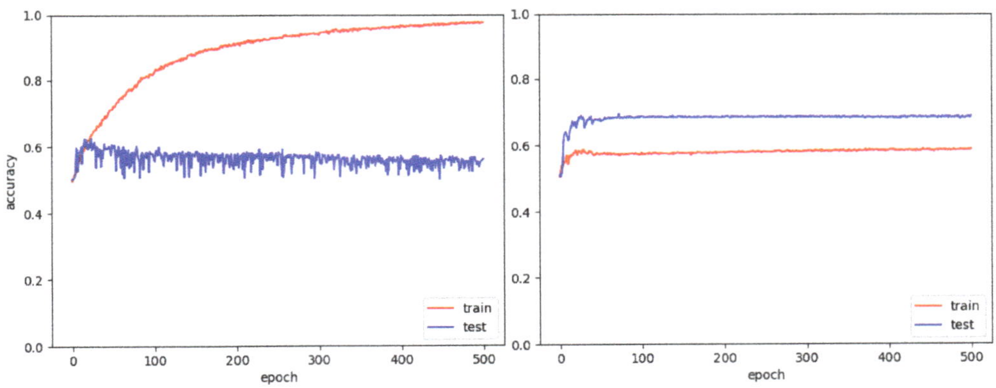

Figure 15. Training and test accuracy for ERM (**left**) and RRM ADH-LP/θ = 0.25 (**right**) on AIRBUS with 40% contamination.

Table 14 reports the u-vector evolution of u-optimization across 3336 mislabeled images and 5004 correctly labeled images, when $\theta = 0.25$.

Table 14. Evolution of u-vector across ADH-LP/SGD in AIRBUS (40% contamination; θ = 0.25).

Nominal Probability (1/N) 12.0·10^{-5}	**Iteration Number**					
	$i = 1$		$i = 2$		$i = 49$	
u_i-Values	**Mislabeled Images**	**Correct Labeled Images**	**Mislabeled Images**	**Correct Labeled Images**	**Mislabeled Images**	**Correct Labeled Images**
>>0	1	0	1	0	1	0
≈0	418	1257	312	1117	1192	3591
$-3.0 \cdot 10^{-5}$	0	0	142	590	7	1
$-6.0 \cdot 10^{-5}$	2917	3747	106	140	2	2
$-9.0 \cdot 10^{-5}$	0	0	2775	3157	5	6
$-12.0 \cdot 10^{-5}$	0	0	0	0	2129	1404
Total of images	3336	5004	3336	5004	3336	5004

During the initial iteration, a considerable number of images from both label categories receives the u_i-value of $-6.0 \cdot 10^{-5}$, 2917 of 3336 in the mislabeled portion and 3747 of

5004 in the correct labeled portion; the rest retain their initial u_i-value of zero. In the second update, until the run reaches 50 epochs, we have a picture of how the u-optimization process of assigning more mislabeled images with lower and lower u-values iteratively coincides with iterative improvements in test accuracy. Looking at Figure 15, this appears to manifest as "sawtooth" jumps improvements in test accuracy early iterations.

At the final update, the algorithm identifies 2129 out of 3336 incorrectly labeled images, 144 more than when $\theta = 0.15$. These images are assigned a u_i-value of $-12 \cdot 10^{-5}$, and they are effectively excluded of the NN learning process. Otherwise, in the correctly labeled image column, the number of images retaining zero u_i-value increases to 3591, which guarantees 456 more correctly labeled images from the NN training process.

4. Conclusions and Future Work

4.1. Conclusions

In this study, we propose an alternative strategy for training neural networks (NNs) that deviates from the traditional empirical risk minimization (ERM) approach. Our method, known as Rockafellian Risk Minimization (RRM), provides a more robust model that may outperform ERM, especially in cases where data contains label noise.

We apply our methodology to two distinct datasets, MASATI and AIRBUS. A suitable NN architecture is selected for both, and we conduct analyses under two optimizer configurations (Adam and SGD). In each case, we compare ERM with RRM ADH-LP.

In the MASATI dataset using the Adam optimizer, the ADH-LP configuration benefits significantly from RRM, with optimal θ values ranging from 0.15 to 0.25, outperforming or matching ERM. When switching to the SGD optimizer, RRM with ADH-LP performs especially well in handling data corruption levels from 20–40%. The ideal θ values are approximately 0.20 to 0.30, highlighting ADH-LP's resilience in high data corruption contexts.

Across both optimizers (Adam and SGD), ADH-LP demonstrates superior performance with the RRM method. Notably, under the SGD optimizer, ADH-LP maintains high-performance levels even as data corruption increases, showcasing a remarkable ability to withstand noise without significant performance loss. This resilience in the face of data corruption positions ADH-LP as the preferred algorithm for both Adam and SGD contexts.

For the AIRBUS dataset, RRM using Adam and ADH-LP consistently achieves competitive or superior results compared to ERM across all corruption levels, with the optimal θ around 0.25. This shows RRM's robustness in uncovering complex hidden data patterns and handling label noise. With the SGD optimizer, RRM again performs well under ADH-LP, particularly under corruption scenarios of 20%, 30%, and 40%, with a notable θ = 0.30 yielding improvements of up to 12.6%.

In cases without label contamination in the AIRBUS dataset, ADH-LP achieves the highest accuracy across architectures, suggesting RRM's effectiveness in managing complex dataset patterns, allowing for it to exceed ERM.

To conclude, RRM has proven to be a robust method for handling label corruption in the AIRBUS dataset. Across all configurations, the ADH-LP architecture under the SGD optimizer stands out for its ability to sustain high performance levels despite increasing corruption, establishing it as the top choice.

Overall, our analysis of the MASATI and AIRBUS datasets, using different optimizers and data corruption levels, consistently demonstrates ADH-LP as the superior performer. ADH-LP's strength lies in its balance between managing data corruption and optimizing performance, showing remarkable resilience against label noise while maintaining accuracy. While RRM ADH-LP requires slightly longer processing time than traditional methods, the test accuracy and robustness make ADH-LP the optimal choice in both Adam and SGD contexts, confirming its resilience and robustness as the preferred algorithm.

4.2. Future Work

Despite the notable improvements in accuracy and robustness achieved through the implementation of the RRM method, there remains significant potential for further enhancement. A key area for exploration is the u-optimization process. Currently, u-optimization effectively excludes up to 64% of mislabeled images in experiments where RRM outperforms ERM. However, by incorporating novel techniques into the u-optimization procedure, this exclusion rate could be further improved, leading to even greater model accuracy and resilience.

Advancements in computer vision, driven by enhanced AI and machine learning algorithms, have unlocked unprecedented capabilities in automating complex tasks that traditionally required human oversight. Tools such as convolutional neural networks (CNNs), generative adversarial networks (GANs), and deep learning frameworks are now integral to computer vision tasks. These tools enable more refined feature extraction and pattern recognition, which are particularly valuable in high-stakes environments like maritime surveillance. When applied to ship detection, these models can be tuned to distinguish between various objects in challenging environments, such as open waters, where factors like lighting, sea state, and occlusion can complicate detection.

For maritime surveillance systems, robust AI-driven algorithms like the enhanced ADH-LP could offer transformative benefits. Unmanned aerial and underwater vehicles equipped with high-precision computer vision models could autonomously detect, track, and analyze vessels. This capability would bolster maritime security, improve illegal fishing detection, and assist in search-and-rescue operations. By integrating RRM-based models, which are resilient to label noise, these systems can maintain high accuracy even when faced with inconsistent or imperfect data a common issue in real-world surveillance data.

Moreover, the adaptive nature of machine learning algorithms enables continuous improvement in performance as they encounter diverse datasets, enhancing the ability of these models to generalize across different maritime conditions and vessel types. This adaptability is especially valuable in monitoring vast, complex marine environments, where a single model must manage varied inputs. Advanced models using techniques like RRM combined with adaptive learning strategies could help create surveillance systems capable of making precise and timely decisions with minimal human intervention.

As AI and machine learning continue to evolve, their integration into computer vision systems like the enhanced ADH-LP holds the potential to revolutionize maritime operations. These advancements provide not only heightened accuracy and robustness but also operational scalability, transforming how we approach surveillance and security across extensive, dynamic environments such as open waters.

Author Contributions: Conceptualization, G.C.R., V.B.A.d.S.A. and E.C.E.; Methodology, G.C.R. and V.B.A.d.S.A.; Software, G.C.R. and V.B.A.d.S.A.; Validation, G.C.R. and V.B.A.d.S.A.; Formal analysis, G.C.R. and V.B.A.d.S.A.; Investigation, G.C.R. and V.B.A.d.S.A.; Resources, G.C.R. and V.B.A.d.S.A.; Data curation, G.C.R., V.B.A.d.S.A., I.P.d.A.C., M.Â.L.M. and A.P.d.A.C.; Writing—original draft, G.C.R., V.B.A.d.S.A. and E.C.E.; Writing—review & editing, G.C.R., V.B.A.d.S.A., I.P.d.A.C., M.Â.L.M., A.P.d.A.C. and E.C.E.; Visualization, I.P.d.A.C., M.Â.L.M., A.P.d.A.C. and M.d.S.; Supervision, I.P.d.A.C., M.Â.L.M., A.P.d.A.C., M.d.S. and E.C.E.; Project administration, I.P.d.A.C., M.Â.L.M., A.P.d.A.C. and M.d.S.; Funding acquisition, I.P.d.A.C. All authors have read and agreed to the published version of the manuscript.

Funding: This research received no external funding.

Data Availability Statement: The original contributions presented in this study are included in the article. Further inquiries can be directed to the corresponding author.

Conflicts of Interest: The authors declare no conflict of interest.

References

1. Wiesebron, M. Blue Amazon: Thinking the Defense of Brazilian Maritime Territory. *Austral Braz. J. Strategy Int. Relat.* **2013**, 2, 101–124. Available online: https://www.files.ethz.ch/isn/166009/37107-147326-1-PB.pdf#page=102 (accessed on 12 June 2023).
2. Brazilian Navy Command. Brazilian Navy Naval Policy. Available online: https://www.naval.com.br/blog/wp-content/uploads/2019/04/PoliticaNavalMB.pdf (accessed on 9 April 2023).
3. Interdisciplinary Observatory on Climate Change. Last Frontier at Sea. Available online: https://obsinterclima.eco.br/mapas/ultima-fronteira-no-mar/ (accessed on 7 April 2023).
4. Brazilian Navy Social Communication Center. Blue Amazon—The Heritage Brazilian at Sea. *Villegagnon Journal Supplement—VI Academic Congress on National Defense.* Available online: http://www.redebim.dphdm.mar.mil.br/vinculos/000006/00000600.pdf (accessed on 9 April 2023).
5. de Oliveira Andrade, I.; da Rocha, A.J.R.; Franco, L.G.A. *DP 0261—Blue Amazon Management System (SisGAAz): Sovereignty, Surveillance and Defense of the Brazilian Jurisdictional Waters*; Discussion Paper; Instituto de Pesquisa Economica Aplicada—IPEA: Brasília, Brazil, 2021; 35p. [CrossRef]
6. de Oliveira Andrade, I.; Franco, L.G.A.; Hillebrand, G.R.L. *DP 2471—Science, Technology and Innovation in The Brazilian Navy's Strategic Programs*; Discussion Paper; Instituto de Pesquisa Economica Aplicada—IPEA: Brasília, Brazil, 2019. Available online: https://www.researchgate.net/publication/335925079_Ciencia_Tecnologia_e_Inovacao_nos_Programas_Estrategicos_da_Marinha_do_Brasil (accessed on 12 June 2023).
7. Murphy, K.P. *Machine Learning: A Probabilistic Perspective*; Adaptive Computation and Machine Learning Series; MIT Press: Cambridge, MA, USA, 2012.
8. Gewalt, R. Supervised vs. Unsupervised vs. Reinforcement Learning—The Fundamental Differences, Fly Spaceships with Your Mind. Available online: https://www.opit.com/magazine/supervised-vs-unsupervised-learning/ (accessed on 10 April 2023).
9. Sohil, F.; Sohali, M.U.; Shabbir, J. *An Introduction to Statistical Learning with Applications in R, Statistical Theory and Related Fields*; Informa UK Limited: London, UK, 2021; Volume 6. [CrossRef]
10. Petercour, Machine Learning Classification vs. Regression, DEV Community. Available online: https://dev.to/petercour/machine-learning-classification-vs-regression-1gn (accessed on 10 April 2023).
11. Gunjal, S. Logistic Regression from Scratch with Python, Quality Tech Tutorials. Available online: https://satishgunjal.com/binary_lr/ (accessed on 24 April 2023).
12. Goodfellow, I.; Bengio, Y.; Courville, A. *Deep Learning*; MIT Press: Cambridge, MA, USA, 2016; Available online: http://www.deeplearningbook.org (accessed on 17 June 2023).
13. Burkov, A. The Hundred-Page Machine Learning Book. 2019. Available online: http://ema.cri-info.cm/wp-content/uploads/2019/07/2019BurkovTheHundred-pageMachineLearning.pdf (accessed on 22 June 2023).
14. Haykin, S.S. *Neural Networks and Learning Machines*, 3rd ed.; Prentice-Hall: New York, NY, USA, 2009.
15. Banoula, M. What is Perceptron? A Beginner's Guide [updated]: Simplilearn. Available online: https://www.simplilearn.com/tutorials/deep-learning-tutorial/perceptron (accessed on 24 April 2023).
16. Shanmugamani, R. *Deep Learning for Computer Vision: Expert Techniques to Train Advanced Neural Networks Using TensorFlow and Keras*; Packt Publishing: Birmingham, UK, 2018.
17. Activation Function—AI Wiki. Available online: https://machine-learning.paperspace.com/wiki/activation-function (accessed on 24 April 2023).
18. Szeliski, R. *Computer Vision: Algorithms and Applications*, 2nd ed.; Springer Ltd.: London, UK, 2022.
19. Varghese, L.J.; Jacob, S.S.; Sundar, C.; Raglend, J. Design and Implementation of a Machine Learning Assisted Smart Wheelchair in an IoT Environment. Research Square Platform LLC: Durham, NC, USA, 2021. [CrossRef]
20. Khan, S.; Rahmani, H.; Shah, S.A.A.; Bennamoun, M. *A Guide to Convolutional Neural Networks for Computer Vision*; Springer International Publishing: Berlin/Heidelberg, Germany, 2018. [CrossRef]
21. Chen, L.; Li, S.; Bai, Q.; Yang, J.; Jiang, S.; Miao, Y. Review of Image Classification Algorithms Based on Convolutional Neural Networks. *Remote Sens.* **2021**, *13*, 4712. [CrossRef]
22. Torén, R. Comparing CNN Methods for Detection and Tracking of Ships in Satellite Images. Master's Thesis, Department of Computer and Information Science, Linköping University, Linköping, Sweden, 2020.
23. TensorFlow Core. Introduction to Automatic Encoders. Available online: https://www.tensorflow.org/tutorials/generative/autoencoder?hl=pt-br (accessed on 28 April 2023).
24. CS 230—Convolutional Neural Networks Cheatsheet. Available online: https://stanford.edu/~shervine/teaching/cs-230/cheatsheet-convolutional-neural-networks (accessed on 28 April 2023).
25. Royset, J.O.; Chen, L.L.; Eckstrand, E. Rockafellian Relaxation in Optimization under Uncertainty: Asymptotically Exact Formulations. *arXiv* **2022**, arXiv:2204.04762. Available online: https://arxiv.org/abs/2204.04762 (accessed on 27 June 2023).
26. Maron, M.E. Automatic Indexing: An Experimental Inquiry. *J. ACM* **1961**, *8*, 404–417. [CrossRef]
27. Cover, T.; Hart, P. Nearest neighbor pattern classification. *IEEE Trans. Inf. Theory* **1967**, *13*, 21–27. [CrossRef]

28. Quinlan, J.R. Induction of Decision Trees. *Mach. Lang.* **1986**, *1*, 81–106. [CrossRef]
29. Cortes, C.; Vapnik, V. Support-vector networks. *Mach. Learn.* **1995**, *20*, 273–297. [CrossRef]
30. Freund, Y.; Schapire, R.E. A Decision-Theoretic Generalization of On-Line Learning and an Application to Boosting. *J. Comput. Syst. Sci.* **1997**, *55*, 119–139. [CrossRef]
31. Breiman, L. Random Forests. *Mach. Lang.* **2001**, *45*, 5–32. [CrossRef]
32. Hinton, G.; Osindero, S.; Teh, Y.-W. A Fast Learning Algorithm for Deep Belief Nets. *Neural Comput.* **2006**, *18*, 1527–1554. [CrossRef] [PubMed]
33. McCulloch, W.; Pitts, W. A logical calculus of the ideas immanent in nervous activity. *Bull. Math. Biophys.* **1943**, *5*, 49–50. [CrossRef]
34. Rosenblatt, F. The perceptron: A probabilistic model for information storage and organization in the brain; American Psychological Association (APA). *Psychol. Rev.* **1958**, *65*, 386–408. [CrossRef]
35. Steinbuch, K.; Widrow, B. A Critical Comparison of Two Kinds of Adaptive Classification Networks. *IEEE Trans. Electron. Comput.* **1965**, *EC-14*, 737–740. [CrossRef]
36. Rumelhart, D.E.; Hinton, G.E.; Williams, R.J. Learning representations by back-propagating errors. *Nature* **1986**, *323*, 6088. [CrossRef]
37. Lecun, Y.; Bottou, L.; Bengio, Y.; Haffner, P. Gradient-based learning applied to document recognition. *Proc. IEEE* **1998**, *86*, 2278–2324. [CrossRef]
38. Glorot, X.; Bengio, Y. Understanding the difficulty of training deep feedforward neural networks. In Proceedings of the Thirteenth International Conference on Artificial Intelligence and Statistics; JMLR Workshop and Conference Proceedings, Sardinia, Italy, 13–15 May 2010; pp. 249–256. Available online: https://proceedings.mlr.press/v9/glorot10a.html (accessed on 8 May 2023).
39. Krizhevsky, A.; Sutskever, I.; Hinton, G.E. ImageNet Classification with Deep Convolutional Neural Networks. In *Advances in Neural Information Processing Systems*; Curran Associates, Inc.: New York, NY, USA, 2012. Available online: https://papers.nips.cc/paper_files/paper/2012/hash/c399862d3b9d6b76c8436e924a68c45b-Abstract.html (accessed on 8 May 2023).
40. Szegedy, C.; Liu, W.; Jia, Y.; Sermanet, P.; Reed, S.; Anguelov, D.; Erhan, D.; Vanhoucke, V.; Rabinovich, A. Going Deeper with Convolutions. *arXiv* **2014**, arXiv:1409.4842. Available online: http://arxiv.org/abs/1409.4842 (accessed on 8 May 2023).
41. Simonyan, K.; Zisserman, A. Very Deep Convolutional Networks for Large-Scale Image Recognition. *arXiv* **2015**, arXiv:1409.1556. Available online: http://arxiv.org/abs/1409.1556 (accessed on 8 May 2023).
42. Shadeed, G.A.; Tawfeeq, M.A.; Mahmoud, S.M. Automatic Medical Images Segmentation Based on Deep Learning Networks. *IOP Conf. Ser. Mater. Sci. Eng.* **2020**, *870*, 012117. [CrossRef]
43. He, K.; Zhang, X.; Ren, S.; Sun, J. Deep Residual Learning for Image Recognition. In Proceedings of the 2016 IEEE Conference on Computer Vision and Pattern Recognition (CVPR), Las Vegas, NV, USA, 26 June–1 July 2016; pp. 770–778. [CrossRef]
44. Fortunati, V. Deep Learning Applications in Radiology: A Deep Dive on Classification. Available online: https://www.quantib.com/blog/deep-learning-applications-in-radiology/classification (accessed on 9 May 2023).
45. Veit, A.; Wilber, M.J.; Belongie, S. Residual Networks Behave Like Ensembles of Relatively Shallow Networks. In *Advances in Neural Information Processing Systems*; Curran Associates, Inc.: New York, NY, USA, 2016. Available online: https://proceedings.neurips.cc/paper_files/paper/2016/hash/37bc2f75bf1bcfe8450a1a41c200364c-Abstract.html (accessed on 7 June 2023).
46. Hanin, B. Which Neural Net Architectures Give Rise to Exploding and Vanishing Gradients? *arXiv* **2018**, arXiv:1801.03744. Available online: http://arxiv.org/abs/1801.03744 (accessed on 7 June 2023).
47. Tetreault, B.J. Use of the Automatic Identification System (AIS) for maritime domain awareness (MDA). In Proceedings of the OCEANS 2005 MTS/IEEE, Washington, DC, USA 17–23 September 2005; Volume 2, pp. 1590–1594. [CrossRef]
48. Zardoua, Y.; Astito, A.; Boulaala, M. A Comparison of AIS, X-Band Marine Radar Systems and Camera Surveillance Systems in the Collection of Tracking Data. *arXiv* **2020**, arXiv:2206.12809.
49. Ma, M.; Chen, J.; Liu, W.; Yang, W. Ship Classification and Detection Based on CNN Using GF-3 SAR Images. *Remote Sens.* **2018**, *10*, 2043. [CrossRef]
50. Ødegaard, N.; Knapskog, A.O.; Cochin, C.; Louvigne, J.-C. Classification of ships using real and simulated data in a convolutional neural network. In Proceedings of the 2016 IEEE Radar Conference (RadarConf), Philadelphia, PA, USA, 2–6 May 2016; pp. 1–6. [CrossRef]
51. Bentes, C.; Velotto, D.; Tings, B. Ship Classification in TerraSAR-X Images with Convolutional Neural Networks. *IEEE J. Ocean. Eng.* **2018**, *43*, 258–266. [CrossRef]
52. Gallego, A.-J.; Pertusa, A.; Gil, P. Automatic Ship Classification from Optical Aerial Images with Convolutional Neural Networks. *Remote Sens.* **2018**, *10*, 511. [CrossRef]
53. Fang, H.; Chen, M.; Liu, X.; Yao, S. Infrared Small Target Detection with Total Variation and Reweighted $\ell 1$ Regularization. *Math. Probl. Eng.* **2020**, *2020*, 1529704. [CrossRef]
54. Kanellakis, C.; Nikolakopoulos, G. Survey on Computer Vision for UAVs: Current Developments and Trends. *J. Intell. Robot. Syst.* **2017**, *87*, 141–168. [CrossRef]

55. Cruz, G.; Bernardino, A. Aerial Detection in Maritime Scenarios Using Convolutional Neural Networks. In *Advanced Concepts for Intelligent Vision Systems*; Lecture Notes in Computer Science; Blanc-Talon, J., Distante, C., Philips, W., Popescu, D., Scheunders, P., Eds.; Springer International Publishing: Cham, Switzerland, 2016; Volume 10016, pp. 373–384. [CrossRef]
56. Lo, L.-Y.; Yiu, C.H.; Tang, Y.; Yang, A.-S.; Li, B.; Wen, C.-Y. Dynamic Object Tracking on Autonomous UAV System for Surveillance Applications. *Sensors* **2021**, *21*, 7888. [CrossRef]
57. Lygouras, E.; Santavas, N.; Taitzoglou, A.; Tarchanidis, K.; Mitropoulos, A.; Gasteratos, A. Unsupervised Human Detection with an Embedded Vision System on a Fully Autonomous UAV for Search and Rescue Operations. *Sensors* **2019**, *19*, 3542. [CrossRef]
58. Redmon, J.; Divvala, S.; Girshick, R.; Farhadi, A. You Only Look Once: Unified, Real-Time Object Detection. *arXiv* **2016**, arXiv:1506.02640. Available online: http://arxiv.org/abs/1506.02640 (accessed on 10 May 2023).
59. WACV 2023—Maritime Workshop. Available online: https://seadronessee.cs.uni-tuebingen.de/wacv23 (accessed on 20 May 2023).
60. Hickey, R.J. Noise modelling and evaluating learning from examples. *Artif. Intell.* **1996**, *82*, 157–179. [CrossRef]
61. Song, H.; Kim, M.; Park, D.; Shin, Y.; Lee, J.-G. Learning from Noisy Labels with Deep Neural Networks: A Survey. *arXiv* **2022**, arXiv:2007.08199. Available online: http://arxiv.org/abs/2007.08199 (accessed on 21 May 2023). [CrossRef] [PubMed]
62. Liu, T.; Tao, D. Classification with Noisy Labels by Importance Reweighting. *IEEE Trans. Pattern Anal. Mach. Intell.* **2016**, *38*, 447–461. [CrossRef] [PubMed]
63. Yang, C.; Wu, Q.; Li, H.; Chen, Y. Generative Poisoning Attack Method Against Neural Networks. *arXiv* **2017**, arXiv:1703.01340. Available online: http://arxiv.org/abs/1703.01340 (accessed on 22 May 2023).
64. Ren, M.; Zeng, W.; Yang, B.; Urtasun, R. Learning to Reweight Examples for Robust Deep Learning. In Proceedings of the 35th International Conference on Machine Learning, Stockholm, Sweden, 10–15 July 2018.
65. Thulasidasan, S.; Bhattacharya, T.; Bilmes, J.; Chennupati, G.; Mohd-Yusof, J. Combating Label Noise in Deep Learning Using Abstention. *arXiv* **2019**, arXiv:1905.10964. Available online: http://arxiv.org/abs/1905.10964 (accessed on 21 May 2023).
66. Chen, L.; Huang, N.; Mu, C.; Helm, H.S.; Lytvynets, K.; Yang, W.; Priebe, C.E. Deep Learning with Label Noise: A Hierarchical Approach. *arXiv* **2022**, arXiv:2205.14299. Available online: http://arxiv.org/abs/2205.14299 (accessed on 21 May 2023).
67. Narasimhan, H.; Menon, A.K.; Jitkrittum, W.; Kumar, S. Learning to reject meets OOD detection: Are all abstentions created equal? *arXiv* **2023**, arXiv:2301.12386. Available online: http://arxiv.org/abs/2301.12386 (accessed on 21 May 2023).
68. Ni, C.; Charoenphakdee, N.; Honda, J.; Sugiyama, M. On the Calibration of Multiclass Classification with Rejection. *arXiv* **2019**, arXiv:1901.10655. Available online: http://arxiv.org/abs/1901.10655 (accessed on 22 May 2023).
69. Ramaswamy, H.G.; Tewari, A.; Agarwal, S. Consistent algorithms for multiclass classification with an abstain option. *Electron. J. Stat.* **2018**, *12*, 530–554. [CrossRef]
70. Katz-Samuels, J.; Nakhleh, J.; Nowak, R.; Li, Y. Training OOD Detectors in their Natural Habitats. *arXiv* **2022**, arXiv:2202.03299. Available online: http://arxiv.org/abs/2202.03299 (accessed on 22 May 2023).
71. Royset, J.O.; Wets, R.J.-B. An Optimization Primer. In *Springer Series in Operations Research and Financial Engineering*; Springer International Publishing: Cham, Switzerland, 2021. [CrossRef]
72. Wang, W.; Carreira-Perpiñán, M.Á. Projection onto the probability simplex: An efficient algorithm with a simple proof, and an application. *arXiv* **2013**, arXiv:1309.1541. Available online: http://arxiv.org/abs/1309.1541 (accessed on 22 April 2023).
73. Airbus Ship Detection Challenge. Available online: https://kaggle.com/competitions/airbus-ship-detection (accessed on 30 May 2023).
74. Kingma, D.P.; Ba, J. Adam: A Method for Stochastic Optimization. *arXiv* **2017**, arXiv:1412.6980. Available online: http://arxiv.org/abs/1412.6980 (accessed on 8 May 2023).
75. Bishop, C.M. Neural Networks: A Pattern Recognition Perspective. In *Handbook of Neural Computation*; CRC Press: Boca Raton, FL, USA, 2020; pp. 1–6.

water

MDPI

rticle

Perception of Mangrove Social–Ecological System Governance in Southeastern Cuba

'anet Cruz Portorreal [1,*], Natalie Beenaerts [2], Nico Koedam [2,3,4,5], Orlando Joel Reyes Dominguez [6], 'elene B. Milanes [7,8,*], Farid Dahdouh-Guebas [3,5,9,10] and Ofelia Pérez Montero [1]

1 Center for Multidisciplinary Studies of Coastal Zones, Universidad de Oriente, Ave. Las Américas s/n, Santiago de Cuba 90100, Cuba; ofelia@uo.edu.cu
2 Centre for Environmental Sciences, Hasselt University, Agoralaan z/n, B3590 Diepenbeek, Belgium; natalie.beenaerts@uhasselt.be (N.B.); koedamnico@gmail.com (N.K.)
3 Systems Ecology and Resource Management, Department of Organism Biology, Faculty of Sciences, Université Libre de Bruxelles—ULB, Av. F.D. Roosevelt 50, CPi 264/1, B1050 Brussels, Belgium; farid.dahdouh-guebas@ulb.be or farid.dahdouh-guebas@vub.be
4 Marine Biology Research Group, Universiteit Gent, Krijgslaan 281-S8, B9000 Gent, Belgium
5 Mangrove Specialist Group (MSG), Species Survival Commission (SSC), International Union for the Conservation of Nature (IUCN), 1196 Gland, Switzerland
6 Eastern Center for Ecosystems and Biodiversity (BIOECO), Museo de Historia Natural "Tomás Romay" José A. Saco # 601, Esquina a Barnada, Santiago de Cuba 90100, Cuba; joel@bioeco.cu
7 Facultad de Ingeniería, Universidad del Magdalena, Calle 29H3 No 22-01, Santa Marta 470004, Magdalena, Colombia
8 Estación Científica (Coiba AIP), Gustavo Lara Street, Building 145B, City of Knowledge, Clayton, Panama City 0801, Panama
9 bDIV: Ecology, Evolution & Genetics, Department of Biology, Faculty of Sciences and Bio-Engineering Sciences, Vrije Universiteit Brussel—VUB, Pleinlaan 2, B1050 Brussels, Belgium
10 Interfaculty Institute of Social-Ecological Transitions, Université Libre de Bruxelles—ULB, Av. F.D. Roosevelt 50, B1050 Brussels, Belgium
* Correspondence: yanetc@uo.edu.cu (Y.C.P.); cmilanes@unimagdalena.edu.co (C.B.M.); Tel.: +57-605-4381000 (C.B.M.)

Abstract: This study examined local communities' perceptions of mangroves in coastal southeastern Cuba. A variety of methods were employed, including mixed and structured questionnaire surveys, interviews with key informants, and document reviews. Data were gathered from 334 respondents living in communities adjacent to four mangrove social–ecological systems (SESs). The analysis focused on five variables: community use of mangrove resources, ecosystem services, threats to the ecosystem, management activities, and social–ecological relationships. To qualitatively assess the influence of social–ecological relationships and governance, a matrix was created based on anthropogenic activities identified by respondents and their perceptions of ecosystem services. A Spearman's rank correlation analysis was performed between demographic variables and identified mangrove uses. The Kruskal–Wallis test was used to compare the frequency of mangrove uses and the perception of ecosystem services among the studied areas. The results indicate that, while local people recognise the uses and ecosystem services of mangroves, they do not rely on them for their livelihoods. Perceptions of ecosystem services vary significantly depending on the occupation of the respondents and the locality. They also showed moderate to full awareness of management responsibilities and activities at each site. The most commonly identified threats were climate change, drought, and deforestation. Three types of social–ecological relationships were identified based on the characteristics of the communities, their economic activities, and their impacts on the mangroves: urban–industrial, rural–agricultural, and rural–agricultural/tourism. Based on the results, recommendations are made for ecosystem governance in the southeast of Cuba.

Keywords: ecosystem service; mangroves; ecosystem services; management; climate change; mitigation; adaptation; nature-based solution; citizen participation

1. Introduction

Addressing climate change mitigation requires a comprehensive understanding of natural processes and the impacts of human development on these processes. In this context, the framework of social–ecological systems (SESs) is pertinent. SESs represent complex adaptive systems where human societies are closely integrated with nature [1,2]. Within the SES framework, the social component encompasses all human activities, while the ecological component refers to the biosphere and its associated natural processes [3]. These components are interrelated, and their boundaries depend on the perspective of the analysis. The interaction between these components is dynamic, and their limits are defined according to the analytical perspective adopted. Ecosystem services, defined as the benefits society derives from ecosystems, illustrate this interaction [3]. Analysing the relationships within a defined SES facilitates the search for integrated solutions to problems arising from human activities in specific ecosystems. This integrated approach is essential for developing strategies that not only mitigate the negative effects of human development on natural processes but also increase the resilience and sustainability of human and ecological systems.

Understanding and managing social–ecological relationships is crucial for the sustainability of mangrove socioecological systems, ensuring that both human and ecological needs are addressed in a balanced and sustainable manner. SESs comprising mangroves are characterised by complex social and ecological relationships [2,4–7], including the provision of ecosystem services and human activities [3] affecting them. The ecosystem services encompass supporting, provisioning, regulating, and cultural or spiritual services [8,9]. For example, coastal ecosystems contribute to climate regulation through carbon sequestration. Mangroves, along with marshes and seagrasses, are classified as blue carbon ecosystems due to their significant capacity to sequester atmospheric carbon [10–13]. They can absorb and sequester substantial amounts of carbon through aboveground and belowground biomass [13]. Moreover, the environmental conditions of these ecosystems favour the long-term accumulation of organic matter in the soil [14]. Mangroves provide critical regulatory services to coastal communities by mitigating the impact of waves, wind, and flooding [15–18]. The climate regulation ecosystem service not only addresses the causes of climate change but also mitigates its impacts. They also directly contribute to the subsistence of coastal communities, as some species living in the mangrove ecosystem are an important food resource; thus, mangroves support the local economy or household livelihoods [19–22].

Despite their importance, mangroves are threatened by human activities, mainly those that transform the spaces where these ecosystems develop [17,23,24]. Human activities include overexploitation of resources, land use change, agricultural development, damming of freshwater sources hydrological change in general, and pollution. The reduction or loss of mangroves leads to the reduction or loss of the ecosystem services they provide, thus diminishing the benefits for local communities [16,25,26]. In response, the international scientific community is striving to integrate the value of ecosystem services into decision-making processes regarding human development and conservation efforts [8,27].

Initiatives such as the Global Mangrove Alliance and the International Blue Carbon Initiative recognise mangroves as high-priority ecosystems for climate change mitigation through climate regulation [17,20]. Moreover, the way coastal communities perceive mangrove ecosystem services influences the success of conservation efforts [28–30] and the acceptance or not of management measures. Therefore, the priorities and preferences of these communities should be integrated into the decision-making process [27,31,32]. It is essential to work with both the mangrove ecosystem and the surrounding social system to analyse how society benefits from nature [33,34]. Identifying the reasons for protecting ecosystem services helps determine which services are relevant to stakeholders and informs on the options to consider for management decisions [18]. There is substantial evidence indicating that coastal communities significantly influence the ecological health of mangroves, the quality of ecosystem services [35,36], and their governance structures [37–39]. Inadequate participation hinders, for example, the establishment of governance norms within the existing political framework, which in turn affects the management and conservation of mangroves [39].

In Cuba, mangroves are recognised as vulnerable and significant coastal wetlands [24]. Globally, Cuba ranks 10th in terms of mangrove surface area [40], covering 5.1% of the national territory and constituting 27% of the total forest area. Approximately 35% of mangroves are legally safeguarded by the National System of Protected Areas (SNAP) and are managed through various conservation measures [41]. Within SNAP, environmental education initiatives, such as training sessions, workshops, and festivals, are exclusively conducted in the communities surrounding the protected mangrove areas. The objective of these activities is to foster community participation in mangrove conservation [42]. Although mangroves are primarily valued for their role in protecting coastal communities and infrastructure from the effects of climate change, such as coastal flooding and strong winds [24], efforts to quantify the ecosystem services they provide in terms of provisioning, regulating, and supporting functions are gaining momentum in the country [43].

Despite the limited research on the analysis of perceptions within coastal communities and governance to improve mangrove management in Cuba, existing evidence underscores the local recognition of the importance of coastal ecosystems, particularly mangroves, in sustaining subsistence activities such as fishing, tourism, and agriculture [24,44]. To date, comprehensive studies are lacking in exploring the intricate social, ecological, and governance dynamics within mangrove habitats. This study aims to elucidate the interactions between SESs and governance in the mangroves of southeastern Cuba, focusing on community perceptions of this valuable ecosystem, which are essential for its integrated and sustainable management.

This research contributes to addressing climate change impact in Cuba by highlighting the role of these ecosystems in carbon capture and storage to mitigate greenhouse gases. Additionally, it promotes the sustainable use of terrestrial ecosystems by generating knowledge on how to manage mangroves to maximise carbon retention. Furthermore, it supports the development of sustainable coastal urban communities by revitalising urban green spaces for carbon capture. Finally, this research contributes to sustainable development goals, specifically to SDG 13 "Life of Terrestrial Ecosystems" and 14 "Marine Life" [45].

2. Materials and Methods

2.1. Study Area

The study area in Cuba is part of the Northwest Atlantic Province [46] and experiences significant influence from seasonal tropical cyclones, which bring heavy rainfall that reduces salinity and increases nutrient loads in watersheds draining into mangrove ecosystems [47]. Cuban mangroves typically form narrow strips, which vary from approximately 0.25 to 0.80 m [47]. The extent and structure of these mangroves are influenced by geomorphology, characteristics of rivers and tributaries, as well as local climatic conditions [47]. Mangrove development is more extensive in the southern central part of the island, while the northeastern coast generally shows mangroves of a lower canopy height. Mangrove coverage is notably limited in eastern Cuba. Our study focused on the southeastern region of Cuba, defined as a natural and political entity extending along the eastern part of the island, south of an imaginary line from the centre of Guacanayabo Bay to the northern edge of Banes Bay [48]. This region is bordered to the north by the Central Region, to the northeast by the Atlantic Ocean, to the east by the Paso de los Vientos, to the south by the Caribbean Sea, and to the west by the Gulf of Guacanayabo [48]. The climate in this region is classified as tropical savannah (type AW according to the Köppen classification [46,49]), with annual average temperatures ranging from 25.6 °C in Las Tunas to 26.8 °C in Guantánamo. Annual rainfall varies from 792 mm in Santiago de Cuba to 1130 mm in Las Tunas. Coastal geomorphological and hydrodynamic characteristics create diverse environments, some of which are conducive to mangrove development.

For this study, four mangrove localities were chosen in southeastern Cuba based on specific criteria, including proximity of communities to mangroves and community accessibility sites: Monte Cabaniguán/Ojo de Agua (MCOA) in Las Tunas province, Guamá (GUAM) and San Miguel de Parada (SAM) in Santiago de Cuba province, and Hatibonico (HAT) in Guantánamo province (Figure 2). A social–ecological survey was designed, and a structured questionnaire was administered from March to October 2023 in selected communities within these specific mangrove localities.

Features of the mangrove SES studied are the following:

(1) Ojo de Agua-Monte Cabaniguán (MCOA) (See Table 1)

This mangrove site is a fragment of the Delta del Cauto vegetation [50], under the administration of the Ojo de Agua-Monte Cabaniguán Wildlife Refuge of the National System of Protected Areas of Cuba. It is located between the municipalities of Jobabo and Colombia in the extreme southwest of the province of Las Tunas next to the Gulf of Guacanayabo (Figure 2A). This sector was selected to represent the Cauto Delta Wetland ecosystem, which contains the largest extent of mangroves in the study area and the second largest in Cuba and the Caribbean Archipelago [50]. Located in a coastal deltaic–alluvial plain, these mangroves, together with the marshes, form an extensive vegetation formation formed by tidal activity and contributions from the Cauto River delta (Figure 1) [47]. The area covers 3929.18 ha and represents 57% of the Delta del Cauto Faunal Refuge [51].

This area is predominantly rural [54]. Economic activities include extensive livestock farming, commercial fishing, and charcoal production made of *Dichrostachys cinerea* Wight et Arn, an exotic leguminous terrestrial tree species. Fishing, as well as poaching, are considered the most impactful activities, including the capture and illegal trade of key mangrove-associated species: the American crocodile, *Crocodylus acutus* (Cuvier), the Cuban iguana, *Cyclura nubila* Gray, and hutias (*Capromys pilorides* Say) [51]. Livestock production also has an impact through uncontrolled animal entry into the reserve. It contributes to the spread of exotic species like *Dichrostachys cinerea* and soil compaction. Despite the measures taken to manage the reserve, these problems continue to exist [51]. Finally, the supply of freshwater and sediment to the mangroves has been significantly reduced by the construction of upstream reservoirs.

Figure 1. Assemblage along a tidal creek, with *Rhizophora mangle* and *Avicennia germinans* in Ojo de Agua-Monte Cabaniguán Wildlife Refuge (credits to Nico Koedam).

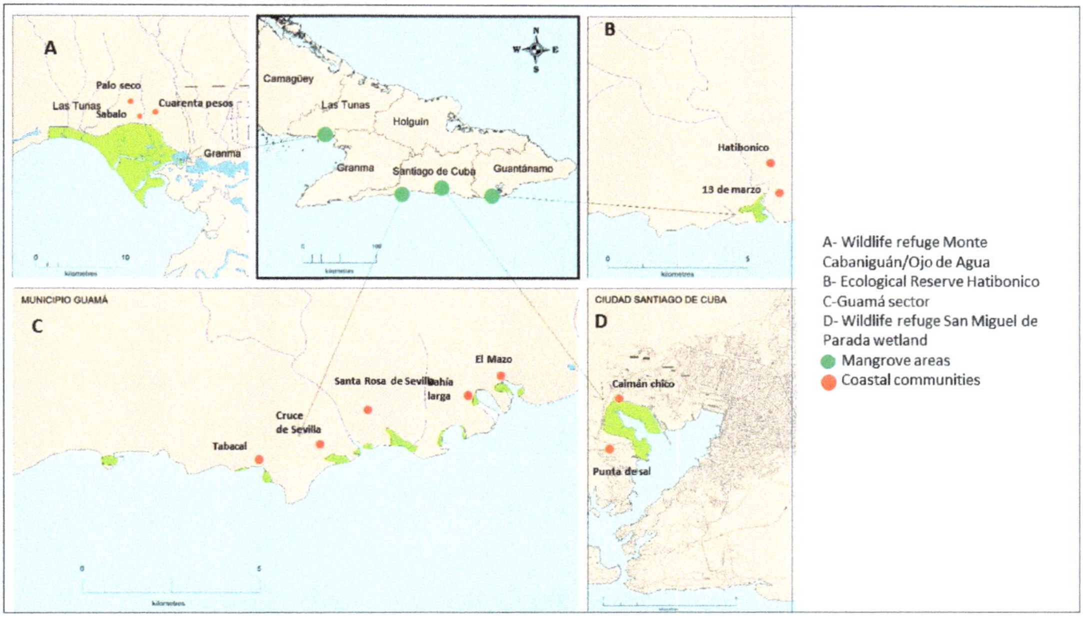

Figure 2. Map of eastern Cuba (inset: Cuba island) with mangrove localities under survey (Maps (**A**)—Wildlife Refuge Monte Cabaniguán/Ojo de Agua, Las Tunas [51]; (**B**)—Ecological Reserve Hatibonico, Guantánamo [52]; (**C**)—Guamá Sector, Santiago de Cuba [20]; and (**D**)—Wildlife Refuge San Miguel de Parada, Santiago de Cuba [53]). Sections with green shades show the mangrove areas, while the red circles indicate the coastal villages that were surveyed.

(2) Guamá Sector (GUAM) (See Table 1)

This sector is located in Guamá municipality to the south of the Sierra Maestra in the province of Santiago de Cuba (Figure 2C) [55–57]. Mangroves in Guamá municipality are not under any type of official management system [20]. They are located along the coastline and are mostly subject to marine influence. However, they can also receive contributions from local terrestrial runoff. Water input comes from land runoff, the Los Lirios stream, and the Sevilla River. The mangroves in the sector from Mazo Bay to Seville were selected because of the concentration of patches, the number and proximity of coastal settlements, and the variety of activities that occur. There are five coastal communities in the vicinity. All four mangrove species reported in Cuba are present there [20], with the highest abundance of *Avicennia germinans* (L.) Stearn. and *Rhizophora mangle* L. (Figure 3). The estimated mangrove area of 162 ha generally occurs as patches [20].

The main activities are agriculture, livestock farming, and forestry. Agricultural impacts include changing land use, such as clearing mangroves to grow crops, which is mainly performed by farmers. Livestock farming is considered one of the most impactful activities, mainly due to soil compaction and erosion. It requires clearing mangroves or other coastal vegetation to plant grasses and fodder. The main activities of the Integrated Forestry Company include the production, collection, and marketing of a range of agricultural products. The company also promotes and manages forests and fruit trees. Mangrove bark production and marketing in local currency and wholesale are also included. Another activity in this area is aquaculture in the Bay of Mazo, where PESCASAN has its El Mazo oyster farm, which covers 30% of the bay. It is exclusively dedicated to the cultivation of oysters (*Crassostrea rhizophorae* Guilding) for the local market.

Figure 3. *Rhizophora mangle* stand in Guamá Sector (credits to Yanet Cruz Portorreal), a patch facing open sea.

(3) San Miguel de Parada Wildlife Refuge (SAM) (See Table 1)

The San Miguel de Parada Wildlife Reserve is situated in the Bay of Santiago de Cuba, surrounded by the bay's industry zone (Figure 2D) [53,58]. Due to environmental changes caused by industry, settlements, and the city of Santiago de Cuba (approx. 500,000 inhabitants) itself at 2 km across the bay, this area is currently considered disturbed. There is no industrial activity within the protected area itself, but there are several industrial and social service infrastructures in the vicinity that have an impact on the functioning of the ecosystem. However, the dumping and leaking of polluting products in the area, such as oil and its derivatives and untreated sewage, is one of the main problems associated with the presence of these facilities. This wetland area covers an area of 55 hectares of contiguous mangroves [53] (Figure 4).

Figure 4. Stands with a low canopy of *Avicennia germinans* in San Miguel de Parada Wildlife Reserve about 2 km from Santiago de Cuba town (credits to Yanet Cruz Portorreal).

(4) Hatibonico Ecological Reserve (HAT) (See Table 1)

The Hatibonico Ecological Reserve lies southwest of Guantánamo Province, between Niceto Pérez and Caimanera (Figure 2B) [52,59,60]. According to the management plan [52], the main function of the local mangrove is "coastal protection", although it is considered important as a forest reserve. Subsistence fishing is allowed and supervised by the Forest Ranger Corps and the Alejandro de Humboldt Environmental Services Unit. Crab fishing of *Ucides cordatus* (Linnaeus, 1763) is a traditional activity that is still allowed in the area for local consumption. There are 64 hectares of low-canopy mangrove forest bordering the mouth of the Hatibonico River (Figure 5).

Figure 5. Low-canopy stands with *Rhizophora mangle*, on the opposite shore *Batis maritima*, a small coastal shrub in Hatibonico Ecological Reserve (credits to Hayler María Pérez Trejo).

2.2. Study Design

2.2.1. Design of the Social–Ecological Study

In this study, we employed a qualitative research approach [61] with exploratory data summary using a structured questionnaire technique [27,32,62–65]. The objective was to gather the perceptions of coastal communities in the mangrove SES regarding the following variables: (1) ecosystem services provided by mangroves, (2) use of mangrove resources, (3) current management practices in place, (4) potential threats posing risks to these ecosystems, and (5) social–ecological relationships existing within the localities. Indeed, for Cuba, and particularly for this region, data are scarce, and there is a need to obtain descriptive data to understand the mangrove social–ecological systems. In a later phase, hypotheses can be put forward regarding the causes of similarities and differences within Cuba, between localities and in comparison, to insights elsewhere.

The questionnaire (Table S1) was structured into three parts, as follows:

Part 1: Community use of mangrove resources.

The following uses were considered: (1) the use of marine species (fish, molluscs, crustaceans) for eating or selling, (2) the use of mangroves bark or wood as dye resource. A three-point Likert scale was used to determine the frequency of use for the following population groups: 1 (never used), 2 (sometimes), and 3 (always, frequently).

Part 2: Perception of ecosystem services.

The perception of ecosystem services was assessed by evaluating the importance of various services to both individual and community life. These included: (i) Supporting

Services, such as providing shelter for the juvenile stages of many species, supporting the feeding and production of a diverse range of marine and terrestrial organisms, and offering habitat for numerous marine and terrestrial fauna; (ii) Provisioning Services, acting as a source of food for consumption and sale and providing natural medicines for various diseases; (iii) Regulatory Services, such as protecting coastal areas from storms, cyclones, and waves; maintaining seawater quality by trapping sediments; and mitigating climate change effects through carbon sequestration; (iv) Cultural Services, serving as recreational and educational sites and representing symbols of local culture and heritage.

Participants rated the importance of these services using the following scale: 1 (not important), 2 (important), and 3 (very important), following the classification proposed by MEA (Millennial Ecosystem Assessment) [8].

Part 3: Perceptions of mangrove ecosystem threats and management framework.

Awareness of mangrove management activities was assessed using a scale ranging from 1 to 3: 1 (unaware), 2 (somewhat aware), and 3 (fully aware). This evaluation focused on several aspects: (i) the existence of a specific management plan or programme for mangroves; (ii) participation in seminars or training sessions related to mangrove management; (iii) awareness of the presence of legal instruments governing mangrove conservation; and (iv) knowledge about the existence of any body, any level, or institution responsible for monitoring and evaluating mangrove resources.

The perception of mangrove managers was gauged by identifying the institutions or agencies responsible for mangrove management within their locality. Additionally, perception of threats to mangroves if existing was elicited through an open-ended question, allowing respondents to freely list events or factors that are impacting the ecosystem in their area, according to them. This structured approach aimed to capture a comprehensive understanding of community awareness, management perceptions, and perceived threats to mangroves in the study area.

2.2.2. Design of the Governance Study

To identify the type of governance present, interviews were conducted involving specialists from each of the three protected areas: MCOA, SAM, and HAT. These experts play active roles in the development and management planning of their respective areas. An unstructured instrument with open-ended questions was used during interview sessions. Key topics explored included: (i) the degree of consultation of communities during the preparation of management plans; (ii) the extent of community involvement in identifying issues, conflicts, and solutions; (iii) the role of communities solely as recipients of management plan actions or as actively involved; (iv) features of management, distinguishing top-down, bottom-up, or co-management systems, with diverse perspectives analysed regarding the specific governance system applied in each study area. The focus group surveys were aimed to gather insights into the governance structures and processes within each protected area, highlighting perspectives from various stakeholders involved in mangrove management.

In addition, we used the interviews to examine the following six variables crucial for understanding mangrove management and decision-making in the study region: (i) *Transparency*: assessing the accessibility and clarity of information related to decision-making processes; (ii) *Knowledge of the Legal Framework*: understanding of the legal provisions concerning mangrove protection and their implementation; (iii) *Accountability*: inquiry into stakeholders' awareness of their responsibility to demonstrate actions related to mangrove management and their ability to assess the consequences of these actions; (iv) *Citizen Participation*: evaluating the extent to which citizens can engage in mangrove decision-making processes and influence public policies; (v) *Equity*: analysis of the distribution of resources and opportunities related to mangrove management; (vi) *Inclusion*: examining the participation of all social groups in mangrove decision-making processes.

To gauge the stakeholders' level of influence and interest, we utilised an influence-interest matrix, more specifically the Mendelow Stakeholder Matrix [66]. This matrix helps to categorise stakeholders based on their level of influence over mangrove management decisions and their level of interest in the outcomes. The interviews and the matrix analysis are intended to provide insights into the dynamics of mangrove governance, stakeholder engagement, and decision-making processes within the study area.

2.2.3. Integrated Matrix Analysis

Finally, to qualitatively assess the impacts of social–ecological and governance interactions, we constructed a matrix incorporating anthropic activities reported by respondents, mangrove ecosystem services, technical reports, public information, and reports from protected area management plans. Each component was evaluated to determine its influence on the overall state of SES. This qualitative weighting or evaluation process involves assessing how the development of specific activities, uses, or ecosystem services affect the SES comprehensively.

↑ High impact: an extensive social–ecological interaction that causes major changes in the SES, thereby affecting its balanced functioning. These are typically very pervasive activities or uses defining the prevalent state of the SES.

→ Medium impact: a strong social–ecological relationship, however, without causing irreversible changes.

↓ Low impact: a weak social–ecological relationship that does not cause irreversible changes. These are typically isolated activities, uses, or services that do not affect the functioning of the mangrove in the SES.

2.3. Sampling

The study baseline was established through a review of the 2021–2025 management plans for the following protected areas: Monte Cabaniguán/Ojo de Agua Wildlife Refuge (MCOA), San Miguel de Parada Wildlife Refuge (SAM), and the Hatibonico Ecological Reserve (HAT). Guamá is a multiple-use area not subject to any official management regime; information was gathered from public population statistics data [67]. Additionally, six specialists from SNAP were interviewed to contrast the information obtained from the above review.

The sample size was determined based on the total population of coastal communities adjacent to each mangrove site. The number of respondents was calculated to ensure a representative sample for each location. A 95% confidence level was chosen with a margin of error of 10% (see Table 1 for details), corresponding to the final number of questionnaires administered per mangrove site. Individuals within these communities were selectively approached, ensuring all participants were over 18 years old. Prior to participation, all individuals were informed of the study's objectives and were requested to provide their consent for involvement.

Social–Demographic Profile of Respondents

Participants from all sites were aged between 18 and 80 years, with a median age of 43 years. The gender balance was nearly equal, with 53.8% male and 46.2% female respondents overall (Table 2).

Regarding the reported occupational backgrounds, the largest proportion of participants was employed in the public sector: 20.5% in MCOA, 41.3% in GUM, 51.4% in SAM, and 60.9% in HAT. These backgrounds included forest rangers, specialists from the Flora and Fauna Company, teachers, and tourism sector workers. Approximately 14.1% of respondents were engaged in the private sector, and an equal percentage (14.1%) were farmers conducting agricultural activities in coastal areas. About 20% identified as housewives, and 8.4% were retired (Table 2).

Table 1. Mangrove SES localities: status, extent, mangrove type, resp. communities, population size, and general appraisal of pollution status.

Mangrove Site	Management Classification (SNAP)	Administration	Extension (ha)	Mangrove Typology [68]	Population (Number of Respondents)
MCOA	Wildlife Refuge	Enterprise for Flora and Faunal Protection Las Tunas	3929.18	Deltaic	596 (83)
GUAM	Multiple use area	No management regime	162	Bay Open coast estuary	1525 (92)
SAM	Wildlife Refuge	Enterprise for Flora and Faunal Protection, Santiago de Cuba	55	Bay	267 (72)
HAT	Ecological Reserve	Ministry of Science Technology and Environment (CITMA)	64	Open coast	1003 (87)
		Number of questionnaires 334			

Table 2. Distribution of the demographic profile of participants (in percentage, n = 334).

Indicators	Percentage			
	MCOA	GUAM	SAM	HAT
Age groups				
18–30	15.7	20.7	30.6	14.9
31–40	28.9	19.6	16.7	25.3
41–50	24.1	18.5	16.7	25.3
51–60	16.9	23.9	16.7	20.7
61–70	7.2	10.9	15.3	6.9
71–80	7.2	6.5	4.2	6.9
Total	100	100	100	100
Gender				
Male	50.6	53.3	54.2	56.3
Female	49.4	46.7	45.8	43.7
Total	100	100	100	100
Occupation *				
Housewife	24.2	20.7	20.8	14.9
State sector 1	-	30.4	44.4	43.7
State sector 2	-	9.8	5.6	13.8
Private Sector	33.7	14.1	16.7	14.9
Farmer	34.9	15.2	-	1.1
Retired	7.2	9.8	12.5	8.2
Student	-	-	-	3.4
Total	100	100	100	100

Note: * State sector 1—nonrelated to conservation, state sector 2—related to conservation.

2.4. Statistical Analysis

The creation of the database and analysis of the frequency tables were conducted using IBM SPSS Statistics for Windows, version 27.0, developed by IBM Corp. in Armonk, NY, USA. This facilitated the identification of patterns and highlighted the notable variable characteristics assessed in the communities for each mangrove site. The participants' profiles, together with the frequency of use and level of awareness of ecosystem services, were tabulated and summarised using frequency distributions and percentages [19].

When dealing with survey data on mangrove management and ecosystem services perceptions, non-parametric tests such as Kruskal–Wallis and Spearman's rank correlation are essential tools. These tests do not assume normal distribution of data, making them robust and reliable for analysing ordinal data and non-normally distributed continuous data, ensuring that the insights drawn are both accurate and meaningful. A Spearman correlation coefficient analysis was conducted to examine the relationship between occupation and frequency of mangrove ecosystem use [19]. Additionally, a Kruskal–Wallis test was employed to compare the frequency of mangrove uses and the perception of ecosystem services among the studied localities [32].

3. Results

3.1. Community Awareness of the Use of Mangrove Ecosystem Resources

The Kruskal–Wallis test revealed significant differences in the frequency of resource consumption as a food source. The results for fish consumption indicated H(3) = 51.9, $p < 0.001$. The post-hoc analysis, conducted using the Games–Howell test, demonstrated that the frequency of fish consumption in GUAM (Mdn = 3) was statistically significantly higher as compared to MCOA (Mdn = 2, $p < 0.001$, 95% CI [0.30, 0.83]), SAM (Mdn = 2, $p < 0.001$, 95% CI [0.60, 1.16]), and HAT (Mdn = 2, $p < 0.001$, 95% CI [0.26, 0.88]) (Table 3).

Table 3. Comparison of the frequency of mangrove uses between the study areas with Mdn—mean, Kruskal–Wallis H—test, *p*—*p* value. Site acronyms given in Section 2.1.

Use of Mangrove Resources	MCOA Mdn (Range)	GUAM Mdn (Range)	SAM Mdn (Range)	HAT Mdn (Range)	H	*p*
Food source (Fish)	2 (2)	3 (2)	2 (2)	2 (2)	51.9	0.001
Food source (Molluscs)	1 (2)	2 (2)	1 (2)	2 (2)	44.0	0.001
Food source (Crustaceans)	1 (2)	2 (2)	1 (2)	2 (2)	44.7	0.001
Traditional medicine	1 (2)	2 (2)	1 (2)	2 (2)	30.2	0.001
Wildlife watching	1 (2)	1 (2)	1 (2)	1 (2)	13.7	0.03
Recreational use	2 (2)	2 (2)	1 (2)	1 (2)	26.1	0.001
Education/Research	1 (2)	1 (2)	1 (2)	1 (2)	9.7	0.021
Spiritual/Religious practices	1 (2)	1 (2)	1 (10)	1 (2)	17.4	0.001

Note: 95% Confidence Interval.

The results of the test for the frequency of mollusc consumption among the localities showed H(3) = 44.0, $p < 0.001$. In GUAM, the consumption of this resource is significantly higher (Mdn = 2) compared to MCOA (Mdn = 1, $p < 0.001$, 95% CI [0.41, 1.05]) and SAM (Mdn = 1, $p < 0.001$, 95% CI [0.44, 1.00]). Differences in crustacean consumption were also examined, resulting in H(3) = 44.7, $p < 0.001$. The analysis indicates that in GUAM, consumption is higher (Mdn = 2) than in MCOA (Mdn = 1, $p < 0.001$, 95% CI [0.41, 1.04]) and SAM (Mdn = 2, $p < 0.001$, 95% CI [0.47, 1.03]), as shown in Table 3. Statistically significant values were also found for traditional medicine use, with H(3) = 30.2, $p < 0.001$. The result was significant for the GUAM locality (Mdn = 2) compared to MCOA (Mdn = 1, $p < 0.001$, 95% CI [0.19, 0.65]), SAM (Mdn = 1, $p < 0.001$, 95% CI [0.27, 0.75]), and HAT (Mdn = 1, $p < 0.001$, 95% CI [0.02, 0.54]). Moreover, significant differences were identified in the use of mangroves for recreation and education, with H(3) = 42.2, $p < 0.001$. This service does not appear to be important for members of the HAT community (Mdn = 1) compared to the communities of MCOA (Mdn = 3, $p < 0.001$, 95% CI [0.36, 0.97]), GUAM (Mdn = 3, $p < 0.001$, 95% CI [0.28, 0.85]), and SAM (Mdn = 3, $p < 0.001$, 95% CI [0.01, 0.61]).

Correlation Analysis of Occupations and Frequency of Mangrove Resource Use in the Study Localities

The Spearman rank correlation analysis (Rho) between demographic variables and mangrove uses in the studied localities indicates generally low to very low correlations (Table S3). In MCOA, age showed significant correlations with fish consumption ($rho_{(81)} = 0.310$, $p = 0.004$) and mollusc consumption, $rho_{(81)} = 0.218$, $p = 0.047$. In HAT, a correlation was found between religious use, $rho_{(85)} = 0.238$, $p = 0.26$) as well as with medicinal use ($rho_{(85)} = 0.229$, $p = 0.33$). In SAM, age correlated with wildlife watching, $rho_{(70)} = 0.340$, $p = 0.004$, and research/education use, $rho_{(70)} = 0.250$, $p = 0.35$. Additionally, in GUAM, age correlated with medicinal use, $rho_{(90)} = 0.281$, $p = 0.007$. In MCOA, length of residence in the community correlated with fish consumption, $rho_{(81)} = 0.354$, $p = 0.001$, mollusc consumption, $rho_{(81)} = 0.704$, $p = 0.001$, and crustacean consumption, $rho_{(81)} = 0.411$, $p = 0.001$. In HAT, time of residence correlated with medicinal use, $rho_{(85)} = 0.229$, $p = 0.33$, and religious practices, $rho_{(85)} = 0.238$, $p = 0.26$.

The rho correlation was also used to identify whether occupation influences the frequency of mangrove use (Table S3). In MCOA, housewives exhibited significant negative correlations with fish consumption, $rho_{(81)} = 0.356$, $p = 0.001$, mollusc consumption, $rho_{(81)} = 0.319$, $p = 0.003$, and crustacean consumption, $rho_{(81)} = 0.279$, $p = 0.011$. This suggests that this occupation relies less on these resources compared to other occupations. In GUAM, both positive and negative correlations were observed, highlighting the influence of housewives and farmers on the utilisation of specific resources. Housewives showed a significant positive correlation with the following uses: fish consumption, $rho_{(90)} = 0.265$, $p = 0.011$), medicinal use, $rho_{(90)} = 0.234$, $p = 0.025$, and dye production, $rho_{(90)} = 0.248$, $p = 0.017$. Farmers correlated significantly with fish consumption, $rho_{(90)} = 0.225$, $p = 0.031$, mollusc consumption, $rho_{(90)} = 0.262$, $p = 0.012$, and medicinal use, $rho_{(90)} = 0.227$, $p = 0.030$.

In SAM, housewives had a significant positive correlation with spiritual use or religious practices, $rho_{(70)} = 0.532$, $p = 0.001$, related to mangroves. The government sector 2 (linked to conservation) correlated with educational practices, $rho_{(70)} = 0.345$, $p = 0.003$. In HAT, housewives exhibited a significant negative correlation with the use of mangroves as a source of molluscs, $rho_{(85)} = 0.236$, $p = 0.028$. Government sector 2 correlated significantly with the consumption of species for food.

3.2. Community Perception of Mangrove Ecosystem Services

Respondents recognised the presence of red mangrove (*Rhizophora mangle*, locally called mangle rojo) and black mangrove (*Avicennia germinans*, locally called mangle prieto) in their area, followed by white mangrove (*Laguncularia racemosa* (L.) C.F. Gaertn, locally called patabán). The least identified (Figure 6) is the button mangrove (*Conocarpus erectus* L., locally called yana). In HAT, between 2.4% and 7.8% of respondents lacked knowledge of any mangrove species.

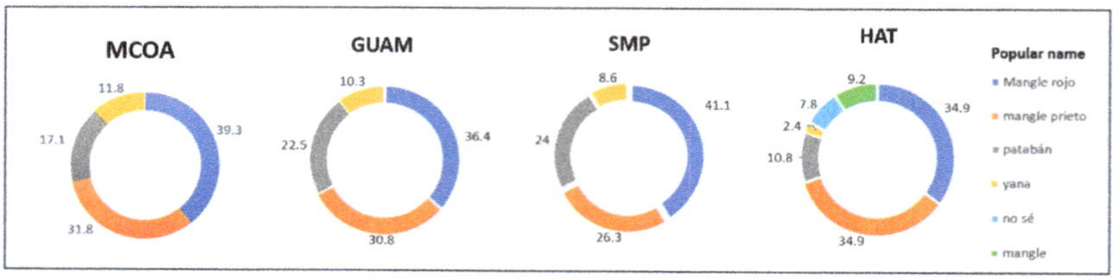

Figure 6. Distribution of responses (in percentage) regarding the knowledge of popular mangrove names across four sites: MCOA, GUAM, SMP, and HAT. The categories include red mangrove, black mangrove, patabán, yana, "I don't know", and mangrove.

Local people were inquired about the importance of ecosystem services provided by mangroves (Table S2, Figure S1). Overall, the perception of ecosystem services was very high in all mangrove communities. Between 43.7% and 100% of participants confirmed that mangroves provide an important supporting function for fauna, including juveniles (nursery function).

Significant differences (p-value < 0.05) were found in perceptions of ecosystem services in mangrove localities (Table 4). The result of the Kruskal–Wallis test for juvenile refuge (nursery) among the localities was $H(3) = 66.9$, $p < 0.001$. The communities in GUAM indicated the highest perception for this service (Mdn = 3) compared to MCOA (Mdn = 3, $p < 0.001$, 95% CI [0.001, 0.26]), SAM (Mdn = 3, $p < 0.001$, 95% CI [0.17, 0.56]), and HAT (Mdn = 2, $p < 0.001$, 95% CI [0.48, 0.97]). Conversely, HAT participants provided the lowest perception compared to the other localities, as shown in Table 4.

Table 4. Comparison of the perception of mangrove ecosystem services between localities studied. Mdn—mean, Krusskal-Wallis H—test, p—p value.

Ecosystem Services	MCOA Mdn (Range)	GUAM Mdn (Range)	SAM Mdn (Range)	HAT Mdn (Range)	H	p
1. Nursery	3 (2)	3 (1)	3 (2)	2 (2)	66.9	0.001
2. Wildlife habitat	3 (0)	3 (0)	3 (2)	2 (2)	97.4	0.001
3. Food source	3 (2)	3 (2)	2 (2)	2 (2)	89.2	0.001
4. Coastal protection	3 (2)	3 (2)	1 (2)	1 (2)	78.0	0.03
5. Water quality	3 (2)	3 (2)	3 (2)	1 (2)	45.5	0.001
6. Recreation/Education	3 (2)	1 (2)	2 (2)	2 (2)	42.2	0.001

Note: 95% Confidence Interval.

In addition, MCOA (Mdn = 3) and GUAM (Mdn = 3) communities identified food sources to be very important as provisioning services. The SAM (Mdn = 2) and HAT (Mdn = 2) communities reported this less frequently as important. We defined three different levels of importance for mangrove regulating services. Coastal protection is considered very important, with between 65.2% and 95.7% of respondents indicating that mangroves help to protect against the high waves and strong winds associated with extreme hydrometeorological events. Significant results were found regarding this service, with H(3) = 78.0. In GUAM, the perception of this service was considered very important (Mdn = 3) compared to MCOA (Mdn = 3, $p < 0.001$, 95% CI [0.7, 0.45]), SAM (Mdn = 1, $p < 0.001$, 95% CI [0.14, 0.56]), and HAT (Mdn = 1, $p < 0.001$, 95% CI [0.66, 1.17]). In HAT, there is less recognition of this service compared to the other localities (Table 4).

Regarding the perception of water quality maintenance, the Kruskal–Wallis test indicated H(3) = 45.5, $p < 0.001$. In HAT, the perception of this service is significantly lower (Mdn = 1) compared to MCOA (Mdn = 3, $p < 0.001$, 95% CI [0.43, 1.00]), GUAM (Mdn = 3, $p < 0.001$, 95% CI [0.29, 0.86]), and SAM (Mdn = 3, $p < 0.001$, 95% CI [0.08, 0.72]).

The role of mangroves in climate change mitigation through carbon sequestration was perceived to be minimal by the participants, with 45.8% in MCOA, 40.2% in GUAM, 41% in SAM, and 43.7% in HAT (See Table S2). The Kruskal–Wallis test indicated no significant differences in the distribution of perception values among localities. This suggests that perceptions of mangroves were similar across the four studied areas.

3.3. Community Perceptions of Mangrove Management Framework

We asked the participants per site if they were aware of the existence of any management activity (Figure 7A). Respondents indicated they were fully aware of the existence of the protected area and management activities in MCOA (61.4%) and SAM (44.4%). In GUAM, between 23.9% and 29.3% were, respectively, aware and fully aware that there was no management plan. However, they referred to the activities of the rangers and the government to protect the mangrove forest (Figure 7D). In HAT, 50.6% of the villagers were moderately aware of management activities.

In the study sites, a significant percentage of respondents in MCOA (63.9%) and SAM (40.3%) indicated that they were very aware of ongoing management training activities, such as seminars, workshops, and training sessions (Figure 7B). In GUAM, where mangroves are not designated as a protected area, perceptions of management activities were linked to the implementation of the COSTASURESTE project, which was concluded in 2012 and is no longer operational. Conversely, in HAT, local awareness of mangrove management activities was reported as moderate percentages.

Across all sites, the communities displayed moderate to high levels of awareness regarding the existence of laws regulating mangrove use, penalties for non-compliance, and the governmental authorities responsible for enforcement (Figure 7C,D).

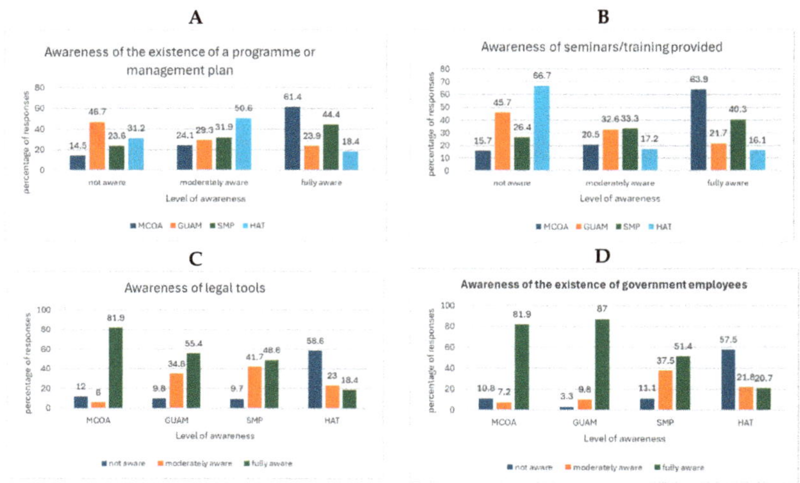

Figure 7. (**A**)—Awareness of mangrove management in the study sites. Perception of management plan existence, (**B**)—awareness of management training activities, (**C**)—awareness of the existence of legal tools regarding mangrove management, and (**D**)—awareness of any body, any level, or institution responsible for monitoring and evaluating mangrove resources in the various areas.

3.4. Community Perception of Potential Threats to Mangroves

During the survey, villagers identified and ranked 15 natural and anthropogenic threats affecting their localities (Table 5). Climate change and drought emerged as the natural threats that raise the most concern across all areas. The villagers of GUAM (16.1%) and SAM (9.0%) also identified tropical cyclones as significant threats. Salinisation was also noted as a relevant concern in MCOA (4.2%), GUAM (12.5%), and SAM (4.2%).

Table 5. Threats to the mangrove perceived in the sites studied (ordered alphabetically).

	Answers (%)			
Mangrove Threats/Localities	**MCOA**	**GUAM**	**SAM**	**HAT**
Agriculture	-	5.1	-	-
Bark and roots removal	-	6.8	-	-
Climate change	24.6	18.5	19.6	31.9
Drought	16.3	5.5	20.1	39.3
Exotic species	5.3	4.1	-	4.3
Fires	14.4	-	-	24.5
Grazing	-	4.5	-	-
Industries	-	-	18.0	-
Logging	24.2	24.7	12.7	-
Pollution/waste dumping	6.4	9.9	11.6	-
Rising sea level	-	1.4	0.5	-
River damming	3.0	-	4.2	-
Road construction	4.8	4.8	-	-
Salinisation	4.2	12.5	4.2	-
Tropical cyclones	-	16.1	9.0	-

Locally, forest fires were seen as a threat by respondents in MCOA (14.4%) and HAT (24.5%). Among direct anthropogenic threats, logging was prominently mentioned in MCOA (24.2%), GUAM (24.7%), and SAM (12.7%). Industries were specifically noted in SAM (18.0%) as contributing to environmental risks.

Road construction was reported as impacting mangroves in MCOA (4.8%) and GUAM (4.8%). Damming of rivers was cited in MCOA (3.0%) and SAM (4.2%). Agricultural expansion was mentioned in GUAM (5.1%) as a replacement of mangrove areas with crops, while 4.5% reported this conversion was into pastures.

Local communities also expressed their concern regarding the extraction of mangrove bark and roots for various purposes. Moreover, the presence and spread of exotic species were identified as hazards in MCOA (5.3%), GUAM (4.1%), and HAT (4.3%) but were not considered hazards in SAM.

We also queried local residents regarding their perceptions of changes and the current state of mangroves near their communities (Figure 8). To conduct this part of the study, we surveyed local communities to assess their awareness and perception of the impact of mangrove degradation on their communities (Figure 8A). In MCOA, 54.2% of respondents acknowledged the severe impact of mangrove degradation, while in GUAM, this figure was notably higher at 77.1%. Conversely, in SAM, 34.7% of respondents perceived mangrove degradation as having minimal or moderate negative consequences. In HAT, most respondents (67.8%) believed that mangrove degradation would either be harmless or would only have a moderate negative impact on their communities.

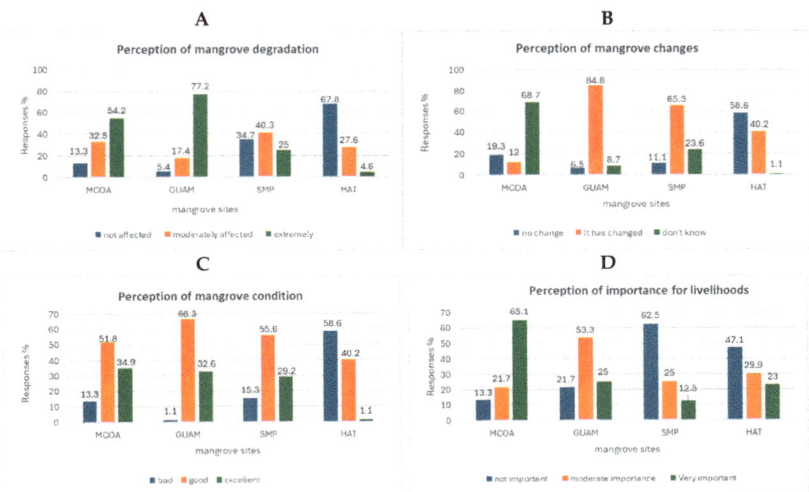

Figure 8. Perception of the studied communities regarding degradation, condition, changes in the mangroves, and importance for their livelihood. (**A**)—Perception of mangrove degradation impact on the community, (**B**)—perception of mangroves changes as a landscape, (**C**)—perception of mangrove condition, (**D**)—perception of the importance of the mangrove for the livelihood of the local people.

Within the same inquiry, the participants of MCOA (68.7%) indicated that they perceived no significant changes in the mangrove landscape, while 51.8% felt that mangroves were in good condition (Figure 8C). Additionally, 65.1% of the MCOA respondents emphasised that mangroves are really important to their livelihoods (Figure 8D). In GUAM, a notable 84.8% of the participating community felt that there were changes in local mangroves; however, 66.3% of them are convinced that these ecosystems are in good condition (Figure 8B). About 53.3% of this community acknowledges that their livelihoods moderately depend on these mangroves, primarily as a food source (Figure 8D). In SAM, 55.6% of participants observed an increase in mangrove areas within the wetland area, and an equal part of the community considered that their mangroves were in good shape. Surprisingly, 62.5% of them stated that they do not depend on mangroves for their livelihoods (Figure 8D). In HAT, 58.6% of respondents reported no noticeable changes in mangroves (Figure 8B), and 40.2% of them assessed the mangroves as being in a healthy state (Figure 8C). Similar to the SAM community, the majority (64.4%) of HAT participants expressed that mangroves are not crucial for their livelihoods (Figure 8D).

3.5. Results of the Governance Analysis

Based on insights gathered from both interviews and management plans, distinct approaches to mangrove management and community involvement were observed across the different sites.

In MCOA, mangrove management plans are formulated exclusively by area specialists without direct community participation or voting rights. The community's role is limited to providing information during stakeholder workshops, where only one community representative is invited.

In SAM, workshops serve as platforms to reconcile diverse interests, gathering inputs from various experts within the protected area. However, community participation here is also restricted to providing information rather than influencing decision-making directly.

HAT employs a more inclusive approach where workshops precede plan development or updates. These sessions involve community participation in defining tasks and reporting on previous plans. Decision-makers and community representatives are integral to these workshops, held locally with support from government entities such as the Ministry of the Revolutionary Armed Forces (MINFAR) and civil defence. Economic challenges sometimes hinder plan implementation, especially for continuity plans, which often rely solely on data updates.

In GUAM, there is no programme or framework for mangrove management, but activities for preservation fall under the responsibility of the Integral Forestal Guamá Enterprise and the State Forestry Service of the Ministry of Agriculture (MINAG), with involvement from the municipal government and the Ministry of Science Technology and Environment (CITMA) in decision-making. However, there is a recognised need that other local institutions and community members should be involved in the future.

The analysis of stakeholder relationships identified that key public entities such as the Cuban State, CITMA, MINAG, provincial governments, and municipal councils have significant influence and interest in mangrove management (Figure S2). In contrast, entities such as the National Hydraulic Institute (INRH) and the Ministry of Food Industry (MINAL) currently exhibit lower influence and interest, resulting in minimal involvement in the management framework.

Local communities and the Ministry of Tourism (MINTUR) are identified as stakeholders who have high levels of interest but low influence in shaping outcomes within the mangrove social–ecological system. This suggests that they are highly affected by the results of actions taken that impact mangroves, yet their ability to influence these outcomes through management decisions is rather limited. Conversely, stakeholders such as MINFAR and the Ministry of the Interior (MININT) possess high influence but exhibit low interest in mangrove-related activities.

3.6. Description of the Observed Social–Ecological Relationships

Our analysis emphasises how human activities (social, economic, and cultural) affect the mangrove ecosystem in each locality analysed. The matrix (Figure 9) synthesises the social–ecological relationships observed, revealing three distinct interaction types: urban/industrial, rural/agricultural, and rural/agricultural/tourist.

In MCOA, significant interaction occurs with the agricultural sector and expanding livestock farming. Probably due to the size of the local mangroves, these activities have minimal impact on the ecosystem. The mangroves primarily provide supporting services, crucial for justifying conservation efforts, with provisioning services playing a secondary role, according to the participants.

GUAM exhibits a high level of interaction between tourism and agriculture activities. Agricultural expansion poses the greatest threat to mangrove areas through land use changes. Tourism, while not directly impacting the mangrove ecosystem, focuses on exploiting the scenic and cultural values of the local forest, thereby fostering strong social relations based on goods supply and cultural services.

SAM is characterised by intense urban–industrial interaction, situated within the industrial zone of Santiago de Cuba's bay. Here, mangroves face significant challenges from pollution and land use changes driven by industrial growth, compromising ecosystem services. Supporting services, particularly the mangrove's role in sheltering migratory birds, dominate despite these pressures.

HAT experiences minimal interaction. The interaction that is happening is primarily centred on agricultural activities and non-productive forestry endeavours. Social interactions thrive without adverse effects on mangroves. This area exhibits low ecological impact, with supporting services being most pronounced and provisioning services playing a moderate role.

Mangrove sites	Anthropic activities				Ecosystem services			
	1	2	3	4	S	A	R	C
MCOA	↑	↓	↓	↑	↑	→	↓	↓
GUAM	↑	↑	↓	↓	↑	↑	↑	↑
SMP	→	↓	↑	↑	↑	↓	→	→
HAT	↑	↓	↑	↑	↑	→	↓	↓

Legend: 1 Agriculture, 2. Industry, 3. Tourism, 4, Conservation, S. Support service, A. Supply service, R. Regulation service, C. Cultural service. Degree of influence:

↓ *High influence* → *Average influence* ↑ *Low influence*

Figure 9. Evaluation matrix of social–ecological interactions in the study sites.

4. Discussion

Our study underscores that knowledge of and exploitation of mangroves are deeply intertwined in the daily lives and cultural practices of southeastern Cuban communities, consistent with studies from elsewhere [7,18,32]. The increased awareness by local communities of ecosystem services and threats such as climate change likely stems from environmental education initiatives integrated into Cuba's Environmental Strategy implementation [69].

Cuba, situated in the West Atlantic–East Pacific mangrove region, naturally exhibits a relatively low mangrove species richness as compared to the Indo-West Pacific and the global number of approximately 70 species. Mangroves in Cuba comprise primarily *Rhizophora mangle* (red mangrove), *Avicennia germinans* (black mangrove), and further *Laguncularia racemosa* (white mangrove), and *Conocarpus erectus* (button mangrove). These species, particularly red and black mangroves, also dominate the mangrove forests across the study sites in southeastern Cuba [20,58], whereby coastal communities mostly recognised red mangroves, in particular [20,58]. Yet, in terms of areal extent, Cuba ranks 10th worldwide [40] and its mangrove SESs are important.

The identification of mangrove species appears, on the one hand, to be influenced by the distribution and composition of the local ecosystem. Due to the environmental conditions of the study area, the presence of *C. erectus* is minimal compared to the other species. On the other hand, the duration of time a participant has resided in the locality also appears to affect their ability to identify the species. Further research should be conducted in this regard.

In this study, we provide descriptive data for the mangrove social–ecological systems of southeastern Cuba in support of a better understanding of issues related to good management and conservation.

4.1. Use of Mangrove Resources by Communities

Fishing and the exploitation of forest products, whether for commercial or subsistence purposes, have historically been primary livelihoods for coastal communities near mangroves [4,6,65]. However, our study indicates a shift in southeastern Cuba where local livelihoods are increasingly diversifying towards inland (terrestrial) activities [55,56] in industrial employment [67] and job opportunities in emerging agricultural and livestock sectors [59,68]. Despite these changes, the use of mangrove resources for food purposes remains prominent, alongside medicinal and recreational uses such as bathing areas. Notably, while mangrove wood is valued elsewhere for fuel and construction [4,64], in this study, participants perceived its use as illegal, which reflects Cuba's stringent regulations against mangrove logging [70]. Studies on mangroves in Mexico, Brazil, Bangladesh, and Madagascar highlight uses such as recreation, tourism, and cultural preservation tied to mangroves [7,71–73]. In southeastern Cuba, local communities also utilise red mangroves for traditional medicine, including treatments for kidney infections and skin diseases. Despite the high scenic value of mangroves, which presents opportunities for nature and ecotourism, such as wildlife observation within protected areas, their use is not widespread in this region. Sustainable practices for these activities remain underdeveloped.

Comparing local occupations with mangrove resource use revealed interesting patterns showing how residents' jobs affect both the frequency and type of mangrove resource utilisation (Supplementary Table S3). The governmental sector is mostly associated with its conservation and research/educational efforts towards the mangrove ecosystem. In SAM, housewives use the mangroves mainly for medicine and religious practices. More research needs to be conducted to understand the significance of this small percentage of religiosity, probably of African origin.

When combining occupation with perceptions of mangrove ecosystem services the respondents' occupation and residency play an important role (Table 5). The social context on the valuation of ecosystem services, management, and conservation activities should take these differences into account.

To better understand the causality behind these observed differences in perception, further research is needed. However, it is crucial to consider these differences when designing conservation and management policies for mangroves, ensuring all community sectors are involved.

4.2. Perception of Ecosystem Services in Communities

Communities near mangroves in southeastern Cuba demonstrate a robust understanding of the ecosystem goods and services provided by mangroves, similar to knowledge found globally, such as in Mauritius [6]. The communities widely recognise their habitat support for fauna, especially juveniles, and the provisioning services such as food and medicine [7,22,74]. In GUAM and MCOA, mangroves play a crucial role in supporting local fisheries, underscoring their significant economic importance [51]. This dependency is more closely related to community livelihood strategies than to demographic factors, unlike other studies [7,74,75]. In GUAM, the proximity of communities to mangroves may contribute to the frequent use of mangrove resources, a pattern also observed in Tanzania [31]. Perception of the regulatory function of climate mitigation services depends on the population's awareness of the protective role of mangroves against extreme hydrometeorological events. This awareness extends beyond the experience of the respondents with national environmental policies [24], which emphasise this ecosystem service [4,32].

While there is a general awareness of mangroves' regulatory role in climate mitigation, understanding of their carbon sequestration capacity remains limited, possibly due to the abstract nature of this service in local contexts [76,77]. This highlights the need for improved environmental education and capacity building to enhance local knowledge and leverage it for future blue carbon initiatives, especially in light of Cuba's position in the world rankings of mangrove areas.

In brief, this study's findings reveal that respondents' occupations in mangrove areas affect their views on the ecosystem services provided by these environments. In several localities, such as GUAM, SAM, and HAT, there were noticeable links between government roles, conservation efforts, and the perception of specific services. Conversely, in MCOA, housewives showed a lower appreciation for services like nursery function, erosion protection, and CO_2 assimilation and carbon retention, while retirees in HAT had a reduced perception of nursery services, erosion protection, consumption and sale, and salinity protection. These variations highlight how different occupational groups perceive the utility and benefits of mangroves differently (Table 5). Overall, these findings underscore the strong impact of occupation on the management and conservation practices related to mangroves.

Additionally, this study reveals that respondents in the four areas studied are unaware of the role of mangroves in carbon sequestration (Table S4). This lack of awareness limits local communities' abilities to value and support this ecosystem's conservation. Mangroves are effective carbon sinks, with a carbon sequestration capacity that surpasses that of tropical forests. Yet this benefit is not widely recognised among local stakeholders and cannot be used as an argument for conservation yet, let alone the establishment of carbon credit programmes. There is an urgent need to enhance awareness and environmental education programmes about the critical role of mangroves in carbon sequestration and climate change adaptation. It is recommended that mangrove management initiatives include these efforts aimed at highlighting the multiple benefits of mangroves. Some examples are integrating relevant information into school curricula, conducting public communication campaigns, and developing materials for key stakeholders such as farmers, fishers, and policymakers. Enhancing this awareness has the potential to foster greater support for mangrove conservation and restoration, thereby maximizing both the benefits for local communities and the environment.

4.2.1. Perception of Mangrove Ecosystem Threats and Management Framework

Coastal residents are usually well aware of ongoing mangrove conservation efforts and the regulatory frameworks governing their use in Cuba. However, clarity on responsible management entities remains a concern despite established legal structures [24]. Apparently involving local governments is essential for successful management programmes [32, 48]. Management plans often prioritise charismatic faunal species over entire mangrove ecosystems, reflecting funding biases. In GUAM, initiatives like the COSTASURESTE project [78] have bolstered local environmental awareness, albeit without specific protected area designations or integrated coastal zone management (ICZM) programmes. Though concluded in 2012, the awareness lingers on.

Governance types around the world could be divided roughly into bottom-up, top-down, and co-management [79], amongst others. In Cuba, the top-down management system is still prevailing [39,80]. Moving forward, embracing bottom-up or co-management models could enhance local participation and governance efficacy, aligning with international best practices and ensuring sustainable mangrove management [79]. Such governance also allows two-directional information flows, a sense of property and belonging and acceptance of constraining measures.

4.2.2. Potential Threats to Mangroves

Mangroves face varied threats, categorised as natural (e.g., drought, cyclones) and anthropogenic (e.g., logging, agricultural expansion, pollution) hazards, each posing different levels of risk [16,81]. Natural forces have caused significant mangrove loss [23]. Climate change has natural and anthropogenic roots, but locals identified it as a natural threat. Climate change and drought are perceived as the most significant natural threats in southeastern Cuba, exacerbated by local climatic conditions and regional forecasts [82].

Cyclones are a major threat, causing area loss and damage, though mangroves can recover [23,83]. Salinisation and forest fires are other concerns expressed, while sea level rise is less perceived due to the elevated eastern coast.

Anthropogenic threats such as logging and agricultural expansion are significant concerns, driven by economic activities that alter land use and hydrology, impacting mangrove ecosystems [16,84–86]. Effective governance and community involvement are crucial for mitigating these threats, aligning policies with local realities and sustainable practices [16,79,80].

4.2.3. New Perspective for Governance and Management of Mangroves in Eastern of Cuba

While Cuba has established institutional mechanisms for environmental governance, their effectiveness in incorporating community voices and ensuring equitable, accountable management of mangroves remains a challenge. Enhancing stakeholder engagement and fostering inclusive governance models could bridge existing gaps, facilitating more sustainable mangrove conservation practices and increasing awareness [38,39]. The proposed model (Figure 10) outlines actionable steps for integrating local knowledge and fostering collaborative management approaches, building on Cuba's existing legal frameworks [24]. The existing legal framework in Cuba supports the development of a scheme (Figure 10) for identifying key stakeholders and their responsibilities towards potential conservation efforts. This provides a solid foundation for improving mangrove governance even though the prevailing approach predominantly follows a top-down model.

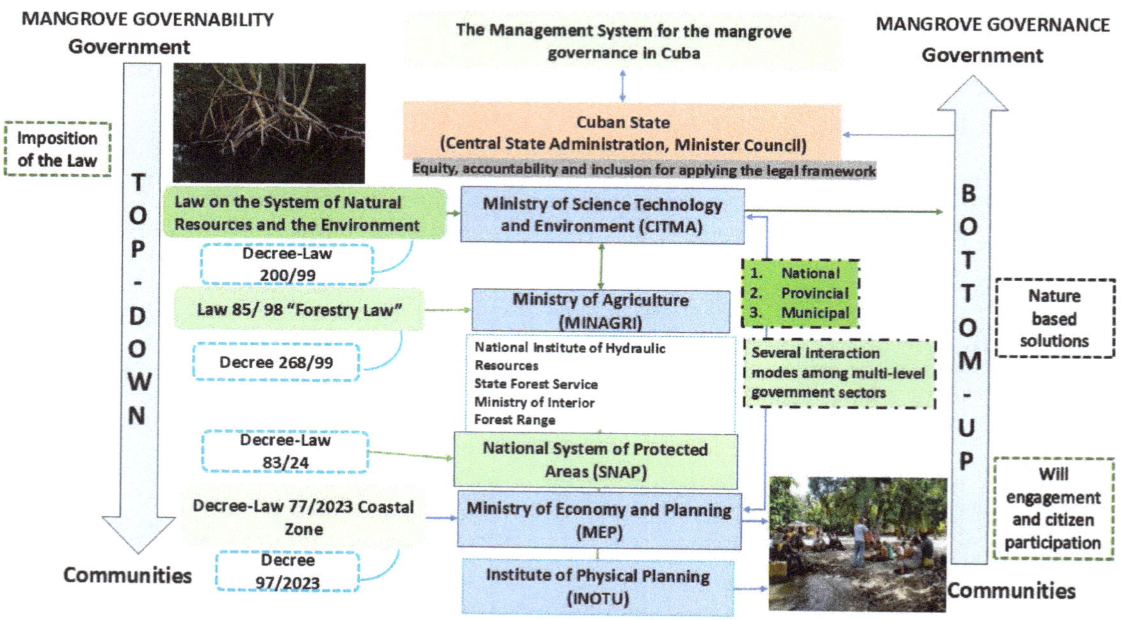

Figure 10. The management system proposed for mangrove governance in Cuba, particularly in southeastern Cuba. Photos credit Nico Koedam ((**top left**), *R. mangle* in MCOA) and Yanet Cruz Portorreal ((**bottom right**), meeting with residents of GUAM).

4.2.4. Analysis of the Observed Social–Ecological Relationships

We conducted a preliminary qualitative and descriptive analysis by integrating the survey results with a structured documentary review. For future research, we recommend incorporating indicators and measurable variables to better analyse these social–environmental interactions. Our study suggests that management programmes for protected areas should align with integrated coastal zone management principles. This alignment will facilitate the development of social–ecological system approaches and increase governance efficacy.

We found gaps in local knowledge on the role of mangrove ecosystems in carbon sequestration. There is a need for training and identifying opportunities for implementing local development projects. Emphasising capacity building could address market issues for ecosystem services, including carbon sequestration, and raise awareness of this intangible service. Understanding the complex international carbon market assessments and realising the ecosystem's potential for this service in Cuba is crucial.

Our study concludes that communities have different social–environmental relationships based on the rural, urban, and coastal characteristics of these settlements. There are also different manifestations of anthropic activities depending on the dominant sector of economic development and its impact on mangrove functioning. There is unexploited potential for understanding and engagement in mangrove SESs of southeastern Cuba if considering the level of public knowledge and awareness. There is a need when summing the contributions of mangrove goods and services to the local economy or subsistence livelihoods, even though these are not similar in each community.

5. Conclusions

This study explored the knowledge, usage patterns, interdependencies, and social-cultural aspects related to social–ecological dynamics between four southeastern Cuban communities and coastal mangrove ecosystems. The four communities and mangroves have different settings and legal frameworks. The expected diversity in human–mangrove relations has been confirmed, with several recurrent features. Understanding these relationships offers valuable insights for decision-makers to promote conservation and highlights gaps in existing management programmes, including the need for community-specific training programmes for local public and private stakeholders.

Our findings emphasise the necessity of a comprehensive approach to mangrove management and conservation, with implications that extend beyond Cuba. Policymakers should recognise NbS as an opportunity for inclusive solutions and forge partnerships for effective mangrove management in southeastern Cuba, even where they have not been called NbS yet in the region. Bridging the gap between NbS and formal governance, addressing diverse opinions, and providing continuous support for community participation is essential. Developing NbS capacities based on regional contexts is crucial for effective management.

The variations in mangrove use frequency and occupational roles highlight the importance of considering the interests of different stakeholder groups in management activities. This also indicates that certain groups may be more susceptible to understanding and remediating mangrove degradation. This should be considered in mangrove restoration practices currently proliferating in the country.

The positive links between the government sector and conservation along with certain educational and research uses underscore the key role of the government in fostering sustainable use. Perceptions of ecosystem services differ by occupation and locality, illustrating the importance of the social and geographical context in the valuation of mangroves.

Overall, the analysis reveals key opportunities and challenges for effective governance by clarifying the complex relationships between human communities and mangrove ecosystems. This research provided valuable scientific information on community perceptions of use, management, and threats to the mangrove ecosystem valid for effective governance of this socio-ecosystem. Upon implementation, decision-makers could promote community participation, identify threats, and develop effective policies that enhance the value placed on these resources. They could better valorise the awareness and knowledge baselines in the communities to the benefit of local economies, livelihoods, and securing coastal safety under environmental threats.

Supplementary Materials: The following supporting information can be downloaded at: https://www.mdpi.com/article/10.3390/w16172495/s1, Table S1: Questionnaire; Table S2: Percentage distribution of respondents at each level of importance of mangrove ecosystem services, Table S3: Correlations between the occupation and frequency of resource utilization; Table S4. Correlations between the occupation and awareness of ecosystem services. Figure S1: Frequency of mangrove resource use in the sites studied, Figure S2: Level of specific stakeholders in the management system for the mangrove governance in Cuba (Mendelow's Stakeholder Matrix).

Author Contributions: Conceptualisation, Y.C.P., N.K., C.B.M., F.D.-G., and O.P.M.; methodology, Y.C.P., N.K., C.B.M., F.D.-G., and O.P.M. software, Y.C.P.; validation, Y.C.P., C.B.M., and O.P.M.; formal analysis, Y.C.P., N.K., N.B., O.J.R.D., C.B.M., and O.P.M.; investigation, Y.C.P.; writing—original draft preparation, Y.C.P.; writing—review and editing, Y.C.P., N.K., N.B., O.J.R.D., C.B.M., F.D.-G., and O.P.M.; visualisation, Y.C.P. and C.B.M.; supervision, N.K., N.B., O.J.R.D., and O.P.M.; funding acquisition, Y.C.P., N.K., N.B., O.J.R.D., and O.P.M. All authors have read and agreed to the published version of the manuscript.

Funding: This research was funded by the Flemish Development Cooperation through the Flemish Interuniversity Council-University Cooperation for Development (VLIR-UOS) as part of an Institutional University Cooperation programme with Universidad de Oriente in Santiago de Cuba, Cuba. YC received a grant from Global Minds, Hasselt University, Belgium. Moreover, we received support from the Science, Technology, and Innovation research project: Adaptive Governance to Climate Change in Coastal Municipalities of Cuba, PN211LH012-018, Associated with the National Local Development Programme in Cuba.

Data Availability Statement: The data presented in this study are available on request from the corresponding author.

Acknowledgments: The authors thank Agustin Garzón for valuable discussions and insights, also Yusneida Alarcón Jorge, Yairén Alonso Jiménez, Ing. Yuri Debros Trutié, Hayler M. Pérez Trejo for assistance with field trips and questionnaire implementation. The fifth author thanks Universidad del Magdalena, Colombia, COIBA-AIP, Panama, and ESPOCH, Ecuador, for supporting her post-doctoral project entitled "Anthropogenic pressures in the Galapagos Islands: pluralistic regulations to mitigate anthropogenic pressures and promote sustainable development." The methodology used in this article is a pivot for qualitatively assessing the influence of social–ecological relationships and governance of mangroves in Galapagos Island.

Conflicts of Interest: The authors declare no conflicts of interest.

References

1. Chapin, F.S.; Kofinas, G.P.; Folke, C. *Principles of Ecosystem Stewardship Resilience-Based Natural Resource Management in a Changing World*; Springer Science+Business Media: New York, NY, USA, 2009; p. 402.
2. Dahdouh-Guebas, F.; Hug, J.; Abuchahla, G.M.O.; Cannicci, S.; Jayatissa, L.P.; Kairo, J.G.; Kodikara, S.; Koedam, N.; Mafaziya, T.W.G.F.; Mukherjee, N.; et al. Coastal and Shelf Science Reconciling nature, people and policy in the mangrove social-ecological system through the adaptive cycle heuristic. *Estuar. Coast. Shelf Sci.* **2021**, *248*, 1–29. [CrossRef]
3. SARAS Institute. Key Concepts: Socio-Ecological Systems. South American Institute for Resilience and Sustainability Studies 2019; p. 2. Available online: https://saras-institute.org/social-ecological-systems/ (accessed on 5 February 2020).
4. Walters, B.B.; Rönnbäckb, P.; Kovacsc, J.M.; Cronab, B.; Hussain, S.A.; Badola, R.; Primavera, J.H.; Barbier, E.B.; Dahdouh-Guebas, F. Ethnobiology, socio-economics and management of mangrove forests: A review. *Aquat. Bot.* **2008**, *89*, 220–236. [CrossRef]
5. Nagendra, H. Reforestation and regrowth in the human-dominated landscapes of South Asia. In *Reforesting Landscapes*; Springer: Berlin/Heidelberg, Germany, 2009; pp. 149–174.
6. Abib, S. Local people and mangroves: Ecosystem perception and valuation on the southwest coast of Mauritius. *West. Indian Ocean. J. Mar. Sci.* **2021**, *20*, 11–19. [CrossRef]
7. Merven, R.; Appadoo, C.; Florens, V.; Iranah, P. Dependency on mangrove ecosystem services is modulated by socioeconomic drivers and socio-ecological changes–insights from an insular biodiversity hotspot. *Res. Sq.* **2023**, 1–18. [CrossRef]
8. MEA (Millennium Ecosystem Assessment). *Ecosystems and Human Well-Being: Biodiversity Synthesis 2005*; World Resources Institute: Washington, DC, USA, 2005.
9. Mitra, A.; Mitra, A. Ecosystem services of mangroves: An overview. In *Mangrove Forests in India: Exploring Ecosystem Services*; Springer: Berlin/Heidelberg, Germany, 2020; pp. 1–32.
10. Nellemann, C.; Corcoran, E.; Duarte, C.M.; Valdés, L.; De Young, C.; Fonseca, L.; Grimsditch, G. *Blue Carbon: A Rapid Response Assessment, Blue Carbon*; United Nations Environment Programme, GRID-Arendal: Arendal, Norway, 2009; Volume 80.

11. Kauffman, J.B.; Adame, M.F.; Arifanti, V.B.; Schile-Beers, L.M.; Bernardino, A.F.; Bhomia, R.K.; Donato, D.C.; Feller, I.C.; Ferreira, T.O.; Garcia, M.C.J.; et al. Total ecosystem carbon stocks of mangroves across broad global environmental and physical gradients. *Ecol. Monogr.* **2020**, *90*, e01405. [CrossRef]
12. Macreadie, P.I.; Anton, A.; Raven, J.A.; Beaumont, N.; Connolly, R.M.; Friess, D.A.; Kelleway, J.J.; Kennedy, H.; Kuwae, T.; Lavery, P.S.; et al. The future of Blue Carbon science. *Nat. Commun.* **2019**, *10*, 3998. [CrossRef] [PubMed]
13. Cummings, A.R.; Shah, M. Mangroves in the global climate and environmental mix. *Geogr. Compass* **2017**, *12*, e12353. [CrossRef]
14. Bouillon, E. Carbon cycle: Storage beneath mangroves. *Nat. Geosci.* **2011**, *4*, 282–283. [CrossRef]
15. Das, S.C.; Das, S.; Tah, J. Mangrove Forests and People's Livelihoods. In *Mangroves: Biodiversity, Livelihoods and Conservation*; Das, S.C., Pullaiah, T., Ashton, E.C., Eds.; Springer Nature: Singapore, 2022; pp. 153–173.
16. Akram, H.; Hussain, S.; Mazumdar, P.; Chua, K.O.; Butt, T.E.; Harikrishna, J.A. Mangrove Health: A Review of Functions, Threats, and Challenges Associated with Mangrove Management Practices. *Forests* **2023**, *14*, 1698. [CrossRef]
17. Friess, D.A.; Yando, E.S.; Alemu, J.B.; Wong, L.W.; Soto, S.D. Ecosystem services and disservices of mangrove forests and salt marshes. In *Oceanography and Marine Biology*; CRC Press: Boca Raton, FL, USA, 2020.
18. Come, J.; Peer, N.; Nhamussua, J.L.; Miranda, N.A.; Macamo, C.C.; Cabral, A.S.; Madivadua, H.; Zacarias, D.; Narciso, J.; Snow, B. A socio-ecological survey in Inhambane Bay mangrove ecosystems: Biodiversity, livelihoods, and conservation. *Ocean Coast. Manag.* **2023**, *244*, 106813. [CrossRef]
19. Gomez, R.G.; Baldago, R.M. Peoples' Resource Utilization of Mangroves and Their Awareness to Its Environmental Importance. *J. Educ. Khon Kaen Univ.* **2016**, *39*, 35–45.
20. Portorreal, Y.C.; Montero, O.P. Evaluación de impactos a la salud del manglar en el municipio Guamá, Santiago de Cuba, Cuba. *Madera Bosques* **2017**, *23*, 27–41. [CrossRef]
21. Alati, V.M.; Olunga, J.; Olendo, M.; Daudi, L.N.; Osuka, K.; Odoli, C.; Tuda, P.; Nordlund, L.M. Mollusc shell fisheries in coastal Kenya: Local ecological knowledge reveals overfishing. *Ocean Coast. Manag.* **2020**, *195*, 17. [CrossRef]
22. Zu Ermgassen, P.S.; Mukherjee, N.; Worthington, T.A.; Acosta, A.; da Rocha Araujo, A.R.; Beitl, C.M.; Castellanos-Galindo, G.A.; Cunha-Lignon, M.; Dahdouh-Guebas, F.; Diele, K.; et al. Fishers who rely on mangroves: Modelling and mapping the global intensity of mangrove-associated fisheries. *Estuar. Coast. Shelf Sci.* **2021**, *248*, 106975. [CrossRef]
23. Goldberg, L.; Lagomasino, D.; Thomas, N.; Fatoyinbo, T. Global declines in human-driven mangrove loss. *Glob. Change Biol.* **2020**, *26*, 5844–5855. [CrossRef]
24. Portorreal, Y.C.; Dominguez, O.J.R.; Cuker, B.; Milanes, C.B.; Montero, O.P. Environmental policy and regulatory framework for managing mangroves as a carbon sink in Cuba. *Water* **2022**, *14*, 3903. [CrossRef]
25. RMunang, T.; Thiaw, I.; Rivington, M. Ecosystem management: Tomorrow's approach to enhancing food security under a changing climate. *Sustainability* **2011**, *3*, 937–954. [CrossRef]
26. Polidoro, B.A.; Carpenter, K.E.; Collins, L.; Duke, N.C.; Ellison, A.M.; Ellison, J.C.; Farnsworth, E.J.; Fernando, E.S.; Kathiresan, K.; Koedam, N.E.; et al. The Loss of Species: Mangrove Extinction Risk and Geographic Areas of Global Concern. *PLoS ONE* **2010**, *5*, 10. [CrossRef] [PubMed]
27. Satyanarayana, B.; Bhanderi, P.; Debry, M.; Maniatis, D.; Foré, F.; Badgie, D.; Jammeh, K.; Vanwing, T.; Farcy, C.; Koedam, N.; et al. A Socio-Ecological Assessment Aiming at Improved Forest Resource Management and Sustainable Ecotourism Development in the Mangroves of Tanbi Wetland National Park, The Gambia, West Africa. *Ambio* **2012**, *41*, 513–526. [CrossRef]
28. Friess, D.A.; Yando, E.S.; Abuchahla, G.M.O.; Adams, J.B.; Cannicci, S.; Canty, S.W.J.; Cavanaugh, K.C.; Connolly, R.M.; Cormier, N.; Diele, K.; et al. Conservation optimism, for now. *Curr. Biol.* **2020**, *30*, R153–R154. [CrossRef] [PubMed]
29. Dahdouh-Guebas, F.; Ajonina, G.N.; Amir, A.A.; Andradi-Brown, D.A.; Aziz, I.; Balke, T.; Barbier, E.B.; Cannicci, S.; Cragg, S.M.; Cunha-Lignon, M.; et al. Public Perceptions of Mangrove Forests Matter for Their Conservation. *Front. Mar. Sci.* **2020**, *7*, 603651. [CrossRef]
30. Nguyen, H.; Harper, R.J.; Dell, B. Examining local community understanding of mangrove carbon mitigation: A case study from Ca Mau province, Mekong River Delta, Vietnam. *Mar. Policy* **2023**, *148*, 10. [CrossRef]
31. Nyangoko, B.P.; Berg, H.; Mangora, M.M.; Gullström, M.; Shalli, M.S. Community Perceptions of Mangrove Ecosystem Services and Their Determinants in the Rufiji Delta, Tanzania. *Sustainability* **2021**, *13*, 63. [CrossRef]
32. Quevedo, J.M.D.; Uchiyama, Y.; Kohsaka, R. Perceptions of local communities on mangrove forests, their services and management: Implications for Eco-DRR and blue carbon management for Eastern Samar, Philippines. *J. For. Res.* **2020**, *25*, 12. [CrossRef]
33. Martín-López, B.; Iniesta-Arandia, I.; García-Llorente, M.; Palomo, I.; Casado-Arzuaga, I.; Amo, D.G.; Gómez-Baggethun, E.; Oteros-Rozas, E.; Palacios-Agundez, I.; Willaarts, B.; et al. Uncovering Ecosystem Service Bundles through Social Preferences. *PLoS ONE* **2012**, *7*, e38970. [CrossRef] [PubMed]
34. Bimrah, K.; Dasgupta, R.; Hashimoto, S.; Saizen, I.; Dhyani, S. Ecosystem Services of Mangroves: A Systematic Review and Synthesis of Contemporary Scientific Literature. *Sustainability* **2022**, *14*, 12051. [CrossRef]
35. Turschwell, M.P.; Tulloch, V.J.D.; Sievers, M.; Pearson, R.M.; Andradi-brown, D.A.; Ahmadia, G.N.; Connolly, R.M.; Bryan-Brown, D.; Lopez-Marcano, S.; Adame, M.F.; et al. Multi-scale estimation of the effects of pressures and drivers on mangrove forest loss globally. *Biol. Conserv.* **2020**, *247*, 11. [CrossRef]
36. Rodríguez-Crespo, G.C.; Domínguez-Junco, O. Servicios ecosistémicos en manglares: Beneficios a: Resiliencia del ecosistema ante cambios climáticos, la comunidad y su desarrollo local. *Rev. Transdiscipl. Estud. Soc. Tecnol.* **2022**, *2*, 5–10. [CrossRef]

37. Mahardika SM, A.; Yulianda, F.; Adrianto, L. Interactive Governance for Mangrove Social-Ecological System in Tangerang Regency: A DPSIR Approach. *Int. J. Adv. Sci. Eng. Inf. Technol.* **2023**, *13*, 1249–1257. [CrossRef]
38. Gayo, L. Local community perception on the State Governance of mangroves in Western Indian coast of Kinondoni and Bagamoyo, Tanzania. *Glob. Ecol. Conserv.* **2022**, *39*, 11. [CrossRef]
39. Ahmed, J.; Kathambi, B.; Kibugi, R. Policy perspective on governance standards setting using community participation for sustainable mangrove management in Lamu Kenya. *Int. J. Conserv. Sci.* **2023**, *14*, 315–326. [CrossRef]
40. Bunting, P.; Rosenqvist, A.; Hilarides, L.; Lucas, R.M.; Thomas, N.; Tadono, T.; Worthington, T.A.; Spalding, M.; Murray, N.J.; Rebelo, L.-M. Global Mangrove Extent Change 1996–2020: Global Mangrove Watch Version 3.0. *Remote Sens.* **2022**, *14*, 3657. [CrossRef]
41. Valderrama, S.P.; Ávila, A.H.; Méndez, J.G.; Martínez, O.M.; Rojas, D.C.; Azcona, H.F.; Hernández, E.M.; Aragón, H.C.; Alcolado, P.M.; Pina-Amargós, F.; et al. Marine protected areas in Cuba. *Bull. Mar. Sci.* **2018**, *94*, 423–442. [CrossRef]
42. García, M.A.V.; González, B.A. Manglar vivo en cuba: Costos y beneficios de las acciones basadas en ecosistemas. Análisis económico-ecológico en las provincias Sur Artemisa y Mayabeque. *Rev. Iberoam. Econ. Ecol.* **2021**, *34*, 86–110.
43. Ferro-Azcona, H.; Espinoza-Tenorio, A.; Calderón-Contreras, R.; Ramenzoni, V.; País, M.G.; Mesa-Jurado, M. Adaptive capacity and social-ecological resilience of coastal areas: A systematic review. *Ocean Coast. Manag.* **2019**, *173*, 36–51. [CrossRef]
44. Perera-Valderrama, S.; Hernández-Ávila, A.; Ferro-Azcona, H.; Cobián-Rojas, D.; González-Méndez, J.; Caballero-Aragón, H.; de la Guardia-Llansó, E.; Ramón-Puebla, A.; Hernández-González, Z.; Espinosa-Pantoja, L.; et al. Increasing marine ecosystems conservation linking marine protected areas and integrated coastal management in southern Cuba. *Ocean. Coast. Manag.* **2020**, *196*, 11. [CrossRef]
45. CITMA. *Estrategia Ambiental Nacional 2021–2025*; CITMA: La Habana, Cuba, 2021; p. 44.
46. Spalding, M.D.; Fox, H.E.; Allen, G.R.; Davidson, N.; Ferdaña, Z.A.; Finlayson, M.; Halpern, B.S.; Jorge, M.A.; Al Lombana, S.A.L.; Martin, K.D.; et al. Marine Ecoregions of the World: A Bioregionalization of Coastal and Shelf Areas. *BioScience* **2007**, *57*, 573–583. [CrossRef]
47. Menéndez, L. The Mangrove Ecosystem in the Cuban Archipelago: Bases for Its Management. Ph.D. Thesis, University of Alicante, San Vicente del Raspeig, Spain, 2013.
48. Martínez Quesada, E. Relación entre morfología foliar de antófitos y factores abióticos en las principales pluvisilvas de la Región Oriental cubana. *Rev. Biol. Trop.* **2009**, *57*, 235–256. [CrossRef]
49. Kottek, M.; Grieser, J.; Beck, C.; Rudolf, B. World map of the Köppen-Geiger climate classification. *Meteorologische Zeitschrift* **2006**, *15*, 59–263. [CrossRef] [PubMed]
50. Reyes, O.J.; Acosta-Cantillo, F. Principales fitocenosis en el humedal del delta del río Cauto, Cuba Oriental. I. Vegetación lacustre y herbazal de humedal. *For. Veracruzana* **2007**, *9*, 15–22.
51. Alonso-Tabet, M.; López-Salcedo, M.; Olano-Labrada, R.; Alarcón Jorge, Y.; Gil Rivero, A.; Alonso Jiménez, Y.; Egard Tamayo, D.; Rodríguez González, O.; Borrero Nieves, Y. *Plan de Manejo 2021–2025 Refugio de Fauna Ojo de Agua-Monte Cabaniguán*; Empresa Flora y Fauna Las Tunas: Las Tunas, Cuba, 2021; Unpublished work.
52. González-Rivera, D.; Manet-Bombú, I.; Aroche-Rodríguez, M.; Frómeta-Barrera, A.; Rodriguez-Munivel, I.; Leiva Cueto, A. *Management Plan 2021–2025 Hatibonico Ecological Reserve*; Unidad de Medio Ambiente-CITMA-Guantánamo: Guantánamo, Cuba, 2021; Unpublished work.
53. Fernández-Rodríguez, I.L.; Hechavarria, S.L.; Mestril Cosme, K.; Bouza Alonso, J.A.; Debros Trutié, Y. *Plan de Manejo Refugio de Fauna San Miguel de Parada 2021–2025*; Empresa Flora y Fauna Santiago de Cuba: Santiago de Cuba, Cuba, 2021; Unpublished work.
54. Cabrera-Díaz, J.P.; Naranjo-Peña, L.A. Estudio de vulnerabilidad de paisajes para la gestión ambiental del área protegida Reserva Ecológica Monte Cabaniguán-Ojo del Agua del municipio Jobabo, Las Tunas. *Estud. Desarro. Soc. Cuba. Am. Lat.* **2019**, *7*, 172–181.
55. Ferrera Bergues, W.V.; Pérez Montero, O.; Soler Nariño, O. Población y vulnerabilidad social ante los efectos del cambio climático en el municipio costero de Guamá. *Noved. Población* **2020**, *16*, 190–217. Available online: http://www.novpob.uh.cu (accessed on 10 February 2024).
56. Pérez, O.; Carbonero, M.A.; Poveda, I.; Gómez, M.; Oliver, M.A. Cuando la mujer migra. Una mirada a las migraciones internas, desde la perspectiva del desarrollo sostenible, en el municipio costero de Guamá, Santiago de Cuba. *Noved. Población* **2018**, *28*, 1–9.
57. Lara, S.C.; Núñez, E.G. Reflexiones desde la relación población-vulnerabilidad en el municipio Guamá, Santiago de Cuba Reflections on the relationship population-vulnerability in the Municipality Guamá, Santiago de Cuba. *Noved. Población* **2020**, *16*, 31–44.
58. Milanés, B.C.; Pérez, M.O. Ordenamiento y Manejo Integrado de la Zona Costera Frente a los Riesgos del Cambio Climático en la Región Suroriental de Cuba. Available online: http://www.revistaccuba.cu/index.php/revacc/article/view/572 (accessed on 10 February 2024).
59. González Rivera, D. Conservación de *Leptocereus nudiflorus* en la Reserva Ecológica Hatibonico, Guantánamo. *Bissea* **2023**, *15*, 1.
60. Manet-Bombus, I.; Barroso-Frometa, L.; González-Rivera, D.; Pérez-Trejo, H.M.; Begué-Quiala, G. Caracterización de la biodiversidad de especies florísticas en la Reserva Ecológica Hatibonico. *Hombre Ciencia y Tecnología* **2020**, *24*, 31–39.
61. RSampieri, H.; Collado, C.F.; Lucio, M.D.P.B. *Metodología de la Investigación*; Mcgraw-Hill: Mexico City, Mexico, 2014; p. 634. ISBN 978-1-4562-2396-0.
62. Quevedo, J.M.D.; Uchiyama, Y.; Kohsaka, R. Local perceptions of blue carbon ecosystem infrastructures in Panay Island, Philippines. *Coast. Eng. J.* **2021**, *63*, 227–247. [CrossRef]

63. Dahdouh-Guebas, F.; Collin, S.; Lo Seen, D.; Rönnbäck, P.; Depommier, D.; Ravishankar, T. Analysing ethnobotanical and fishery-related importance of mangroves of the East-Godavari Delta (Andhra Pradesh, India) for conservation and management purposes. *J. Ethnobiol. Ethnomed.* **2006**, *2*, 24. [CrossRef] [PubMed]
64. Mohamed, M.K.; Adam, E.; Jackson, C. MAssessing the Perception and Contribution of Mangrove Ecosystem Services to the Well-Being of Coastal Communities of Chwaka and Menai Bays, Zanzibar. *Resources* **2024**, *13*, 23. [CrossRef]
65. Asante, F.; Huge, J.; Asare, N.K.; Dahdouh-Guebas, F. Does mangrove vegetation structure reflect human utilisation of ecosystem goods and services? *iScience* **2023**, *26*, 6. [CrossRef]
66. Krkač, K. Stakeholder Mapping. In *Encyclopedia of Sustainable Management*; Idowu, S., Schmidpeter, R., Capaldi, N., Zu, L., Del Baldo, M., Abreu, R., Eds.; Springer: Cham, Switzerland, 2022. [CrossRef]
67. Oficina Nacional de Estadísticas e Información (ONEI). *Anuario Estadístico de Cuba. Capítulo 1: Territorio*; Oficina Nacional de Estadísticas e información República de Cuba: La Habana, Cuba, 2021; p. 15.
68. Worthington, A.; Ermgassen, P.S.E.; Friess, D.A.; Krauss, K.W.; Lovelock, C.E.; Thorley, J.; Tingey, R.; Woodroffe, C.D.; Bunting, P.; Cormier, N.; et al. A global biophysical typology of mangroves and its relevance for ecosystem structure and deforestation. *Sci. Rep.* **2020**, *10*, 14652. [CrossRef] [PubMed]
69. Soto, H. *Priorización de los ODS en Cuba: Articulación de la Agenda 2030 con el Plan Nacional de Desarrollo e Identificación de prioridades de Desarrollo sostenible. Sede Subregional de la CEPAL en México (Estudios e Investigaciones) 49071*; Ciudad de México, México, 2023; 164p.
70. Ley 85 "Ley Forestal". Gaceta Oficial de la República de Cuba. Edición Ordinaria Número No. 46: 773. 1998. Available online: https://www.gacetaoficial.gob.cu/ (accessed on 21 January 2024).
71. Uddin, M.S.; Shah, M.A.R.; Khanom, S.; Nesha, M.K. Climate change impacts on the Sundarbans mangrove ecosystem services and dependent livelihoods in Bangladesh. *Asian J. Conserv. Biol.* **2013**, *2*, 152–156.
72. Souza, L.D.; Rossi, S.; Calvet-Mir, L.; Ruiz-Mallén, I.; García-Betorz, S.; Salvà-prat, J.; Jeovah, A.; Meireles, D.A. Neglected ecosystem services: Highlighting the socio-cultural perception of mangroves in decision-making processes. *Ecosyst. Serv.* **2017**, *26*, 137–145.
73. Reyes-Arroyo, N.; Camacho-Valdez, V.; Saenz-Arroyo, A. Socio-cultural analysis of ecosystem services provided by mangroves in La Encrucijada Biosphere Reserve, southeastern Mexico. *Local Environ.* **2021**, *26*, 86–109. [CrossRef]
74. Afonso, F.; Félix, P.M.; Chainho, P.; Heumüller, J.A.; de Lima, R.F.; Ribeiro, F. Community perceptions about mangrove ecosystem services and threats. *Reg. Stud. Mar. Sci.* **2022**, *49*, 102114. [CrossRef]
75. Mallick, B.; Priodarshini, R.; Kimengsi, J.N.; Biswas, B.; Hausmann, A.E.; Islam, S.; Huq, S.; Vogt, J. Livelihoods dependence on mangrove ecosystems: Empirical evidence from the Sundarbans. *Curr. Res. Environ. Sustain.* **2021**, *3*, 100077. [CrossRef]
76. Warren-Rhodes, K.; Schwarz, A.-M.; Boyle, L.N.; Albert, J.; Agalo, S.S.; Warren, R.; Bana, A.; Paul, C.; Kodosiku, R.; Bosma, W. Mangrove ecosystem services and the potential for carbon revenue programs in the Solomon Islands. *Environ. Conserv.* **2011**, *38*, 485–496. [CrossRef]
77. Song, A.M.; Dressler, W.H.; Satizábal, P.; Fabinyi, M. From conversion to conservation to carbon: The changing policy discourse on mangrove governance and use in the Philippines. *J. Rural. Stud.* **2021**, *82*, 184–195. [CrossRef]
78. Chircop, A. Local Integrated Coastal Zone Management in Cuba (COSTASURESTE Project); S06499; Report to the Department of Foreign Affairs. 2015. Available online: https://ssrn.com/abstract=2757073 (accessed on 10 February 2024).
79. Golebie, E.J.; Aczel, M.; Bukoski, J.J.; Chau, S.; Ramirez-Bullon, N.; Gong, M.; Teller, N. A qualitative systematic review of governance principles for mangrove conservation. *Conserv. Biol.* **2021**, *36*, e13850. [CrossRef]
80. Walker, J.E.; Ankersen, T.; Barchiesi, S.; Meyer, C.K.; Altieri, A.H.; Osborne, T.Z.; Angelini, C. Governance and the mangrove commons: Advancing the cross-scale, nested framework for the global conservation and wise use of mangroves. *J. Environ. Manag.* **2022**, *312*, 114823. [CrossRef] [PubMed]
81. Alongi, D.M. Responses of mangrove ecosystems to climate change in the Anthropocene. In *Mangroves: Ecology, Biodiversity and Management*; Rastogi, R.P., Phulwaria, M., Gupta, D.K., Eds.; Springer: Berlin/Heidelberg, Germany, 2021; pp. 201–224.
82. Iturralde-Vinent, M.; Méndez, H.S. *Peligros y Vulnerabilidades de la Zona Marino-Costera de Cuba: Estado Actual y Perspectivas ante el Cambio Climático Hasta el 2100*; Editorial Academia: La Habana, Cuba, 2015; p. 86. ISBN 978-959-270-338-4.
83. Krauss, K.W.; Osland, M.J. Tropical cyclones and the organization of mangrove forests: A review. *Ann. Bot.* **2020**, *125*, 213–234. [CrossRef] [PubMed]
84. Kissinger, G.; Herold, M.; De Sy, V. *Drivers of Deforestation and Forest Degradation: A Synthesis Report for EDD+ Policymakers*; Lexeme Consulting: Vancouver, BC, Canada, 2012.
85. Madhav, S.; Nazneen, S.; Singh, P. *Coastal Ecosystems Environmental Importance, Current Challenges, and Conservation Measures*; Springer Nature AG: Cham, Switzerland, 2022; p. 392.
86. Friess, D.A.; Adame, M.F.; Adams, J.B.; Lovelock, C.E. Mangrove Forests under Climate Change in a 2 °C World. *Wiley Interdiscip. Rev. Clim. Change* **2022**, *13*, e792. [CrossRef]

water

MDPI

Article

Reflections on How to Reach the "30 by 30" Target: Identification of and Suggestions on Global Priority Marine Areas for Protection

Chang Zhao [1,2], Yuejing Ge [1] and Miaozhuang Zheng [2,*]

1 Faculty of Geographical Science, Beijing Normal University, Beijing 100875, China
2 China Institute for Marine Affairs, Ministry of Nature Resources, Beijing 100860, China
* Correspondence: zhengmiaozhuang@cimamnr.org.cn

Abstract: The establishment of marine protected areas (MPAs) is an important method to ensure marine protection. To protect and conserve global marine biodiversity, with the adoption of the "Kunming-Montreal Global Biodiversity Framework" during the 15th meeting of the Conference of the Parties of Convention on Biodiversity (CBD) in December 2022, the establishment of an effectively managed MPA network by 2030 and the protection of 30% of the world's oceans will be common goals for all countries party to the CBD over the next decade. Based on the distribution of over 150 types of marine species, habitats, ecosystems, and abiotic elements, ArcGIS10.5 and Zonation are used in this study to calculate the marine protection priority levels of coastal, nearshore, open ocean, and deep ocean trench areas, and a plan to reach the "30 by 30" targets is proposed. The suggestions for scientifically identifying and managing MPAs are as follows: first, improve MPA planning and establish a well-connected MPA network in national jurisdictions, then conduct scientific marine investigations to obtain background data on MPA establishment and delimitation.

Keywords: marine protected areas; marine ecosystems; marine biodiversity; global marine governance; 30 by 30 target

1. Introduction

The ocean is the foundation of all life, providing us with space for transportation, trade, and recreation [1]; cultural heritage [2]; and resources, including food [3], drugs [4], and minerals [5]. Meanwhile, the ocean is also the largest carbon reservoir in the Earth system [6], with approximately a quarter of anthropogenic carbon dioxide emissions being absorbed by oceans over the last two decades [7]. With the rapid increases in the human population and material demands, the scope of human utilization of the ocean has expanded from coastal regions [8] to the high seas [9] and international seabeds [10]. Over-exploitation and harvesting, land-based pollution, and marine invasion have created great losses of marine life [8]. Moreover, oceangoing activities such as biological resource surveys and deep-sea mineral exploration and exploitation pose a serious threat to areas beyond national jurisdictions. Establishing marine protected areas (MPAs) is an effective way to systematically address marine environmental problems and resolve the crisis of biodiversity destruction [11,12]. Considering the synergistic effects of marine connectivity, fluidity and population growth, climate change, and other stressors, there is a need to identify and establish MPAs on a global scale. In December 2022, the second phase of the 15th Conference of the Parties (COP) on the Convention on Biological Diversity (CBD) adopted the "Kunming-Montreal Global Biodiversity Framework", which sets a highly ambitious global marine conservation target for the next decade, aiming to "reverse and halt global biodiversity loss and enhance biodiversity and ecosystem functions and services, ecological integrity and connectivity through area-based conservation measures to protect at least 30 percent of global coastal and marine areas by 2030" [13], referred to as the so-called "30 by 30" target. There is a global consensus to designate marine protected areas and protect marine biodiversity in the coming decade.

Since the 10th COP of the CBD in 2010, which set the target of protecting 10% of the world's oceans and seas by 2020, the pace of MPA expansion and new designations has accelerated [14]. However, the science behind new MPA designations and effectiveness needs to be strengthened [15]. By August 2023, a total of 29.59 million square kilometers of MPAs of all types and levels had been established globally, accounting for 8.17% of the global marine area, of which 18.7% was protected in marine areas under national jurisdiction and 1.44% in marine areas beyond national jurisdiction, falling short of the 10% protection target [16,17]. There is also a general lack of scientifically informed and effective management plans and feasible management measures for MPAs, insufficient attention to ecosystem representation and connectivity, and low effectiveness of actual protection efforts [15]. Biodiversity loss has not been effectively mitigated, and the security of critical habitats is still threatened [18,19]. According to the International Union for Conservation of Nature (IUCN), 5510 threatened species (21% of all threatened species) are covered by biodiversity-critical areas globally, of which 13% are entirely within protected areas, while another 31% are only partially covered by marine protected areas [15]. The appropriateness and effectiveness of selecting MPAs already in place must be further explored.

In accordance with current international law, all MPAs established under the United Nations Convention on the Law of the Sea (UNCLOS), except for MPAs in the Antarctic Ocean, are within areas of national jurisdiction and are designated and managed by countries in accordance with domestic laws, with insufficient attention given to the high seas and international seabed areas. To promote marine protection in areas beyond national jurisdictions, since 2018, the United Nations (UN) has initiated negotiations on an intergovernmental legally binding instrument for the conservation and sustainable use of marine biodiversity in areas beyond national jurisdictions, better known by its acronym BBNJ [20]. In June 2023, the BBNJ agreement was adopted, which forms a legal basis for the establishment of area-based management tools, including MPAs beyond national jurisdictions [21]. At the scientific and technical level, many UN and non-governmental initiatives have proposed maps for globally important marine areas using different criteria [22]. Seven criteria, including "uniqueness or rarity, special importance of the life history stage of the species, importance for threatened, endangered or declining species and/or habitats, vulnerability, fragility, susceptibility or slow recovery, biological productivity, biodiversity, naturalness", have been used in the CBD framework to describe ecologically or biologically significant marine areas (EBSAs) in more than 75% of the world's regions. These areas described by CBD have been directly used by the European Union, Japan, and small island countries to establish MPAs. However, the description of EBSAs has scientific gaps, such as a lack of data regarding the deepest parts of the ocean, and many EBSA descriptions were provided more than 10 years ago with no consideration for new data available. Moreover, the coverage of EBSAs is not yet universal, with no EBSAs described in the Southwest Atlantic [23]. If this description is used as the scientific basis for achieving the "30 by 30" initiative on the high seas and international seabed, it will be difficult to effectively protect the global marine ecosystem. Greenpeace [24] and the Pew Charitable Trusts [25] have also made recommendations for identifying 30% of the world's marine conservation priority areas based on global species and habitat distribution data using spatial planning software.

Scientists have used various methods to prioritize marine conservation, including the use of leading and lagging indicators to define scoring criteria for marine ecosystem services [26], weighted overlay analysis of different habitat types [27], presence–absence or density surface models [28], systematic conservation planning [29], and calculating priority through an environmental sensitivity mapping approach for mapping environmentally sensitive assets [30]. Remote sensing imagery combined with habitat mapping and modeling [31] and rarity-weighted richness maps [32] have also been used for marine conservation priority identification. Moreover, systematic conservation planning approaches and optimization algorithms, including Marxan [33], Prioritizr [34], and Zonation [35,36], are commonly used to identify high-priority sites for biodiversity conservation. The research object includes single species such as sea turtles [37], coral reefs [38], seabirds [39], salmon [40], and sea mammals [41]; specific sea areas [42–45]; or Earth as a whole [33,46,47].

Some studies have analyzed the proposed priorities for marine biodiversity protection based on existing important marine areas for conservation [22,33,47] or the distribution of marine species [34,35,48]. In these studies, the dimensions of data and the methods used differ. Fan et al. used mitochondrial DNA barcode sequences and constructed a phylogenetic tree as the basis for a priority proposal. Belote et al. used 1697 marine species and subspecies in the continental US to compare the performance of the two methods. Gownaris et al. quantified the consensus of existing initiatives on MPAs and proposed gaps in globally important marine protection areas. Visalli et al. focused on maritime areas beyond national jurisdictions and used the species richness of fishes, marine mammals, marine invertebrates, seagrass, and benthic habitats for calculation. Compared to previous studies, this study focused on the conservation of rare and endangered marine species, including seabirds, considering the different habitats in the nearshore and deep ocean, as well as the five main catches that affect the livelihoods of fisherfolk on a global scale for priority calculation. The differences in the data and the methods used provide an opportunity to compare the outputs and take an integrated approach to choose priority areas for conservation to realize the target.

As policymakers seek to meet the "30 by 30" target in the global ocean, based on existing studies [22,34,35], this study focuses on typical nearshore, offshore, and deep-sea habitats, as well as the conservation of global rare, endangered, and commercially important marine species. This is achieved by considering abiotic factors, including temperature and salinity, and providing a marine conservation priority map of the global ocean. We also examine the overlap between EBSAs and existing MPAs with our result and conduct several discussions to provide suggestions on the establishment and management of protected area plans to achieve the considered target. Finally, we establish MPAs in areas beyond national jurisdictions under the BBNJ agreement.

2. Materials and Methods

2.1. Data Description and Sources

This study identifies global priority marine areas for protection at three levels: the marine species level, the habitat level, and the ecosystem level. A total of 162 layers of important habitats, rare and endangered species, and ecosystem distribution characteristics were selected as the basis for selecting priority marine areas for protection. All data used, including ocean features and biogeographic province distribution data, were obtained from publicly accessible sources on the Internet.

2.1.1. Surveyed Nearshore, Offshore, and Deep-Sea Habitats

Nearshore coral reefs, mangroves, salt marshes, kelp beds, and seagrass meadows, together with deep-sea seamounts, cold seeps, and hydrothermal vent ecosystems, are typical habitats that require priority attention and protection. Coral reefs are the most diverse marine ecosystem worldwide and are crucial for human survival and development [49]. Mangroves can secure the coastline from erosion, provide habitats for seabirds, fish, and invertebrates, and mitigate climate change [50]. Salt marshes are one of the most productive ecosystems, providing food sources and habitats for marine organisms in estuaries and coastal areas [51]. Kelp beds have high productivity and can help improve water quality, maintain biodiversity, sequester carbon, and prevent coastal erosion. Moreover, kelp beds can contribute to recreation and cultural services [52]. Seagrass meadows may help in mitigating ocean acidification due to their ability to raise pH and decrease DIC and pCO_2 [53]. They also serve as an important blue carbon habitat [54] and various ecosystem goods and services, including biodiversity maintenance, filtering, and sediment enrichment [55]. Cold seeps appear in seafloor cracks caused by tectonic activities, and the surrounding area forms a unique habitat and provides a haven for a variety of unique organisms [56].

Seamounts are also rich in deep-sea species, among which plankton, nekton, and benthos are quite different from those of the surrounding in terms of biomass, abundance, diversity, and so on [57]. Hydrothermal vents are underwater heat leaks formed in active volcanoes and seamounts, most of which are distributed in the continental plate boundary area. Hundreds of species live in and around hydrothermal vent ecosystems, many unique to the vent area [58].

Water depth is also an important factor that affects species distribution over seamounts. Research has shown that the composition of nekton and benthos distributed over seamounts varies with water depth [59], and different species live at different water depths. This study establishes layers based on the division of a range of ocean depths and divides them into four categories from the surface to the seafloor and even the abyss (the categories include shallower than 200 m, from 200 m to 800 m, from 800 m to 2000 m, and deeper than 2000 m) to distinguish different types of seamounts [60,61]. A seamount that is shallower than 200 m represents an area with a protrusion from the summit to the photic zone and hosts a particular community in the shallow depths. A seamount with a depth from 800 m to 2000 m represents an area with the distribution of vertically migrating animals (the scattering layer). A seamount with a depth deeper than 800 m forms the biogeographic area of the deep-sea bottom [60], characterized by the settlement of invertebrates and habitat for bathyal fishes, where the depth from 800 to 2000 m refers to the upper portions of the lower bathyal, with summits of seamounts at fishable depths; depths deeper than 2000 m refer to the lower portions of the lower bathyal [61]. Considering that hydrothermal vent ecosystems are distributed in seamount areas, only cold seeps and seamount data layers were used for deep sea areas in order to avoid duplicate calculations.

2.1.2. Endangered, Rare, and Commercially Valuable Marine Species

The establishment of MPAs considers not only environmental protection and conservation but also social and economic elements, including ensuring fisheries and fishing livelihoods, food security, and cultural values [62]. Focusing on the sustainable use of fishery resources, this study considers endangered species, rare marine species, commercial fish of significant economic value, and a high pelagic fishing volume when determining the selection parameters of priority marine areas for protection. To find a balance between the protection and utilization of living marine resources and ensure the livelihoods of fisherfolk, not only are rare and endangered species considered, but also important economic fish species with high pelagic fishing volume.

Based on "The State of World Fisheries And Aquaculture 2022", Anchoveta, Alaska poll, Skipjack Tuna, Atlantic herring, Yellowfin Tuna, and Jumbo flying squid with a high global catch were selected as representative species [63]. We are also aware that the construction and management of an MPA may have a certain impact on the livelihoods of fisherfolk, and we have proposed targeted recommendations in Section 4.1 [64].

A total of 114 species of rare and endangered marine organisms listed in the "IUCN Red List of Threatened Species" were included in the model [33], including 44 species of fish, 25 species of reptiles, and 45 species of marine mammals [65]. Furthermore, as the distribution of seabirds may vary with breeding sites, age, time of day, and the seasons, and as some seabirds are opportunistic feeders that disperse widely, it is hard to define specific areas or habitats for marine protection [39]. We used the Important Bird and Biodiversity Areas (IBAs) established by BirdLife International [66] to represent the spatial distribution of seabirds that require consideration. IBAs in the ocean are described by BirdLife International based on satellite tracking data, marine surveys, and the research literature. There are currently 2621 seabird areas, 38% of which include a significant marine component, and others are coastal areas such as seabird breeding colonies [67]. In addition, this study considers global ship activity using Automatic Identification System (AIS) data of global vessels for the whole year of 2023 to delineate hotspots that are highly influenced by human activity.

2.1.3. Abiotic Elements

In addition to the above data layers, the current scope and technical limitations of deep-sea surveys were considered, given that the species distribution in existing survey data cannot fully reflect the actual situation of the deep sea. To better reflect the impact of the spatial distribution of abiotic factors in the ocean, such as cold water masses, hypoxic areas, temperature, salinity, dissolved oxygen, and ocean currents, this study drew inspiration from the classification of global marine habitats reported by Sutton et al., which incorporates ecosystem classification layers into the model [68], thus representing different types of marine ecosystems worldwide. This could meet the predetermined protection goals of various types of habitats and compensate for the lack of deep-sea surveys in the current survey and the absence of deep-sea species on the IUCN Red List. Global marine ecosystems were divided into 33 categories, including 20 types of coastal and nearshore ecosystems and 13 types of far-reaching marine ecosystems [68].

2.2. Methodology

2.2.1. Research Area and Data Pre-Processing

Referring to the scope of the "30 by 30" target, the research scope of this study covers all parts of the ocean from the surface to the floor, including territorial waters, exclusive economic zones (EEZs), the outer continental shelf, high seas, and international seabed areas. To facilitate layer comparison and subsequent calculations, all data were processed into grid layers of the same extent and grid size, with 4377 × 1996 grid units (grid size 0.083° × 0.083°) and the same coordinate system, "the 1984 World Geodetic Coordinate System." Specifically, layers of point formats were interpolated as a separate raster map layer using the kernel density estimation function, then normalized from 0 to 1 using the Raster Calculator in ArcGIS 10.5. Layers of surface formats were converted to raster layers, with rasters with data being assigned a value of 1 and the rest assigned 0. To reduce the repetitive time and effort to perform each layer, the iterator of ModelBuilder was used for some level of automation.

2.2.2. Priority for Protection Calculation

This study used ArcGIS 10.5 and Zonation software (version 4.0.0) for priority calculation. Zonation is land-planning software based on ecosystem analysis and conservation prioritization at large spatial scales [36]. It is widely used to establish terrestrial habitats, species conservation areas [69], and marine protected areas [70,71]. The Zonation algorithm makes it possible to find a balance among multiple categories of biodiversity features—such as different species, habitats, and ecosystems—by dividing the study area evenly into several smaller units and calculating and ranking the priority order of the elements in each unit to produce a hierarchical prioritization of the conservation value of a study area. The results were achieved by repeatedly removing the units in the remaining unit set that caused the least marginal loss to the overall 'conservation value' and ranking all units.

After several attempts, the additive benefit function was chosen for this work, and the distortion factor was set to 100 to achieve a balance between the speed of computation and reproducibility of the results; the weights of each type of parameter took the value of 1. The "Additive benefit function" enables convergence to a solution to find a proportional coverage solution for data in order to identify prioritized areas containing multiple species or habitat layers, rather than considering only areas where a particular species or habitat is very rich to find a minimum set of solutions covering all data categories [51]. The distortion factor used in the model indicates the number of grid cells removed in each iteration.

2.2.3. Visualization and Comparison

After calculation, the conservation priority order was reclassified, filtered, and ranked using ArcGIS 10.5. To consider the range from coastal and nearshore areas to the open ocean and deep ocean trenches in the adoption of global marine protection priority calculations,

30% of the global marine protection priorities were selected within and beyond national jurisdictions using the mask function in ArcGIS 10.5 in the case of data gap of different areas. Both within and beyond national jurisdictions, priority areas were reclassified and ranked in order of protection priority using the reclassify function and Raster Calculator to form a global marine conservation priority distribution map.

In order to compare the result with existing MPAs and EBSAs, the MPAs and EBSAs layer was first converted to the same raster size as the results above and assigned values of 1 separately. The blank area was assigned 0 and then analyzed by overlaying the two layers with the Raster Calculator in ArcGIS 10.5, then calculating the distribution promotion of existing MPAs in priority areas via the number of grids with different values.

3. Results

3.1. Description of Global Priority Marine Areas for Protection Layout

As shown in Figure 1, the conservation priority of coastal and nearshore areas is much higher than that of the open ocean. Considering that global marine protection and conservation will revolve around the "30 by 30" initiative in the next decade, this article proposes the top 30% of global priority marine areas for protection by analyzing the model's results.

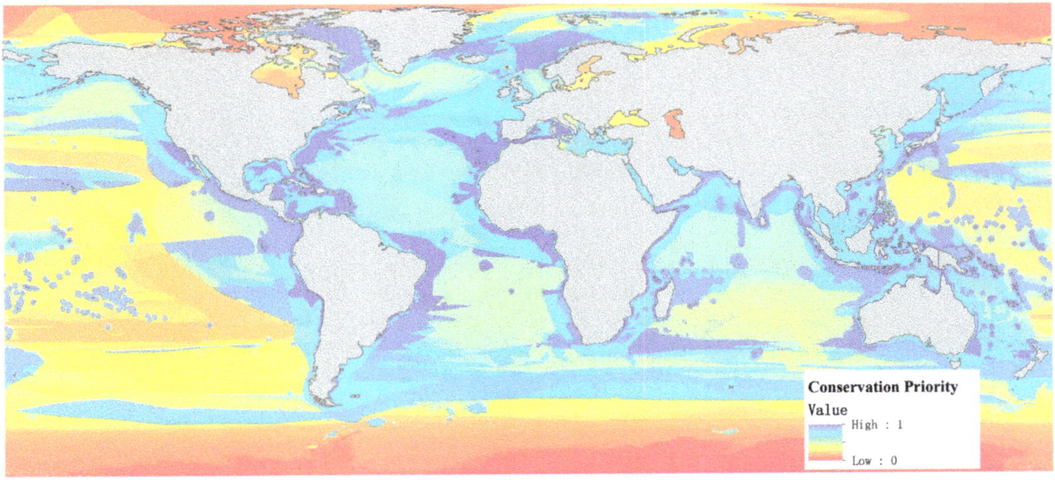

Figure 1. The conservation priority of the global ocean. The output from the model is shown in this map. The conservation priority ranges from 0 to 1. Blue and green indicate higher-priority areas for protection, and conversely, red and yellow indicate lower-priority areas. Gray indicates the land.

The top 30% of the global marine protection priorities mainly include the Arabian Sea, the Chagos–Laccadive Ridge, the marine areas around Madagascar and the Southwest Indian Ocean region, the Western Pacific seamount area and the Eastern Pacific seamount area, the Emperor seamount chain, the Tasman Sea, the North Pacific transitional zone, the Southeast Pacific Ocean, the coral reef delta, and the hydrothermal region of the Midwestern ridge of the North Pacific Ocean. It also includes the Argentine Basin, the Sargasso Sea, the hydrothermal region of the Midwestern ridge of the North Pacific Ocean, the equatorial high productivity area of the Atlantic Ocean, and other sea areas in the Atlantic Ocean. The top 5% of the most protected areas are mainly distributed along the Atlantic and Indian Ocean coasts, the Mediterranean, and the Western Pacific seamount region. Among them, the priority of marine protection in territorial waters and EEZs is much higher than in areas beyond national jurisdictions (Figure 2).

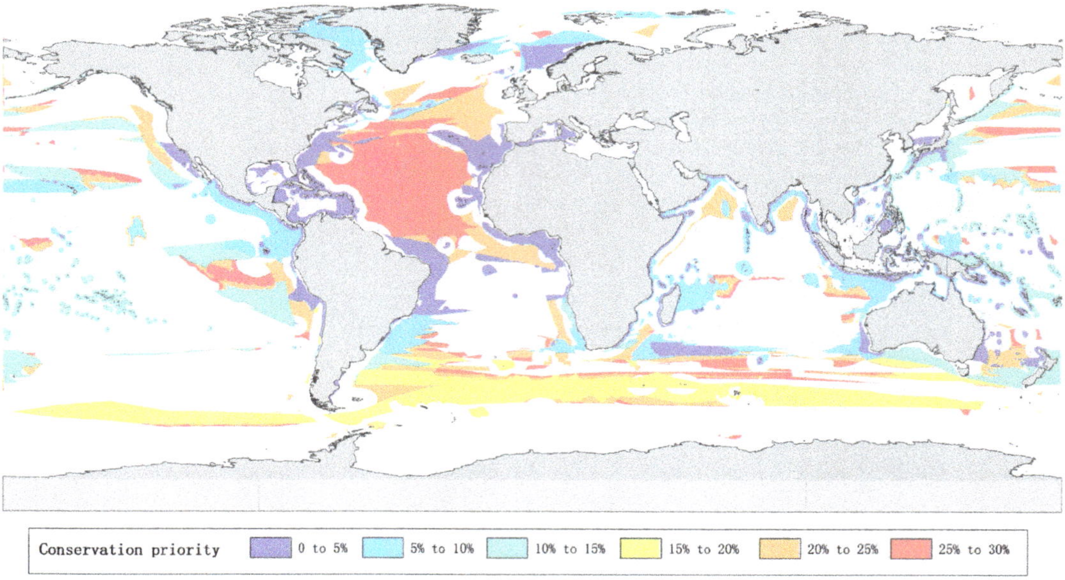

Figure 2. The layout of 30% of global priority marine areas for conservation. Purple indicates the top 5% of areas with protection priority. Light blue, lake blue, yellow, orange, and pink indicate areas with corresponding protection priorities. Gray indicates the land.

3.2. Distribution of Priority Marine Areas for Protection from the Coastline to Deep Ocean Trenches

In the top 30% of global marine protection priorities, the priority of marine protection and conservation in territorial waters and EEZs is much higher than that in marine areas beyond national jurisdictions. Among the top 5% of areas with priority protection, 82.42% are located in territorial waters and EEZs, while 17.58% are located in high seas and international seabed areas beyond national jurisdictions (see Table 1). However, over 95% of the areas with protection priority of 20% to 30% are beyond national jurisdictions. On the one hand, this is due to the richer biodiversity in the coastal and nearshore waters, which are habitats for numerous marine mammals, reptiles, fish, and seabirds, as well as important habitats such as mangroves, kelp beds, and coral reefs with high conservation value. On the other hand, the data on marine species, habitats, and ocean features used in this study mainly rely on open-source data released after global marine scientific surveys. Due to the abundance of surveys in the offshore area compared to the high sea and international seabed, it is easier to be more accurate in determining the areas that need to be protected in the coastal and nearshore marine areas.

Table 1. The distribution of protection priority areas. The table indicates the proportions of the 6 categories of maritime areas with different priorities within and beyond national jurisdiction, respectively.

Order of Priority Marine Areas for Protection	Percentage of Territorial Sea and Exclusive Economic Zone	Percentage of High Seas and International Seabed Areas
Top 5%	82.42%	17.58%
From 5% to 10%	71.27%	28.73%
From 10% to 15%	53.16%	46.85%
From 15% to 20%	15.09%	84.91%
From 20% to 25%	0.63%	98.47%
From 25% to 30%	0.16%	99.84%

The protection priority of coastal and nearshore areas is higher. In contrast, in the high sea and international seabed, due to the lack of systematic scientific surveys, the selection of protected areas largely relies on available scientific cognition and model speculation, resulting in a lower priority of calculated protection. To more scientifically lay out marine protected areas globally, it is more feasible to first start from the waters under national jurisdiction, and each country should prioritize the establishment of protected areas in its own territorial waters and EEZs in accordance with domestic laws in order to establish a scientific and reasonable system of marine protected areas. Simultaneously, countries could actively cooperate in conducting scientific investigations in marine areas beyond national jurisdictions, for instance, through international cooperation of big science programs under the Ocean Decade. When the background data are sufficient, countries can scientifically determine the priority order for protecting high seas and international seabed areas.

4. Discussion

4.1. Comparison with Marine Protected Areas and Ecologically or Biologically Significant Marine Areas

At present, there are 18,431 MPAs worldwide, representing 8.17% of the global marine area. In contrast, all established MPAs are located within marine areas of national jurisdiction, except those in the Antarctic Ocean (including the Ross Sea region Marine Protected Area and the South Orkney Islands Southern Shelf Marine Protected Area). A total of 27.96% of the world's MPAs are located in the top 30% of the effective order of protection when compared to the priority areas for marine protection within the jurisdiction, with 12.42% of the MPAs located in the top 10% of the effective order of protection (Table 2). Clearly, the reason for such a large difference in results is the difference in the scope of MPAs and the method and data used for their selection (Figure 3). As the current MPAs are almost all selected within national jurisdictions based on the domestic laws of each country and a large part of the data used to select MPAs in each country is not readily available online and in open-source forums, our findings do not overlap well with existing protected areas.

Table 2. The distribution proportion of existing marine protected areas (MPAs) and ecologically or biologically significant marine areas (EBSAs) in priority areas. The table indicates the proportion of MPA and EBSA areas overlapping with sea areas of different protection priority orders to the total MPA and EBSA area.

Order of Priority Marine Areas for Protection	Proportion of Existing Marine Protected Areas	Proportion of Ecologically or Biologically Significant Marine Areas
Top 5%	5.26%	6.59%
From 5% to 10%	7.16%	7.12%
From 10% to 15%	7.29%	7.27%
From 15% to 20%	5.28%	2.29%
From 20% to 25%	1.75%	10.54%
From 25% to 30%	1.22%	8.31%
Sum	27.96%	42.11%

An EBSA, described under the CBD regional expert workshops, is a type of marine ecosystem area that has special ecological importance and biological value, such as marine species breeding grounds, spawning grounds, migration corridors, or food resources. Due to efforts by experts for more than a decade, more than 300 EBSAs have been described both within and beyond national jurisdictions at a variety of depths from the surface to the deep sea [23,72]. The comparison showed that approximately 42.11% of the described EBSAs overlapped with the top 30% of priority protected areas, with 13.71% of the EBSAs overlapping with the top 10% of priority protected areas (Table 2) (Figure 4).

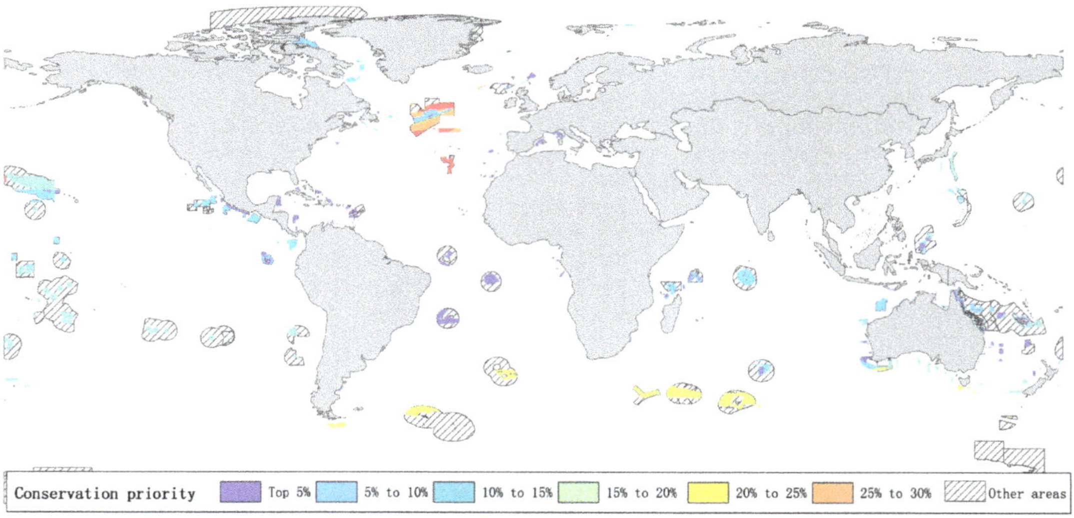

Figure 3. The overlay of priority marine areas for protection with marine protected areas. This map compares the layout of existing MPAs with the result of priority marine areas for protection. Purple indicates MPAs that overlap with the top 5% of priority sea areas. Correspondingly, light blue, lake blue, green, yellow, and orange indicate MPAs that overlap with the corresponding protection priority domains, respectively. The slash indicates MPAs that do not overlap with the top 30% of priority sea areas. Gray indicates the land.

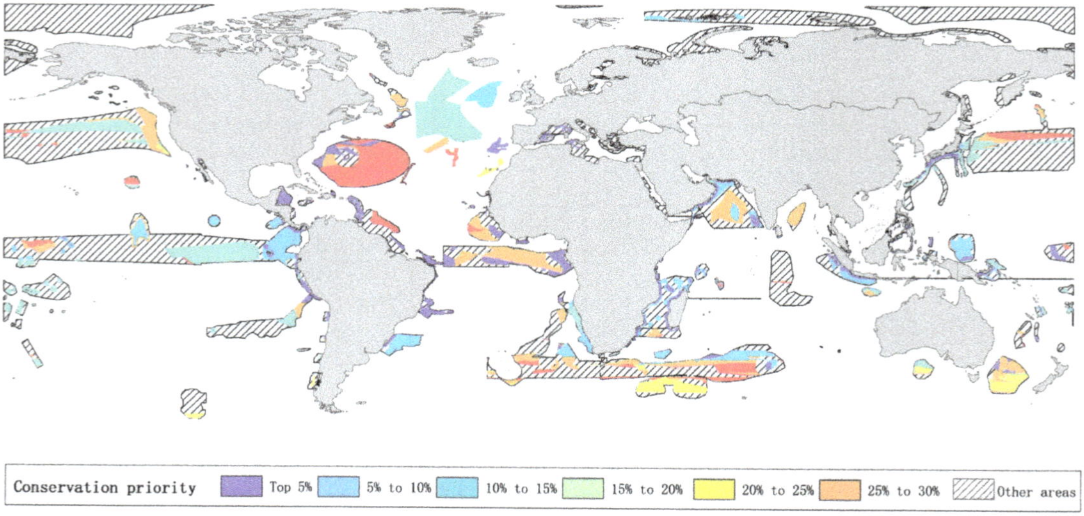

Figure 4. The overlay of priority marine areas for protection with ecologically or biologically significant marine areas (EBSAs). This map compares the layout of EBSAs with the result of priority marine areas for protection. Purple indicates EBSAs that overlap with the top 5% of priority sea areas. Correspondingly, light blue, lake blue, green, yellow, and orange indicate EBSAs that overlap with the corresponding protection priority domains, respectively. The slash indicates EBSAs that do not overlap with the top 30% of priority sea areas. Gray indicates the land.

The difference in results between the two kinds of areas is due to differences in the used identification methods, as described by EBSAs through regional expert workshops held under the CBD. Only relevant countries, regional organizations, non-governmental organizations, experts, and scholars within a region are involved in the description, with technical support provided by the Commonwealth Scientific and Industrial Research Organization of Australia and the Marine Geospatial Ecology Laboratory of Duke University [23]. Using the Central Indian Ocean Basin as an example, this area was described as an EBSA, as it is a foraging area during the non-breeding season for four seabird species based on past research [73]. However, the data we used to assess the distribution of seabirds came from the IBA, and the Central Indian Ocean Basin was not identified as an IBA because of different identification methods [67], which led to different results. Considering that multi-species approaches using overall enclosure or diversity specifications are currently the main method for selecting MPAs [39,74] for seabirds, due to the variation in seabird distribution depending on their breeding site, age, or the time of year [39,75], it seems more reasonable not to identify this area as an MPA. Furthermore, many EBSAs have been described for more than 10 years, and objects of protection, such as marine migratory species or deep-sea ecosystems, were not considered in the description. Compared to the analysis methodology that only relies on available scientific information and models in this study, the description of EBSAs had a focus that is more in line with stakeholder participation [76]. This method is more feasible within a region and has high operability but entails strong expert subjectivity. Local support from stakeholders is crucial for the planning and ongoing management of MPAs, as it could make the rules of MPAs more acceptable, as well as increase trust and satisfaction [77]. In the future, it is recommended that the use of available scientific information and evidence support, stakeholder participation, and expert evaluation in the selection and planning of high seas protected areas be combined to improve the scientific nature of protected area planning.

4.2. Discussion on Management of Priority Marine Areas for Protection

Based on open-source data regarding species, habitats, and ocean features and using spatial analysis methods, this study takes global marine regions as the research object and identifies marine regions that need global priority protection. The mere proposal of MPAs does not guarantee actual ocean protection; effective management measures, enforcement [78], surveillance, and compliance with regulations are also decisive for MPAs [79]. The measures of an MPA vary widely from strict protection, with all extractive activities being prohibited, to areas that permit certain levels of extraction and human activities [80]. The different marine areas in Figure 1 should be protected in different ways. Although the protection priority order of different sea areas might be roughly at the same level, the specific protection objects and objectives are different, and establishing each MPA is a different task. It is necessary to adopt a "case-by-case" strategy when establishing an MPA and develop targeted management measures as well as monitoring, surveillance, and control tools that are based on the protection objectives and objects in different regions to protect the marine environment while considering normal marine activities and resource development and utilization.

Management measures of MPAs include routine measures related to human activities monitoring, fishing activity regulation, the establishment of biological rest periods, and the regulation of marine and terrestrial spaces like zoning or buffers [81]. Referring to the "Guidelines for Applying the IUCN Protected Area Management Categories to Marine Protected Areas", MPAs are always divided into six categories (including Ia, strict nature reserve; Ib, wilderness area; II, national park; III, natural monument or feature; IV, habitat/species management area; V, protected landscape or seascape; and VI, protected areas the with sustainable use of natural resources) based on their management objectives [82], and different management measures from exclusionary protection to sustainable utilization are implemented for different categories of MPAs [83].

For example, in protected areas with a priority of 5% to 10%, the protection objects of the Southwestern Indian Ocean are unique seamounts and shoals, with rare and fragile species habitats [84] and some threatened seabird species [85]. An MPA should be established to protect seamount ecosystems, biodiversity hotspots near the seamount [86], and the migration routes of seabirds. Seamount habitats and ecosystems are put at risk by seabed mining and bottom trawling, which are the main human activities in this region [84]. Some deep-sea corals and sponges near the seamount are particularly sensitive to the impacts of bottom fishing [87]. Therefore, the management measures of MPAs in this region should focus on deep-sea mineral exploration, commercial mining, bottom fishing prohibition, catches or footprint limitation, bycatch mitigation, and complete observer coverage on all commercial fishing vessels. The protected objects in the Arabian Basin are the feeding area of petrels [88]. The region is also home to threatened species, including baleen whales, sea turtles, and dolphins [89]. As the main human activities in this area include fishing and shipping, measures of MPA here could focus on improving the environmental protection standards for shipping and fishery and implementing management measures such as vessel noise reduction, speed reduction, marine litter discharge reduction, bycatch mitigation, and low-carbon clean vessel fuels [89].

Management measures should be designed to balance normal marine activities and resource exploitation and should not be generalized to all protected areas. Excessively strict protection measures would greatly increase the cost of MPAs and affect the livelihoods of local communities, such as fisherfolk [90]. In particular, some activities have little impact on the environment, such as maritime scientific investigations and submarine cable laying [91]. Targeted management measures with time bounds could be stratified and graded according to the protected species. Moreover, management measures should consider the livelihoods of fisherfolk, including proposing alternative livelihoods, involving all stakeholders and the wider community of fisherfolk in MPA management [64,92].

4.3. Limitations

There were limitations in this study. We extensively used data on marine species, habitats, ocean features, and biogeographic provinces to describe priority areas for marine conservation, but all data were acquired from open sources. Spatial and temporal gaps of available data are a general limitation for biodiversity conservation [93], and there were uneven data on the distribution of specifications and habitats in different sea areas [94], such as in the South Atlantic [95], which likely influenced our results. Meanwhile, we also note that research technologies regarding marine biodiversity and habitats are emerging, such as the use of eDNA sampling and remote sensing techniques, which could help fill the gaps in data used for MPA identification [34,96]. In addition, the use of data and models with the participation of experts and the public, in line with the description of EBSAs, will undoubtedly be beneficial in a better "30 by 30" proposal.

4.4. Recommendations

Two recommendations are made in this paper to better contribute to the global construction of MPAs and other area-based management tools for marine biodiversity conservation and protection.

First, considering that laws and guidelines for establishing MPAs beyond national jurisdiction are still being developed [97], priority is given to promoting networks of marine protected areas within national jurisdictions. At present, the area of marine protection in waters under the jurisdiction of many countries is seriously inadequate. In China, for example, the total area of marine protected areas only accounts for 4.1% of the area of marine areas under the country's jurisdiction, far short of the intended protection target. Most protected areas are in the territorial seas, with a serious lack of marine protection in EEZs [98]. To achieve effective protection of the oceans, it is first necessary to conduct scientific surveys around China's Bohai, Yellow, East, and South China Seas to obtain basic data for designating MPAs or other area-based protection or management measures.

By combining data on the specific environmental and species distribution status and human activity distribution of the four sea areas, the selection and management plan of MPAs within the exclusive economy can be determined, starting with the establishment of well-designed, well-managed, and strongly enforced MPAs in EEZs based on solid scientific data.

Second, to promote the construction of global protected areas more quickly and at the same time guarantee the right to participation for low- and middle-income countries, it is necessary to promote systematic joint marine scientific surveys in the high seas and international seabed areas, with the participation of all countries, in order to map the distribution of species and habitats in the high seas at an early stage. It is recommended that international scientific programs such as the "2021–2030 United Nations Decade of Ocean Science for Sustainable Development" and the "Global Ocean Observing Programme" be used as a means of scientific capacity building and public awareness raising in low- and middle-income countries and that countries cooperate in conducting scientific surveys. Scientific data should be obtained to support the selection of protected areas in the high seas and international seabed areas, priority areas for global marine biodiversity conservation should be identified as soon as possible, and relevant proposals should be made.

5. Conclusions

The systematic identification of global conservation priority considering nearshore and offshore species, habitats, and abiotic elements, as well as the sustainable use of marine resources, plays an important role in identifying and managing MPAs at the global scale to ensure marine protection and conservation. This study provides a global-scale proposal for the selection of priority areas for marine protection under the scenario of achieving the "30 by 30" target. Feasible recommendations to identify MPAs were proposed based on comprehensive scientific evidence and stakeholder participation in the future. Furthermore, in order to improve the effectiveness of marine protection, scientific delineation of the boundaries of protected areas is necessary, as well as the development of targeted management measures based on the specific objects to be protected and the threats affecting them.

Supplementary Materials: The following supporting information can be downloaded at https://www.mdpi.com/article/10.3390/w16162293/s1, Table S1: Properties and sources of data layers used in this study. References [99–111] are cited in the Supplementary Materials.

Author Contributions: Conceptualization, C.Z. and M.Z.; Methodology, C.Z.; Software, C.Z.; Resources, Y.G.; Data curation, Y.G.; Writing—original draft, C.Z.; Funding acquisition, M.Z. All authors have read and agreed to the published version of the manuscript.

Funding: This research was funded by National Key R&D Program of China, grant number 2023YFC2808802, and Major Project of the Key Research Base for Humanities and Social Sciences of the Ministry of Education, grant number 22JJD790028.

Data Availability Statement: Data are contained within the article and Supplementary Materials.

Conflicts of Interest: The authors declare no conflict of interest.

References

1. NOAA. 'Why Should We Care about the Ocean?', National Ocean Service National Oceanic and Atmospheric Administration. Available online: https://oceanservice.noaa.gov/facts/why-care-about-ocean.html (accessed on 17 August 2023).
2. Fischer, M.; Maxwell, K.; Nuunoq; Pedersen, H.; Greeno, D.; Jingwas, N.; Graham Blair, J.; Hugu, S.; Mustonen, T.; Murtomäki, E.; et al. Empowering her guardians to nurture our Ocean's future. *Rev. Fish Biol. Fish.* **2022**, *32*, 271–296. [CrossRef]
3. Tigchelaar, M.; Leape, J.; Micheli, F.; Allison, E.H.; Basurto, X.; Bennett, A.; Bush, S.R.; Cao, L.; Cheung, W.W.L.; Crona, B.; et al. The vital roles of blue foods in the global food system. *Glob. Food Secur.* **2022**, *33*, 100637. [CrossRef]
4. Newman, D.; Cragg, G. Drugs and Drug Candidates from Marine Sources: An Assessment of the Current "State of Play". *Planta Med.* **2016**, *82*, 775–789. [CrossRef] [PubMed]
5. Blanchard, C.; Harrould-Kolieb, E.; Jones, E.; Taylor, M.L. The current status of deep-sea mining governance at the International Seabed Authority. *Mar. Policy* **2023**, *147*, 105396. [CrossRef]

6. Sala, E.; Mayorga, J.; Bradley, D.; Cabral, R.B.; Atwood, T.B.; Auber, A.; Cheung, W.; Costello, C.; Ferretti, F.; Friedlander, A.M.; et al. Protecting the global ocean for biodiversity, food and climate. *Nature* **2021**, *592*, 397–402. [CrossRef] [PubMed]
7. Quéré, C.L.; Moriarty, R.; Andrew, R.M.; Peters, G.P.; Ciais, P.; Friedlingstein, P.; Jones, S.D.; Sitch, S.; Tans, P.; Arneth, A.; et al. Global Carbon Budget 2014. *Earth Syst. Sci. Data* **2015**, *7*, 47–85. [CrossRef]
8. Vikas, M.; Dwarakish, G.S. Coastal Pollution: A Review. *Aquat. Procedia* **2015**, *4*, 381–388. [CrossRef]
9. Estes, M.; Anderson, C.; Appeltans, W.; Bax, N.; Bednaršek, N.; Canonico, G.; Djavidnia, S.; Escobar, E.; Fietzek, P.; Gregoire, M.; et al. Enhanced monitoring of life in the sea is a critical component of conservation management and sustainable economic growth. *Mar. Policy* **2021**, *132*, 104699. [CrossRef]
10. Lodge, M. The International Seabed Authority and Deep Seabed Mining, United Nations. Available online: https://www.un.org/en/chronicle/article/international-seabed-authority-and-deep-seabed-mining#:~:text=Terrestrial%20mineral%20deposits%20are%20coming%20under%20increasing%20pressure,Easily%20mined,%20high-grade%20ore%20deposits%20are%20quickly%20declining (accessed on 17 August 2023).
11. Butchart, S.; Scharlemann, J.; Evans, M.; Quader, S.; Aricò, S.; Arinaitwe, J.; Balman, M.; Bennun, L.; Bertzky, B.; Besançon, C.; et al. Protecting Important Sites for Biodiversity Contributes to Meeting Global Conservation Targets. *PLoS ONE* **2012**, *7*, e32529. [CrossRef]
12. Lester, S.; Halpern, B. Biological responses in marine no-take reserves versus partially protected areas. *Mar. Ecol. Prog. Ser.* **2008**, *367*, 49–56. [CrossRef]
13. CBD 15/4. Kunming-Montreal Global Biodiversity Framework. Available online: https://www.cbd.int/doc/decisions/cop-15/cop-15-dec-04-en.pdf (accessed on 12 March 2023).
14. CBD. Decision Adopted by the Conference of the Parties to the Convention on Biological Diversity at Its Tenth Meeting, October 2010. Available online: https://www.cbd.int/doc/decisions/cop-10/cop-10-dec-02-en.pdf (accessed on 17 August 2023).
15. Protected Planet. Protected Planet Report 2018. Available online: https://livereport.protectedplanet.net/ (accessed on 12 March 2023).
16. Protected Planet. Marine Protected Areas. Available online: https://www.protectedplanet.net/en/thematic-areas/marine-protected-areas (accessed on 12 March 2023).
17. Protected Planet. Discover the World's Protected and Conserved Areas. Available online: https://www.protectedplanet.net/en (accessed on 10 March 2024).
18. Dinerstein, E.; Vynne, C.; Sala, E.; Joshi, A.R.; Fernando, S.; Lovejoy, T.E.; Mayorga, J.; Olson, D.; Asner, G.P.; Baillie, J.E.M.; et al. A Global Deal For Nature: Guiding principles, milestones, and targets. *Sci. Adv.* **2019**, *5*, eaaw2869. [CrossRef] [PubMed]
19. Un Environment (Ed.) *Global Environment Outlook–GEO-6: Healthy Planet, Healthy People*, 1st ed.; Cambridge University Press: Cambridge, UK, 2019; ISBN 978-1-108-62714-6.
20. United Nations. *UNCITRAL Expedited Arbitration Rules 2021: UNCITRAL Rules on Transparency in Treaty-Based Investor-State Arbitration*; United Nations: New York, NY, USA, 2022. Available online: https://uncitral.un.org/sites/uncitral.un.org/files/media-documents/uncitral/en/rules-on-transparency-e.pdf (accessed on 17 February 2023). [CrossRef]
21. United Nations. UN Delegates Reach Historic Agreement on Protecting Marine Biodiversity in International Waters. UN News Global Perspective Human Stories. Available online: https://news.un.org/en/story/2023/03/1134157 (accessed on 15 March 2023).
22. Gownaris, N.J.; Santora, C.M.; Davis, J.B.; Pikitch, E.K. Gaps in Protection of Important Ocean Areas: A Spatial Meta-Analysis of Ten Global Mapping Initiatives. *Front. Mar. Sci.* **2019**, *6*, 650. [CrossRef]
23. Johnson, D.; Gunn, V.; Bax, N.; Dunn, D. *Special Places in the Ocean: A Decade of Describing Ecologically or Biologically Significant Marine Areas (EBSAs)*; Secretariat of the Convention on Biological Diversity: Montreal, QC, Canada, 2021. Available online: https://www.cbd.int/marine/ebsa/booklet-ebsa-impact-en.pdf (accessed on 17 August 2023).
24. Greenpeace. 30 × 30: A Blueprint for Ocean Protection. Available online: https://www.greenpeace.org/international/publication/21604/30x30-a-blueprint-for-ocean-protection/ (accessed on 12 March 2023).
25. Pew. A Path to Creating the First Generation of High Seas Protected Areas Science-Based Method Highlights 10 Sites That Would Help Safeguard Biodiversity beyond National Waters. Available online: https://www.pewtrusts.org/en/research-and-analysis/reports/2020/03/a-path-to-creating-the-first-generation-of-high-seas-protected-areas (accessed on 12 March 2023).
26. Werner, S.R.; Spurgeon, J.P.G.; Isaksen, G.H.; Smith, J.P.; Springer, N.K.; Gettleson, D.A.; N'Guessan, L.; Dupont, J.M. Rapid prioritization of marine ecosystem services and ecosystem indicators. *Mar. Policy* **2014**, *50*, 178–189. [CrossRef]
27. Ortiz Cajica, A.K.; Hinojosa-Arango, G.; Garza-Pérez, J.R.; Rioja-Nieto, R. Seascape metrics, spatio-temporal change, and intensity of use for the spatial conservation prioritization of a Caribbean marine protected area. *Ocean Coast. Manag.* **2020**, *194*, 105265. [CrossRef]
28. Winiarski, K.J.; Miller, D.L.; Paton, P.W.C.; McWilliams, S.R. A spatial conservation prioritization approach for protecting marine birds given proposed offshore wind energy development. *Biol. Conserv.* **2014**, *169*, 79–88. [CrossRef]
29. Holness, S.D.; Harris, L.R.; Chalmers, R.; De Vos, D.; Goodall, V.; Truter, H.; Oosthuizen, A.; Bernard, A.T.F.; Cowley, P.D.; Da Silva, C.; et al. Using systematic conservation planning to align priority areas for biodiversity and nature-based activities in marine spatial planning: A real-world application in contested marine space. *Biol. Conserv.* **2022**, *271*, 109574. [CrossRef]
30. Pruckner, S.; Bedford, J.; Murphy, L.; Turner, J.A.; Mills, J. Adapting to heatwave-induced seagrass loss: Prioritizing management areas through environmental sensitivity mapping. *Estuar. Coast. Shelf Sci.* **2022**, *272*, 107857. [CrossRef] [PubMed]

31. Hogg, O.; Huvenne, V.; Griffiths, H.; Linse, K. On the ecological relevance of landscape mapping and its application in the spatial planning of very large marine protected areas. *Sci. Total Environ.* **2018**, *626*, 384–398. [CrossRef]
32. Jenkins, C.N.; Houtan, K.S.V.; Pimm, S.L.; Sexton, J.O. US protected lands mismatch biodiversity priorities. *Proc. Natl. Acad. Sci. USA* **2015**, *112*, 5081–5086. [CrossRef]
33. Davidson, L.N.K.; Dulvy, N.K. Global marine protected areas to prevent extinctions. *Nat. Ecol. Evol.* **2017**, *1*, 0040. [CrossRef]
34. Visalli, M.E.; Best, B.D.; Cabral, R.B.; Cheung, W.W.L.; Clark, N.A.; Garilao, C.; Kaschner, K.; Kesner-Reyes, K.; Lam, V.W.Y.; Maxwell, S.M.; et al. Data-driven approach for highlighting priority areas for protection in marine areas beyond national jurisdiction. *Mar. Policy* **2020**, *122*, 103927. [CrossRef]
35. Belote, R.T.; Barnett, K.; Dietz, M.S.; Burkle, L.; Jenkins, C.N.; Dreiss, L.; Aycrigg, J.L.; Aplet, G.H. Options for prioritizing sites for biodiversity conservation with implications for "30 by 30". *Biol. Conserv.* **2021**, *264*, 109378. [CrossRef]
36. Lehtomäki, J.; Moilanen, A. Methods and workflow for spatial conservation prioritization using Zonation. *Environ. Model. Softw.* **2013**, *47*, 128–137. [CrossRef]
37. Wallace, B.; DiMatteo, A.; Hurley, B.; Finkbeiner, E.; Bolten, A.B.; Chaloupka, M.Y.; Hutchinson, B.J.; Abreu-Grobois, F.A.; Amorocho, D.; Bjorndal, K.A.; et al. Regional Management Units for Marine Turtles: A Novel Framework for Prioritizing Conservation and Research across Multiple Scales. *PLoS ONE* **2010**, *5*, e15465. [CrossRef]
38. Wagner, D.; Friedlander, A.M.; Pyle, R.L.; Brooks, C.M.; Gjerde, K.M.; Wilhelm, T. 'Aulani Coral Reefs of the High Seas: Hidden Biodiversity Hotspots in Need of Protection. *Front. Mar. Sci.* **2020**, *7*, 567428. [CrossRef]
39. Lascelles, B.G.; Langham, G.M.; Ronconi, R.A.; Reid, J.B. From hotspots to site protection: Identifying Marine Protected Areas for seabirds around the globe. *Biol. Conserv.* **2012**, *156*, 5–14. [CrossRef]
40. Walsh, J.; Connors, K.; Hertz, E.; Kehoe, L.; Martin, T.; Connors, B.; Bradford, M.; Freshwater, C.; Frid, A.; Halverson, J.; et al. Prioritizing conservation actions for Pacific salmon in Canada. *J. Appl. Ecol.* **2020**, *57*, 1688–1699. [CrossRef]
41. Williams, R.; Grand, J.; Hooker, S.; Buckland, S.; Reeves, R.R.; Rojas-Bracho, L.; Sandilands, D.; Kaschner, K. Prioritizing global marine mammal habitats using density maps in place of range maps. *Ecography* **2014**, *37*, 212–220. [CrossRef]
42. Tardin, R.; Maciel, I.; Espécie, M.; Melo-Santos, G.; Simão, S.; Alves, M. Modelling habitat use by the Guiana dolphin, *Sotalia guianensis*, in south-eastern Brazil: Effects of environmental and anthropogenic variables, and the adequacy of current management measures. *Aquat. Conserv Mar. Freshw. Ecosyst.* **2020**, *30*, 775–786. [CrossRef]
43. Villa, F.; Tunesi, L.; Agardy, T. Zoning Marine Protected Areas through Spatial Multiple-Criteria Analysis: The Case of the Asinara Island National Marine Reserve of Italy. *Conserv. Biol.* **2002**, *16*, 515–526. [CrossRef]
44. Dunn, D.C.; Van Dover, C.L.; Etter, R.J.; Smith, C.R.; Levin, L.A.; Morato, T.; Colaço, A.; Dale, A.C.; Gebruk, A.V.; Gjerde, K.M.; et al. A strategy for the conservation of biodiversity on mid-ocean ridges from deep-sea mining. *Sci. Adv.* **2018**, *4*, eaar4313. [CrossRef] [PubMed]
45. Yamakita, T.; Yamamoto, H.; Nakaoka, M.; Yamano, H.; Fujikura, K.; Hidaka, K.; Hirota, Y.; Ichikawa, T.; Kakehi, S.; Kameda, T.; et al. Identification of important marine areas around the Japanese Archipelago: Establishment of a protocol for evaluating a broad area using ecologically and biologically significant areas selection criteria. *Mar. Policy* **2015**, *51*, 136–147. [CrossRef]
46. Klein, C.J.; Brown, C.J.; Halpern, B.S.; Segan, D.B.; McGowan, J.; Beger, M.; Watson, J.E.M. Shortfalls in the global protected area network at representing marine biodiversity. *Sci. Rep.* **2015**, *5*, 17539. [CrossRef] [PubMed]
47. Jenkins, C.; Van Houtan, K. Global and regional priorities for marine biodiversity protection. *Biol. Conserv.* **2016**, *204*, 333–339. [CrossRef]
48. Fan, H.; Huang, M.; Chen, Y.; Zhou, W.; Hu, Y.; Wei, F. Conservation priorities for global marine biodiversity across multiple dimensions. *Natl. Sci. Rev.* **2023**, *10*, nwac241. [CrossRef] [PubMed]
49. Dodge, R.; Birkeland, C.; Hatziolos, M.; Kleypas, J.; Palumbi, S.; Hoegh-Guldberg, O.; Van Woesik, R.; Ogden, J.; Aronson, R.; Causey, B.; et al. A Call to Action for Coral Reefs. *Science* **2008**, *322*, 189–190. [CrossRef] [PubMed]
50. Zhao, S.; Hong, H.S.; Zhang, L.P.; Chen, W.Q. Emergy Value of Mangrove Ecosystem Services in China. *Resour. Sci.* **2007**, *29*, 147–154.
51. Moilanen, A.; Franco, A.; Early, R.; Fox, R.; Wintle, B.; Thomas, C. Prioritizing multiple-use landscapes for conservation: Methods for large multi-species planning problems. *Proc. R. Soc. B* **2005**, *272*, 1885–1891. [CrossRef] [PubMed]
52. Ota, T.; Takao, K.; Uehara, T.; Mineo, K.; Obata, N.; Nakagami, K.; Yoshioka, T.; Sakurai, R.; Hidaka, T.; Seino, S. Chapter 5-Environmental Economics, Culture, and Negotiation in the Coastal Sea. In *Integrated Coastal Management in the Japanese Satoumi-Restoring Estuaries and Bays*; Elsevier: Amsterdam, The Netherlands, 2019; pp. 131–193.
53. Job, S.; Sekadende, B.; Yona, G.; George, R.; Lugendo, B.R.; Kimirei, I.A. Effect of seagrass cover loss on seawater carbonate chemistry: Implications for the potential of seagrass meadows to mitigate ocean acidification. *Reg. Stud. Mar. Sci.* **2023**, *60*, 102816. [CrossRef]
54. Macreadie, P.I.; Costa, M.D.P.; Atwood, T.B.; Friess, D.A.; Kelleway, J.J.; Kennedy, H.; Lovelock, C.E.; Serrano, O.; Duarte, C.M. Blue carbon as a natural climate solution. *Nat. Rev. Earth Environ.* **2021**, *2*, 826–839. [CrossRef]
55. Garmendia, J.M.; Rodríguez, J.G.; Borja, Á.; Pouso, S.; Del Campo, A.; Galparsoro, I.; Fernandes-Salvador, J.A. Restoring seagrass meadows in Basque estuaries: Nature-based solution for successful management. *Nat.-Based Solut.* **2023**, *4*, 100084. [CrossRef]

56. Joseph, A. Chapter 6-Seafloor Hot Chimneys and Cold Seeps: Mysterious Life around Them. In *Investigating Seafloors and Oceans-from Mud Volcanoes to Giant Squid*; Elsevier: Amsterdam, The Netherlands, 2017; pp. 307–375. Available online: https://www.sciencedirect.com/science/article/abs/pii/B9780128093573000060 (accessed on 12 December 2023).
57. Rogers, A.D. The Biology of Seamounts. *Adv. Mar. Biol.* **1994**, *30*, 305–350. [CrossRef]
58. Stone, K.; Fenner, D.; LeBlanc, D.; Vaisey, B.; Purcell, I.; Eliason, B. Chapter 30-Tonga. In *World Seas: An Environmental Evaluation (Second Edition)-Volume II: The Indian Ocean to the Pacific*; Academic Press: Cambridge, MA, USA, 2019; pp. 661–678.
59. Gage, J.; Tyler, P. *Deep-Sea Biology: A Natural History of Organisms at the Deep-Sea Floor*; Cambridge University Press: Cambridge, UK, 1991.
60. Clark, M.R.; Bennun, L.; Rowden, A.; Guinotte, J.; Smith, C. A global seamount classification to aid the scientific design of marine protected area networks. *Ocean Coast. Manag.* **2011**, *54*, 19–36. [CrossRef]
61. CBD. Global Open Oceans and Deep Seabed (Goods)-Biogeographic Classification, April 2010. Available online: https://www.cbd.int/doc/meetings/sbstta/sbstta-14/information/sbstta-14-inf-10-en.pdf (accessed on 17 March 2023).
62. Charles, A.; Westlund, L.; Bartley, D.; Fletcher, W.; Garcia, S.; Govan, H.; Sanders, J. Fishing livelihoods as key to marine protected areas: Insights from the World Parks Congress: Fishing Livelihoods and Marine Protected Areas. *Aquat. Conserv Mar. Freshw. Ecosyst.* **2016**, *26*, 165–184. [CrossRef]
63. FAO. *The State of World Fisheries and Aquaculture 2022. Towards Blue Transformation*; FAO: Rome, Italy, 2022. [CrossRef]
64. Siddique, M.R.H.; Hossain, M.; Rashid, A.Z.M.M. The dilemma of prioritizing conservation over livelihoods: Assessing the impact of fishing restriction to the fishermen of the Sundarbans. *Trees For. People* **2023**, *11*, 100366. [CrossRef]
65. IUCN. The IUCN Red List of Threatened Species. Version 2022-2. Available online: https://www.iucnredlist.org (accessed on 12 March 2023).
66. Birdlife International. Data Zone. Available online: http://datazone.birdlife.org/site/ibasindanger (accessed on 12 January 2023).
67. Donald, P.; Fishpool, L.; Ajagbe, A.; Bennun, L.; Bunting, G.; Burfield, I.; Butchart, S.; Capellan, S.; Crosby, M.; Dias, M.; et al. Important Bird and Biodiversity Areas (IBAs): The development and characteristics of a global inventory of key sites for biodiversity. *Bird. Conserv. Int.* **2019**, *29*, 177–198. [CrossRef]
68. Sutton, T.; Clark, M.; Dunn, D.; Halpin, P.; Rogers, A.D.; Guinotte, J.; Bograd, S.; Angel, M.; Perez, J.; Wishner, K.; et al. A global biogeographic classification of the mesopelagic zone. *Deep Sea Res. Part I* **2017**, *126*, 85–102. [CrossRef]
69. Mikkonen, N.; Moilanen, A. Identification of top priority areas and management landscapes from a national Natura 2000 network. *Environ. Sci. Policy* **2013**, *27*, 11–20. [CrossRef]
70. Leathwick, J.; Moilanen, A.; Francis, M.; Elith, J.; Taylor, P.; Julian, K.; Hastie, T.; Duffy, C. Novel methods for the design and evaluation of marine protected areas in offshore waters: Designing offshore MPAs. *Conserv. Lett.* **2008**, *1*, 91–102. [CrossRef]
71. Virtanen, E.; Viitasalo, M.; Lappalainen, J.; Moilanen, A. Evaluation, Gap Analysis, and Potential Expansion of the Finnish Marine Protected Area Network. *Front. Mar. Sci.* **2018**, *5*, 402. [CrossRef]
72. Dunn, D.C.; Ardron, J.; Bax, N.; Bernal, P.; Cleary, J.; Cresswell, I.; Donnelly, B.; Dunstan, P.; Gjerde, K.; Johnson, D.; et al. The Convention on Biological Diversity's Ecologically or Biologically Significant Areas: Origins, Development, and Current Status. *Mar. Policy* **2014**, *49*, 137–145. [CrossRef]
73. CBD. Report of the Southern Indian Ocean Regional Workshop to Facilitate the Description of Ecologically or Biologically Significant Marine Areas. Mauritius, June 2013. Available online: https://www.cbd.int/doc/meetings/mar/ebsa-sio-01/official/ebsa-sio-01-04-en.pdf (accessed on 17 March 2023).
74. Nur, N.; Jahncke, J.; Herzog, M.P.; Howar, J.; Hyrenbach, K.D.; Zamon, J.E.; Ainley, D.G.; Wiens, J.A.; Morgan, K.; Ballance, L.T.; et al. Where the wild things are: Predicting hotspots of seabird aggregations in the California Current System. *Ecol. Appl.* **2011**, *21*, 2241–2257. [CrossRef]
75. González-Solís, J.; Croxall, J.P.; Afanasyev, V. Offshore spatial segregation in giant petrelsMacronectes spp.: Differences between species, sexes and seasons. *Aquat. Conserv. Mar. Freshw. Ecosyst.* **2007**, *17*, S22–S36. [CrossRef]
76. Johnson, D.E.; Barrio Froján, C.; Turner, P.J.; Weaver, P.; Gunn, V.; Dunn, D.C.; Halpin, P.; Bax, N.J.; Dunstan, P.K. Reviewing the EBSA process: Improving on success. *Mar. Policy* **2018**, *88*, 75–85. [CrossRef]
77. Katikiro, R.E.; Kweka, O.L.; Minja, R.; Namkesa, F.; Ponte, S. Stakeholder engagement and conservation outcomes in marine protected areas: Lessons from the Mnazi Bay-Ruvuma Estuary Marine Park (MBREMP) in Tanzania. *Ocean Coast. Manag.* **2021**, *202*, 105502. [CrossRef]
78. Agardy, T.; Di Sciara, G.N.; Christie, P. Mind the gap: Addressing the shortcomings of marine protected areas through large scale marine spatial planning. *Mar. Policy* **2011**, *35*, 226–232. [CrossRef]
79. Horta, E.; Costa, B.; Claudet, J.; Franco, G.; Erzini, K.; Caro, A.; Gonçalves, E.J. A regulation-based classification system for Marine Protected Areas (MPAs). *Mar. Policy* **2016**, *72*, 192–198. [CrossRef]
80. Rife, A.N.; Erisman, B.; Sanchez, A.; Aburto-Oropeza, O. When good intentions are not enough . . . Insights on networks of "paper park" marine protected areas: Concerns regarding marine "paper parks". *Conserv. Lett.* **2013**, *6*, 200–212. [CrossRef]
81. Failler, P.; Touron-Gardic, G.; Drakeford, B.; Sadio, O.; Traoré, M.-S. Perception of threats and related management measures: The case of 32 marine protected areas in West Africa. *Mar. Policy* **2020**, *117*, 103936. [CrossRef]
82. Day, J.; Dudley, N.; Hockings, M.; Holmes, G.; Laffoley, D.; Stolton, S.; Wells, S. *Guidelines for Applying the IUCN Protected Area Management Categories to Marine Protected Areas*; IUCN: Gland, Switzerland, 2012. Available online: https://portals.iucn.org/library/node/10201 (accessed on 15 February 2024).

3. Vimal, R.; Navarro, L.M.; Jones, Y.; Wolf, F.; Le Moguédec, G.; Réjou-Méchain, M. The global distribution of protected areas management strategies and their complementarity for biodiversity conservation. *Biol. Conserv.* **2021**, *256*, 109014. [CrossRef]
4. Marsac, F.; Galletti, F.; Ternon, J.-F.; Romanov, E.V.; Demarcq, H.; Corbari, L.; Bouchet, P.; Roest, W.R.; Jorry, S.J.; Olu, K.; et al. Seamounts, plateaus and governance issues in the southwestern Indian Ocean, with emphasis on fisheries management and marine conservation, using the Walters Shoal as a case study for implementing a protection framework. *Deep Sea Res. Part II* **2020**, *176*, 104715. [CrossRef]
5. Heerah, K.; Dias, M.P.; Delord, K.; Oppel, S.; Barbraud, C.; Weimerskirch, H.; Bost, C.A. Important areas and conservation sites for a community of globally threatened marine predators of the Southern Indian Ocean. *Biol. Conserv.* **2019**, *234*, 192–201. [CrossRef]
6. Clark, M.R.; Rowden, A.A.; Schlacher, T.; Williams, A.; Consalvey, M.; Stocks, K.I.; Rogers, A.D.; O'Hara, T.D.; White, M.; Shank, T.M.; et al. The Ecology of Seamounts: Structure, Function, and Human Impacts. *Annu. Rev. Mar. Sci.* **2010**, *2*, 253–278. [CrossRef]
7. FAO. *International Guidelines for the Management of Deep-Sea Fisheries in the High Seas*; FAO: Roma, Italy, 2009.
8. Brown, R.M.; Jordan, W.C. Characterization of polymorphic microsatellite loci from Round Island Petrels (Pterodroma arminjoniana) and their utility in other seabird species. *J. Ornithol.* **2009**, *150*, 925–929. [CrossRef]
9. CBD. Report of the North-West Indian Ocean and Adjacent Gulf Areas Regional Workshop to Facilitate the Description of Ecologically or Biologically Significant Marine Areas. 2016. Available online: https://www.cbd.int/doc/meetings/sbstta/sbstta-20/information/sbstta-20-inf-23-en.pdf (accessed on 17 March 2023).
0. Jones, P.J.S.; De Santo, E.M. Viewpoint–Is the race for remote, very large marine protected areas (VLMPAs) taking us down the wrong track? *Mar. Policy* **2016**, *73*, 231–234. [CrossRef]
1. International Cable Protection Committee. International Cable Protection Committee Welcomes New Marine Biodiversity Treaty, Calls on All Parties to Promote Regulatory Certainty and Network Resilience. Available online: https://www.iscpc.org/news/ (accessed on 17 August 2023).
2. Dutka-Gianelli, J.; Crandall, C.; Garlock, T.; Camp, E.; Lorenzen, K. Effects of short educational workshops on stakeholder knowledge and attitudes on coastal fish stocking programmes. *Fish. Manag. Ecol.* **2019**, *26*, 306–309. [CrossRef]
3. Gilman, E.; Dunn, D.; Read, A.; Hyrenbach, K.D.; Warner, R. Designing criteria suites to identify discrete and networked sites of high value across manifestations of biodiversity. *Biodivers. Conserv.* **2011**, *20*, 3363–3383. [CrossRef]
4. Bridges, A.E.H.; Barnes, D.K.A.; Bell, J.B.; Ross, R.E.; Voges, L.; Howell, K.L. Filling the data gaps: Transferring models from data-rich to data-poor deep-sea areas to support spatial management. *J. Environ. Manag.* **2023**, *345*, 118325. [CrossRef] [PubMed]
5. Howell, K.L.; Hilário, A.; Allcock, A.L.; Bailey, D.M.; Baker, M.; Clark, M.R.; Colaço, A.; Copley, J.; Cordes, E.E.; Danovaro, R.; et al. A Blueprint for an Inclusive, Global Deep-Sea Ocean Decade Field Program. *Front. Mar. Sci.* **2020**, *7*, 584861. [CrossRef]
6. Stat, M.; Huggett, M.J.; Bernasconi, R.; DiBattista, J.D.; Berry, T.E.; Newman, S.J.; Harvey, E.S.; Bunce, M. Ecosystem biomonitoring with eDNA: Metabarcoding across the tree of life in a tropical marine environment. *Sci. Rep.* **2017**, *7*, 12240. [CrossRef]
7. Deasy, K. What we know about the new High Seas Treaty. *NPJ Ocean Sustain.* **2023**, *2*, 7. [CrossRef]
8. Zhao, C. The Development History and Prospects of Marine Protected Areas in China. *Green Technol.* **2022**, *24*, 207–211. [CrossRef]
9. BirdLife International. *Important Bird and Biodiversity Area (IBA) Digital Boundaries: September 2020 Version*; BirdLife International: Cambridge, UK, 2020.
100. Bunting, P.; Rosenqvist, A.; Lucas, R.M.; Rebelo, L.M.; Hilarides, L.; Thomas, N.; Hardy, A.; Itoh, T.; Shimada, M.; Finlayson, C.M. The Global Mangrove Watch—A New 2010 Global Baseline of Mangrove Extent. *Remote Sens.* **2018**, *10*, 1669. [CrossRef]
101. EBSAs. Ecologically or Biologically Significant Marine Areas. 2023. Available online: https://www.cbd.int/ebsa/ (accessed on 15 February 2024).
102. Flanders Marine Institute. Union of the ESRI Country Shapefile and the Exclusive Economic Zones (version 3). Available online: https://www.marineregions.org/ (accessed on 15 February 2024).
103. GBSCO. The GEBCO_2021 Grid. 2023. Available online: https://www.gebco.net/data_and_products/gridded_bathymetry_data/gebco_2021/ (accessed on 15 February 2024).
104. German, C.R.; Ramirez-Llodra, E.; Baker, M.C.; Tyler, P.A.; ChEss Scientific Steering Committee. Deep-Water Chemosynthetic Ecosystem Research during the Census of Marine Life Decade and Beyond: A Proposed Deep-Ocean Road Map. *PLoS ONE* **2011**, *6*, e23259. [CrossRef] [PubMed]
105. Jayathilake, D.R.M.; Costello, M.J. A modelled global distribution of the kelp biome. *Biol. Conserv.* **2020**, *252*, 108815. [CrossRef]
106. Mcowen, C.; Weatherdon, L.V.; Bochove, J.; Sullivan, E.; Blyth, S.; Zockler, C.; Stanwell-Smith, D.; Kingston, N.; Martin, C.S.; Spalding, M.; et al. A global map of saltmarshes (v6.1). *Biodivers. Data J.* **2017**, *5*, e11764. [CrossRef]
107. OBIS. OBIS The OBIS Web Portal Search Interface (2015). 2022. Available online: http://iobis.org/mapper/ (accessed on 15 February 2024).
108. Sayre, R.; Noble, S.; Hamann, S.; Smith, R.; Wright, D.; Breyer, S.; Butler, K.; Van Graafeiland, K.; Frye, C.; Karagulle, D.; et al. A new 30 meter resolution global shoreline vector and associated global islands database for the development of standardized ecological coastal units. *J. Oper. Oceanogr.* **2019**, *12*, S47–S56. [CrossRef]
109. UNEP-WCMC; IUCN. Protected Planet: The World Database on Protected Areas (WDPA). 2023. Available online: www.protectedplanet.net (accessed on 15 February 2024).

110. UNEP-WCMC. *Short FT. 2021. Global Distribution of Seagrasses (Version 7.1). Seventh Update to the Data Layer Used in Green an Short*; UN Environment World Conservation Monitoring Centre: Cambridge, UK, 2003.
111. UNEP-WCMC; WorldFish Centre; WRI; TNC. *Global Distribution of Warm-Water Coral Reefs, Compiled from Multiple Source Including the Millennium Coral Reef Mapping Project. Version 4.1. Includes contributions from IMaRS-USF and IRD (2005), IMaRS-US (2005) and Spalding et al. (2001)*; UN Environment World Conservation Monitoring Centre: Cambridge, UK, 2021.

water

MDPI

Article

Urban Governance, Economic Transformation, and Land Use: A Case Study on the Jimei Peninsula, Xiamen, China, 1936–2023

Shu-Chen Tsai [1], Xue-Fang Zhang [1,2], Su-Hsin Lee [3] and Hui Wang [1,*]

[1] College of Arts and Design, Jimei University, Xiamen 361021, China; tsaishuchen@jmu.edu.cn (S.-C.T.); u23092110359@cityu.edu.mo (X.-F.Z.)
[2] Faculty of Innovation and Design, City University of Macau, Macau, China
[3] Department of Geography, National Taiwan Normal University, Taipei 10644, Taiwan; shlee@ntnu.edu.tw
* Correspondence: huiwang@jmu.edu.cn

Abstract: The purpose of this study was to explain how the heterogeneous elements embedded in the Jimei Peninsula affect the transformation of the production landscape into a consumption landscape and the connection between urban governance and economic transformation. The study took a qualitative approach, utilizing historical literature analysis, a field investigation, and in-depth interviews to explore the driving forces and impacts of coastal-zone functional transformation. A total of 26 residents were interviewed individually or collectively, the current situation in the coastal zone with a length of about 16.1 km was recorded in detail, and all the collected elements were divided into six landscape categories for analysis. The results indicate that urban positioning, economic development, policies, and residents are the main factors driving the continuous advancement of the Jimei Peninsula zone. The coast has completed the functional transformation from meeting the residents' survival needs to tourists' sightseeing needs. The traditional fishing culture in this area is slowly disappearing with the tide of time, and navigation technology is being passed down through the Jimei School Village. This study reveals the dynamic process of the transformation of coastal functions in representative coastal tourism cities in China, bringing attention to coastal ecology and local fishing culture, and raising people's awareness of cautious coastal development and sustainable blue-economy development.

Keywords: heterogeneous; embeddedness; moments; commodity; consumption landscape

1. Introduction

Coastal zones are the most productive areas [1], but the dynamic nature of the interface between land and sea [2] makes coastal areas particularly vulnerable to the effects of human activities. Although the consequences of changes in coastal areas are local or regional, they are likely to affect human society within a decade and cause irreversible changes [3]. Economic globalization has led to the economic growth of coastal port cities, which requires the expansion of hinterlands into the sea.

By the end of 2020, the cargo and container throughputs of Chinese coastal ports (which were 48.5-fold higher than those in 1978) were ranked first, globally [4].

For China, the economic transformation over the past half century has been driven by coastal cities. Most of the prototypes of these cities are fishing villages of varying sizes. Fishing is the main form of livelihood in these villages, which encouraged the development of shipbuilding, and in turn, shipbuilding technology promotes human trade, migration, and exploration. Thus, it can be said that fishers and their related landscapes gave birth to the modern world [1].

However, most of the fishing village landscapes along China's coast have disappeared with the rapid economic growth, and there is a lack of historical records of this process in the literature. This study attempts to document how the fishing village landscapes that relied on the sea for survival disappeared due to economic growth, including the reasons for their disappearance and how the changing coastal landscape accelerates economic transformation.

1.1. Background

China's reform and opening up began in 1978, and every decade since then the coastal zone has undergone drastic changes. In the first decade, the continuing increase in population coupled with economic growth, rapid urbanization, and infrastructure development resulted in the disappearance of space/resources in the coastal zone, including salt pans, breeding ponds, and agricultural land [5]. In the second decade, more than 70% of large Chinese cities were located in the coastal areas, accounting for 55% of China's gross domestic product (GDP) [6]. Furthermore, in 1998 alone, the occurrence of diseases in maricultural animals reached 30%, resulting in high economic losses [7,8]. In the third decade, with rapid economic growth and urbanization in China's coastal zone, new issues emerged, such as increased coastal pollution, ecosystem degradation, water resource depletion, and coastal erosion. From the fourth decade to the present, the coastal areas accounted for approximately 70% of the GDP in China [9,10] and contributed to China's rapid urbanization from 2000 to 2020 [11]; however, there were land use conflicts (cultivated land, construction land, and the ecological spatial layout in coastal zones) in all the coastal provinces of China [12]. Human activities have increased the exposure and vulnerability of China's marine and coastal ecosystems, causing a reduction in coastal wetland areas, as well as degradation of biodiversity and ecosystem stability [13]. The research of Li, Sun [14] showed that in the past 30 years, overall, the landscape in China's coastal areas has rapidly fragmented, with reduced aggregation, complexity, and irregular edge shapes [14].

In general, in each decade of the past half century, China's coastal zones have lost their space/resources, achieved economic growth, and experienced environmental decline.

The coastline of China is approximately 18,000 km long. The comprehensive governance of coastal zones in its true sense began in 1994 with the cooperation between China and the United Nations Development Program (UNDP). It was only in 2004 that the country started considering integrated land–sea governance as a national strategy [15]. However, prior to this, China's coastal provinces had already been the fastest developing regions in the country. By 2017, the per capita GDP in coastal-zone areas was close to that of high-income countries [15,16]. On the other hand, the high intensity of land development and population density in these coastal zones have resulted in significant social and environmental changes, resource pressures, coastal pollution, a continuous decline in ecosystem health, and increased environmental risks [17–21]. These issues indicate that the economic benefits of coastal zones are not proportionate to the governance of their ecological environment.

1.2. The Value of the Jimei Peninsula as a Case Study

Xiamen was a pilot city for the United Nations Partnerships in the Environmental Management for the Seas of East Asia (PEMSEA) in 1994 [22]. Earlier, in the 19th century, Amoy (Xiamen) was one of China's foreign treaty ports [23], which were the instruments that forced China to open up to the world, and also the places that could more readily adapt to globalization [24]. The difference between Xiamen and other treaty ports is that it is an island. Since it was a trading base for the Portuguese, Spanish, and Dutch from the 16th to the 19th century, it was an island on the "must have" list for foreigners [23]. Although it developed relatively slowly in the 19th century, it was also an "attractively provincial" port [23]. The open nature of Xiamen Island means it easily accepts foreign things. The surrounding waters of Xiamen Island are very rich in fishery resources [25], and it has a long history of development of fishing and hunting, fishing ports, and fisheries [26], and even as a treaty port; fishery production continues to be maintained. In 1980, Xiamen Island immediately became a special economic zone in China, and the commercialization of the city was rapidly promoted. In 2001, Xiamen was positioned as a tourist city and began to renovate the surrounding sea areas [27]. At this time, all aquaculture in Xiamen

Island was withdrawn from the coastline [25], and since then, the fishing village landscape of Xiamen Island has completely disappeared.

Tourism became a development priority during the transition from a planned economy to a market economy [28]. Coastal tourism that has undergone the transition from a planned economy to a market economy contributed to the rapid development of the coastal zone in China and it has been one of the major marine industries since 1993 [29]. China's economic growth and the functional transformation of coastal spaces are mutually exclusive, and the coastal zone of the Xiamen Special Economic Zone has a certain degree of representativeness, which can be used to infer the changes in the spatial distribution of fishing villages in southeastern China. At present, only the coastal zone of the Jimei Peninsula can be used to study this process.

1.3. The Issue and Purpose of This Study

Coastal tourism was seldom studied under the framework of the coastal-zone management considering the influence of China's economic and political reform [29]. The coastal function of the Jimei Peninsula has transformed the urban positioning of Xiamen in different periods. The process of how a coastal zone transforms from a production space to a consumption space has rarely been discussed. We posit that the function of a coastal zone changes with the positioning of the city and is affected by the dynamic influence of heterogeneous elements in the process of change. Geographical studies can highlight the appearance of local reproduction for consumption practices and how consumption processes and practices create special geographical features [30].

However, although the Jimei Peninsula in the north of Xiamen Island is closely related to the development of Xiamen and belongs to the Xiamen Special Economic Zone, it has a very different landscape. The Jimei Peninsula is characterized by its long coastline, which still maintains a fishing village landscape, and is also affected by the Xiamen Treaty Port. The peninsula is studded with fishing villages, ports, schools and other elements, still maintaining the landscape of the early 20th century. However, the latest research from Yu-Yan Zhang, Zhe Zou [31] and Zou, Zhang [32] points out that the spatial pattern of the Jimei Peninsula coastal zone generally maintains the mosaic pattern of fishing villages and educational spaces [33]. However, after 1994, the spatial governance model of the coastal zone completely changed. The change in this model came with the economic transformation of the city. Among these changes, the "Coastal Exclusion Policy (CEP)" has had a significant impact on the fishing villages around Xiamen Island, as fishers were forced to withdraw from coastal areas [32]. This policy means that coastal communities and fishing villages will also lose their unique production patterns and landscapes, but this important point was not mentioned in the aforementioned article. This study will continue to explore the current situation of coastal-zone changes after the policy's implementation and all the impacts brought by the changes.

This study takes the coastal land-use changes on the Jimei Peninsula as the object; the purpose of this study was to explain how the heterogeneous elements embedded in the peninsula affect the transformation of the production landscape into a consumption landscape and the connection between urban governance and economic transformation. The coastal zone is a place where people closely interact with the environment, and it is also the most developed. However, focusing only on the economic advantages of a single industry is likely causing us to ignore more important issues, for example, the decline of the ecological environment and the irrecoverable fishery culture, as these important issues are hidden behind the changes in the coastal landscape. The aim of this study is to increase the awareness of sustainability in the current blue economic boom.

2. Methods and Materials

2.1. Study Framework and Design

Today, China's consumption model has also shifted from material goods to more experiential consumption [34]; Chinese consumers have entered an era of cloud consumption, where the start- and end-point of the commodity cannot be seen during the consumption process. Moreover, the concept of a linear or one-way commodity chain is no longer enough to explain the process of production and consumption [30] because different locations and nodes in the commodity chain are bound to be connected [35]. A non-linear research approach transforms commodities into social practices, relationships, and policies, which are conceived as different aspects that are connected in different ways [36]. This study treated these as heterogeneous mosaics and identified the coastal landscape through on-site surveys and coding.

This study adopted two research approaches, namely historical geography and post-structuralism geography; the former can explain the regional dynamics in the long-term process of change, while the latter can explain the investment of political and economic energy in specific areas, making the space appear open, and all endogenous and exogenous factors can freely enter and exit. Spaces refer to newly created areas in the ever-changing dynamics [37,38]. The coastal zone is the junction between the sea and land, with the characteristics of openness, freedom, and imagination. The above characteristics coincide with those of post-structuralism, such as openness and conflict, difference, transcendence, loosening, and freedom.

This study uses the concept of an "underground rhizome" to reflect the spatial fragmentation of the coastal zone. We believe that, hidden under the spatial fragmentation, there are close relationships. It uses the point-to-point rhizome concept [39] to focus on the dynamics of coastal land changes, with particular attention to the relationship between heterogeneous elements and changes in coastal landscapes. Only by constructing a network among all spatial elements in the coastal zone, including culture, education, tourism, ecology, and policies, can we identify the various relationships hidden in the handover and transformation.

2.2. Study Steps and Contents

This study was conducted from October 2022 to February 2024. A total of 26 residents were interviewed individually or collectively, including fishers, retail store owners, coastal residents, community administrators, Jimei University teachers, Jimei University students, and police officers (Table 1). Furthermore, the current situation in the coastal zone with a length of about 16.1 km was recorded in detail, including tourist attractions, public facilities, settlements, and historical sites. All the collected landscape elements were divided into six categories for analysis.

Table 1. List of interviewees.

No.	Sample Coding	Sample Coding	Age	Background	Interview Time and Place
1	S1	M	Above 60	Grandpa's main job is as an employee in the logistics department of a construction institute, and his side job is as a fisher.	15 April 2023, online interview; 28 October 2023, grandpa's home.
2	S2	M	20–35	Fisher	16 April 2023, telephone interview
3	S3	S3-A:M S3-B:F	36–60	Vegetable retailer (husband and wife)	20 April 2023, Jimei Dashe 20 April 2023, Jimei Dashe
4	S4	F	Above 60	Clam-digging residents	20 April 2023, Jimei Dashe
5	S5	F	36–60	Nearby residents	29 May 2023, Jimei Dashe
6	S6	M	36–60	Nearby residents	9 April 2023, East coast of Xinglin Bay Landscape Belt
7	S7	S7-A:M S7-B:F S7-C:F S7-D:F S7-E:F	20–35 36–60 20–35 36–60 36–60	Employees of Dashe Neighborhood Committee	30 May 2023, community neighborhood committee
8	S8	M	36–60	Photographer	30 May 2023, online interview
9	S9	M	36–60	University teacher	31 May 2023, Jimei University College of Art and Design
10	S10	M	20–35	Sailing instructor	31 May 2023, online interview
11	S11	M	Above 60	Clothing store owner; digs oysters and sells them to surrounding businesses.	4 June 2023, Jimei Dashe

Table 1. *Cont.*

No.	Sample Coding	Sample Coding	Age	Background	Interview Time and Place
12	S12	S12-A:F S12-B:M	Above 60	Manager of Chen Clan Association	4 June 2023, Chen clan ancestral hall
13	S13	F	36–60	Fishmonger	4 June 2023, Jimei Dashe
14	S14	M	Above 60	Retired high school teacher	4 June 2023, Jimei Dashe
15	S15	S15-A:F	36–60	Jimei Dashe small shop owner	23 November 2023, Jimei Dashe
		S15-B:M	36–60	Residents of Dashe	23 November 2023, Jimei Dashe
16	S16	F	Above 60	Fishmonger	24 November 2023, Jimei Dashe
17	S17	S17-A:F	Above 60	Street vendors at Jimei Dashe	24 November 2023, Jimei Dashe
		S17-B:M	Above 60	Retired grandfather who lives in Jimei Dashe	24 November 2023, Jimei Dashe
18	S18	F	36–60	Owner of a retail store next to Ao Garden	24 November 2023, in the store
19	S19	F	36–60	Specialty shop owner next to Ao Garden	24 November 2023, in the shop
20	S20	F	36–60	Specialty shop owner next to Ao Garden	24 November 2023, in the shop
21	S21	F	36–60	Specialty shop owner next to Ao Garden	24 November 2023, in the shop
22	S22	F	20–35	Street vendors	24 November 2023, dragon boat pool
23	S23	M	36–60	Police officer	9 November 2023, Shi Li Chang Di
24	S24	F	30	Student (with part-time job selling bouquets).	21 January 2024, online interview
25	S25	F	20–35	Full-time flower shop studio owner.	21 January 2024, online interview
26	S26	F	20–35	Student (with part-time job selling bouquets).	22 January 2024, online interview

Note: The age of the interviewees is based on the year of the interview. The personal information of the interviewees was as complete as possible. In accordance with research ethics, the interviewees are represented by codes. The code "F" represents female, "M" represents male, "A:F" represents female A, and " B:M " represents male B.

In response to the heterogeneous elements in the coastal zone, a variety of data collection methods must be used. The data used in this study include historical documents, official documents (including laws, regulations, propaganda, and announcements), news media reports, publications, and social media content. The above data correspond to the nonlinear research framework, and triangulation was used to determine the correctness of the content. Study methods including case studies (including Dashe Fishing Village, Shi Li Chang Di), secondary literature analysis, fieldwork, and interviews (a total of 26 samples, Table 1) were used. These study methods were selected based on the results of exploratory research and the data of the study cases, which show the main characteristics of fragmentation and heterogeneity.

Changes in coastal fisheries were the focus of the interview, and the topics included the "Coastal Exclusion Policy (CEP) and the differences before and after CEP", land reclamation, and the impact on ecology, culture and life. The interview location was the interviewee's home, pier, workplace, coastal zone, etc. Please see the Supplementary Information for interview time and summaries (Supplementary Table S1).

The numbers and illustrations used in this study are defined as follows: ① refers to the site number shown in Figure 2 and (S1, S2, S4) refer to the interviewees' sample number. The sample numbers and interview summaries can be found in Supplementary Table S1.

2.3. Study Space and Time Scopes and Definitions

This study focused on the Jimei Peninsula (shown in Figure 1), located at 117.57° E to 118.04° E and 24°25′ to 24°46′ N. Its land area is approximately 275.79 square kilometers [40], and is located at the junction of the eastern and western sea areas of Xiamen city. The coastal zone at the southern tip of the Jimei Peninsula is one of the few areas in Xiamen that remains relatively unchanged, and it is also one of the few living areas that still retain the culture of southern Fujian. Administratively, it falls under the jurisdiction of Xiamen City, in the period of 1936–2023. The coastline on the Jimei Peninsula in 1936 still maintained the natural terrain, but 1949 was the starting point for coastline changes. From 1978 to 2000, the coastal zone underwent the most dramatic changes, and from 2020 to 2023, tourism represented the main function of the coastal zone.

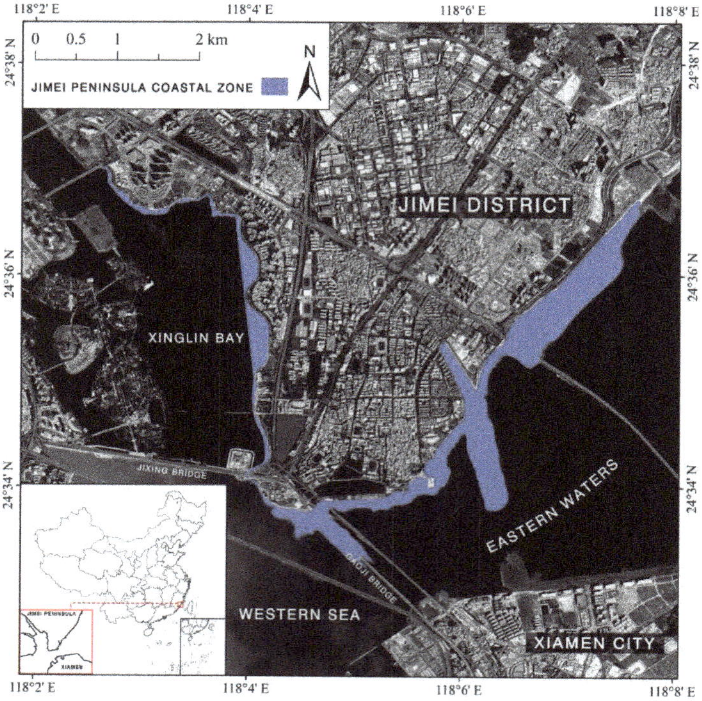

Figure 1. Study area.

The Jimei Peninsula features a combination of various geographical features, including low mountains, hills, plateaus, plains, and coastal tidal flats, as it is situated in a transitional zone between land and sea. This has contributed to the formation of a semi-fishing and semi-agricultural society [41]. The maritime area of the Jimei Peninsula is approximately 40 square kilometers. Due to natural siltation, land reclamation, and the construction of reservoirs along the coast, the coastline and tidal flats have gradually been reduced, and some areas have experienced varying degrees of pollution. Before 1956, the coastline of the Jimei District consisted of the coastlines of the Jimei Peninsula, Xinglin Bay, Xinglin Peninsula, Maluan Bay, and islands in Xinglin Bay, which were relatively long.

However, with the construction of the Xinglin Dam (Xinglin Reservoir) and Xinglin Bay Reservoir, as well as extensive land reclamation, the coastline has gradually shortened. The area of this study was defined as between Jimei Bridge and Shi Li Chang Di, and Antarctic Ao Garden and the end of Xinglin Bay Greenway.

3. Results

3.1. Spatial Distribution of Heterogeneous Landscapes in Coastal Zones

3.1.1. Production Landscape

When the coastal currents from Zhejiang and Fujian carry sediments deep into Xiamen Bay, tidal flats are formed around the Jimei Peninsula [13]. The low temperature, low salinity, and nutrient-rich characteristics of ocean currents create favorable conditions for the reproduction of organisms in the coastal zone, especially in autumn and winter [42].

The coastline of the Jimei Peninsula is winding, and there are clams, oysters, crabs, mantis shrimps, conches, sandworms, octopuses, seaweed, and other coastal creatures on the tidal flats [43]. These stable biomasses have attracted coastal residents for hundreds of years. In autumn and winter, the residents have used many simple tools, such as bamboo baskets of different shapes, and shovels, to set traps, search for food, and collect these organisms, in a process which is called Tao Xiaohai (which means "foraging behavior in the intertidal zone") [43,44]. On the tidal flats of the Jimei Peninsula, as long as you understand the tides, you can collect mollusks in large quantities; thus, residents in the coastal zone have a predictable food source in autumn and winter. A tide sign is still erected in this area (⑧). The location around this signboard is the only place where the collection activities can be seen on the Jimei Peninsula. The tidal flats have a stable productivity, encouraging the cultivation of fields on tidal flats and forming a system of sea fields (S1, S2, S4, S8, S11, S12, S15, S16, and S17). The sea fields of the Jimei Peninsula are distributed at the southernmost end (①, ②, ③, and ④). The sea fields have enabled the fishing village to have mature social organizations, economic activities, and a fishery culture. Coastal residents have a tradition of using the coastal tidal flats to grow food or breed livestock, and reclamation is carried out based on economic needs. For example, the reclamation on the west coast of the peninsula turned 40,000 acres of tidal flats into land for growing rice (⑫) [45,46]. The original intention of reclamation was farming or salting, but the economic benefits of aquaculture are much higher than those of farming or salting, so most of them were converted to aquaculture.

Shellfish excavated on the tidal flats can be easily boiled or grilled, and eaten, but their weight limits the distance they can be transported, and so they must be processed or eaten locally. Therefore, women can be seen everywhere in fishing villages shucking oysters, which are sold fresh or dried (⑪) (S15-B, S16). There are also fishers living on the coast who harvest clams and distribute them to the market (S4, S11). Most residents of Dashe Fishing Village make a living from fishing or aquaculture, and in its heyday there were about a hundred boats (S2) [47]. There is a boat repair shop in Dashe Fishing Village. Fishing equipment such as fishing nets are repaired by the fishers themselves ⑪). Tao Xiaohai, processing, purchasing, selling, and repairing fishing gear constitute the primary production landscape of the Jimei Peninsula coastal zone.

In 2007, the western sea area of the peninsula began to be regulated: aquaculture and fishing were prohibited, and the scale of aquaculture dropped sharply (S1, S2). Due to natural siltation and land reclamation (S1, S6, S7, S10), sea reclamation, and reservoir construction, the area of coastlines and tidal flats gradually shrunk, and the sea areas were also polluted, to varying degrees (S1, S2, S4, S9, S16). So far, the visible production landscape of the fishing village consists of only very sporadic fishing and oyster-shucking activities.

3.1.2. Cultural Landscape

Carefully observing the tides on the shore is the first step before fishers start production. The understanding of nearshore fish drives humans to build rafts to go out to sea for fishing. There were at least four piers on the Jimei Peninsula, and currently, only the ruins of the Dragon King's Palace pier still exist (⑤) [48,49]. However, fishery production is a high-risk labor industry. Due to safety needs and spiritual sustenance, many beliefs and taboos have arisen [48]. For example, the fishers of Dashe Fishing Village go to Tianfei Palace and Dragon King's Palace together before going to sea for the first time in the new year [32]. The same beliefs and practices strengthen the cohesion of local fisher organizations [47,50].

The unpredictability of going to sea is the essence of fishers' traditional beliefs. Therefore, there are many specific landscape elements connected to beliefs on the coast that are exclusive to the fishing culture. Therefore, the Tianfei Palace was built on the rocks on the east side of the peninsula (⑧) [51,52], and the Dragon King's Palace was built on the small island on the west side (⑤). The purpose was to seek safety at sea with the power of faith [51,53]. The landscape pattern also extends to the more exclusive fishing culture. For example, when a fishing boat returns to port, the whole family will greet it and hold a "peace" ceremony. When returning to port, if something unlucky happens, you are not allowed to drive past the Dragon King's Palace [54,55]. During the Lantern Festival, fishers will go to Tianfei Palace and Dragon King Palace to worship (S11), and fishers burn incense twice a month to worship those who died at sea (S1). The elaborate rituals, customs, and economy are all related to fishery production. Before the implementation of the CEP, various clans in Dashe Fishing Village would still have activities in the sixth and seventh lunar months and at the Lantern Festival (S11).

Tianfei Palace was destroyed by artillery fire during World War II, and Ao Garden was built on the original site. The fishing piers have also disappeared, due to land reclamation. Only the Dragon King's Palace pier still has a temple, and is the only fishery cultural landscape (The numbers and illustrations used in this study are defined as follows: ① refers to the site number shown in Figure 2).

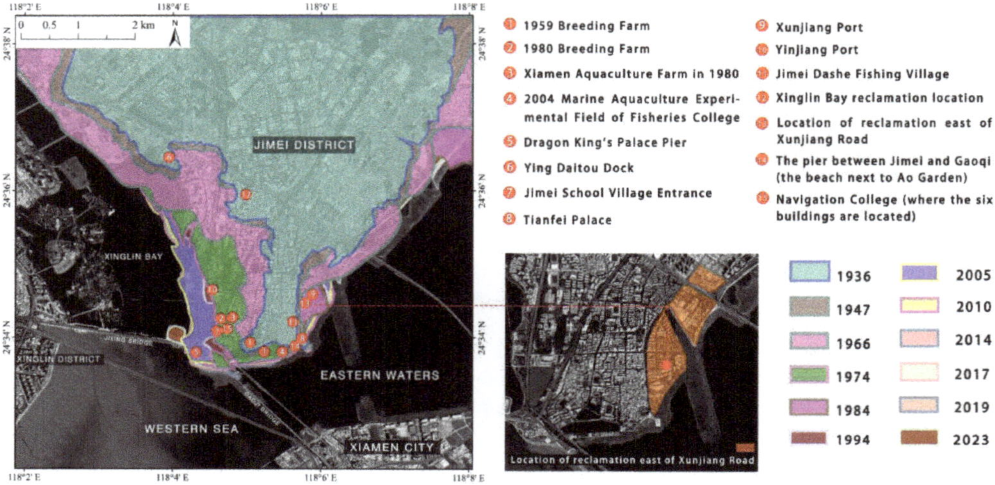

Figure 2. Coastal line changes and landscape elements of the Jimei Peninsula.

3.1.3. Educational Landscape of Maritime Technology

For scholars, illiterate fishing communities are elusive because fishers keep knowledge within their communities, passing it on from generation to generation and rarely allowing it to leave the community [1]. For the same reason, until the early 20th century, education in the fishing villages of the Jimei Peninsula was still poor [45,56]. Because fishers have been cultivating the sea, and using animal husbandry and fishing practices for a long time, children learn to fish on the tidal flats at the age of six or seven and go to sea with their family elders when they are more than ten years old. All knowledge is accumulated and transmitted at sea, and the content is related to fishery and navigation. The new generation of fishers enjoys the experience passed down by their predecessors, has accumulated many industrial habits that adapt to and promote productivity, is proficient in the fishing season and flood season, has sufficient practical experience, and can "observe the mountain situation", judge the direction, and avoid dangers (S1, S2, S3). Regardless of fishery or trade, the livelihood of fishing villages is derived from community inheritance, and education is not widely available in fishing villages.

> *"When they did not have a living skill, they went to Southeast Asia, Singapore, and Malaysia to make money. It was easy to make money with clans, so they usually stopped studying after they reached junior high school. Even if they went to college, they would not necessarily earn more than they did. At that time, the Chen family not many of our children can go to high school."* (S3-a)

Therefore, the education level in coastal areas has not improved.

However, there is the famous Jimei School Village, which is a fishing village at the southern end of the Jimei Peninsula. It consists of a mosaic of villages and schools, which are famous for their knowledge and navigation technology courses [33]. The reproduction of the fishing community is attributed to the inheritance of navigation technology, which can be used to develop fisheries and easily expand the transnational trade space. Tan Kah

Kee (also known as Chen JiaGeng, the founder of Jimei School) of the Chan clan in Dashe Fishing Village is an important kin leader because of his prosperous overseas trade and his allocation of assets for education [56]. He recognized the importance of revitalizing navigation and fisheries, and allocated wealth to navigation technology education [57]. Jimei Primary School, Normal School, Fisheries, Navigation, and other schools are collectively referred to as the Jimei Schools (⑦) [58]. The six buildings at the core of the campus bear witness to the history of maritime education in Jimei School Village. The fifteen-story Nanxun Tower is a landmark on the coast of the Jimei Peninsula. When Tan Kah Kee built it, it was also intended to protect fishery production and provide directions and the time for fishers traveling at night [59]. Jimei School Village is a collection of fishing villages, sailing education and leaders, and other elements, leaving a special educational landscape on the Jimei Peninsula.

3.1.4. Governance Landscape

This study overlaid onto each other the 1936–2023 maps for the Jimei Peninsula, which showed that in the past 90 years land reclamation has expanded the Jimei Peninsula's land area by 42.54% (Figure 2). The area of tidal flats in the coastal zone has also been reduced and artificialized.

From 1955 to 1965 alone, the reclamation area of Xinglin Bay was as high as 2000 hm^2 (⑩) [60]. During this period, Jimei built the Gaoji and Jixing seawalls, with the main purpose of making Jimei a transportation hub connecting the island with the outside world. From 1966 to 1968, large-scale reclamation was carried out in Xinglin Bay. The main purpose at this time was aquaculture [61]. However, although land reclamation has improved the economy, it has led to a reduction in the area of tidal flats. In the late 1970s, it also caused the number of oysters in a large area to decrease [45]. The year 1984 was a turning point in Xiamen's reclamation activities. Before that, the reclaimed areas were mainly used for aquaculture, but after 1984 the reclamation projects were mainly used for port construction, industry, commerce, trade, and urban construction [62]. After the 1990s, a large number of tidal flats in Jimei were developed and built, but the area of tidal flats continued to shrink. The east side of the Jimei Peninsula now consists almost completely of reclaimed land (⑬).

> *"Jimei used to be very small, just a small fishing village. The Jiageng Memorial Hall used to be located on the tidal flats. Now the east of Xunjiang Road is filled in by the government"* (S1, S2, S3, S6, S7, S9, S10)

Since 1994, Xiamen has implemented coastal management for the surrounding sea areas by regulating the coastline [63–66]. The year 2002 marked a turning point in coastal-zone governance. The "Comprehensive Rehabilitation of Aquaculture in the Western Sea Area" policy, commonly known as the "Coastal Exclusion Policy (CEP)", was implemented [32]. In 2006, the Jimei Peninsula began to implement the CEP in an attempt to restore the ecology. It banned tidal flat farming and punished fishers who violated the regulations [67]. By the end of May 2020, the Jimei Peninsula had fully completed the work of clearing the sea area for breeding.

The governance landscape of the Jimei Peninsula coastal zone is mainly land reclamation and artificialization, which is in line with the pace of economic development in China as a whole [68]. The Jimei Peninsula is a historical microcosm of the entire southeastern coast of China.

3.1.5. Tourism Landscape

Xiamen city was approved as a scenic tourism city by the central government in 2001 [27]. The subsequent coastal governance measures were all aimed at promoting the development of tourism in line with the development goals of scenic tourism cities [69]. In 2002, Xiamen promulgated a CEP to build coastal tourism facilities on the coast to increase the economic benefits [70]. Beginning in 2006, fishers on the Jimei Peninsula began to withdraw from the coastal zone; the withdrawal was completed in 2020, thereby meeting the marine environment requirements of the tourist scenic area [71]. The government has also successively proposed projects for the Jimei coastal zone [72], including beautification projects, the planting of mangroves, the construction of sea plank roads and landscape platforms, etc., to cope with the development of tourism [73].

The development of tourism on the Jimei Peninsula has attracted a large number of foreign tourists [74]. The coastline of the peninsula is already dotted with tourist spots (Figure 3), including landscape green belts, bicycle paths, parks, and hotels. These facilities were initially concentrated in the north corner of the southeastern coast of the peninsula before spreading to the entire coastal zone (⑧). The Dragon King Palace, a cultural fishery landmark, is now a thousand-year-old temple, and the surrounding green belt formed by land reclamation was repackaged into the "Shi Li Chang Di" (which means 10 km-long seawall). The intersection of Gaoji Bridge and Jixing Bridge has also been developed in order to boost the economic vitality of the coastal zone [75] (Figure 3).

Figure 3. Tourism landscape elements.

3.1.6. Ecological Landscape

The tidal flats formed by the Zhejiang and Fujian coastal currents around Xiamen Bay are habitats that nurture coastal ecology [13] and provide important stopover, breeding, and wintering habitats for many endangered water bird species [76]. The creatures and migratory birds on the tidal flats form the ecological landscape of the coastal zone [77]. However, since 1964, land reclamation has expanded the area of the peninsula and the length of the coastline, completely changing the original appearance of the coastal zone. Initially, only the coastline at the southern end of the peninsula had minimal changes [32], but it was completely artificial by the end of 2023.

Over the past few decades, the degree of human-induced artificialization along the coast has increased significantly, with the construction of ports, docks, and bridges, which occupy a large portion of the coastline, leading to a significant decrease in fishery resources [78,79]. In 1955, after the construction of two seawalls on the east and west sides of the Jimei Peninsula, the natural nurseries in the coastal zone disappeared, resulting in a decrease in shrimp production. The area of the three major fishing grounds nearby also shrank year by year, the currents were not smooth, and the fish and shrimp resources declined [45]. The interviewees also mentioned that land reclamation and the construction of artificial facilities have caused the disappearance of the tidal flats and the native common fish and shellfish on the coast of the Jimei Peninsula (S1, S2, S3, S7, S8).

The western waters of the Jimei Peninsula are a nature reserve for the Chinese white dolphin [13]. However, the habitats of Chinese white dolphins in the sea area overlap extensively with economic development zones and shipping-intensive areas. The degradation and loss of suitable shallow-sea habitats caused by reclamation and other projects have caused the disappearance of Chinese white dolphin habitats [80].

As Xiamen is a pilot city of the United Nations, the governance of its coastal zone must focus on the protection and use of marine resources [63,64,81–83]; for example, the restoration of fishery resources and marine species [84]. However, the interviewees said that the artificialization of the coastal zone has left organisms with no place to hide. Although the water quality is gradually improving, the number of organisms has not increased (S2). The ecological landscape of the Jimei Peninsula has declined due to the prevalence of land reclamation, industry, and fish farming. However, the interviewees (S2, S5, S6, S9) also pointed out that the government's long-term control has shown signs of easing the degradation trend of the ecological environment [85]. Whether it is an illusion caused by environmental beautification or a true ecological restoration is an issue worthy of continued observation.

3.2. Economic Benefits of Coastal-Zone Management and Tourism Services

In 1994, Xiamen became a demonstration site for coastal-zone governance in China. [71]. Xiamen's GDP of that year was CNY 18.7 billion [86] and it grew to CNY 780.27 billion in 2022 [87]. Most of this comes from cultural tourism (or tourism service industry) income [88]. In 2006, the Jimei District's GDP was CNY 14.274 billion, but by 2022 Jimei's GDP had grown to CNY 95.658 billion, with the service industry accounting for 49.1% [89]. Taking Ao Garden, the main scenic area in Jimei, as an example, the number of tourists increased by 101.5% from 2012 to 2019. Although affected by the COVID-19 epidemic in 2020, the number of tourists in 2023 reached a new high [90–100]. According to recent data, Xiamen has completed its transformation from a commercial port city to a tourist city, and its tourism popularity index and tourist satisfaction have always ranked among the top in China [101].

After the implementation of the CEP on the Jimei Peninsula, the coastal zone has transformed from traditional fishery to tourism service industry [102]; among the sites, "Shi Li Chang Di" and Dashe Fishing Village are the most representative.

In 1955, two seawalls were built to connect the Jimei Peninsula to the land on the east and west sides, with a total length of 5032 m, collectively known as the "Xiamen Seawall". Half a century later, both seawalls have been transformed into cross-sea bridges connecting the coast. The strip is nicknamed the "Shi Li Chang Di" and was promoted as a tourist attraction by coastal-zone governance.

The "Shi Li Chang Di" was built around the Dragon King's Palace pier site. It is one of the most popular tourist attractions on the Jimei Peninsula. Now, the coastal zone has been occupied by various consumer and leisure spaces (Figure 4 (left)). At present, the Dragon King's Palace pier has lost its original function of fishery production and transportation and now serves to promote the economic development of the coastal zone in the form of a tourist attraction [75]. The government manages the surrounding aquaculture facilities and fishing boats [103]. There are only a few fishing boats left (S23).

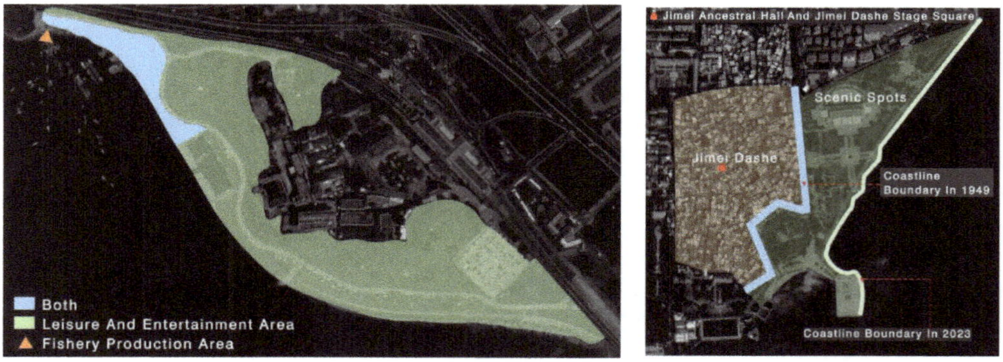

Figure 4. Space configuration of Shi Li Chang Di (**left**) and Coastal zone changes in Dashe Fishing Village (**right**).

Dashe Fishing Village (Figure 4 (right)) was originally a fishing village near the sea. Later, due to the reclamation and construction of Ao-Yuan Garden, the fishing village became separated from the sea [104]. Dashe Fishing Village is the ancestral home of the Tan Kah Kee clan. Therefore, with the help of the Jimei District Government and universities, it has been transformed into a cultural and creative village that promotes Kah Kee's spirit. Now, various shops extend outwards from the Jimei Ancestral Hall. The fishing village industry has been transformed into the tourism service industry. Currently, only one or two households in Dashe are still engaged in fishery production (S5, S7).

3.3. Changes in Coastal Consumption Patterns

From 2010 to 2021, the Jimei Peninsula carried out reclamation and marketization of sea area resources in the coastal zone, outsourcing sea area-use rights through bidding, auctions, listings, etc. In 2021, regulations began to formally transform the coastline of the Jimei Peninsula into a tourist and recreational area [32]. After decades of changes in the coastal zone of the Jimei Peninsula, the consumption theme has changed from the traditional agriculture and fishery economy to the tourism economy.

The ferry terminal next to Ao Garden is the main space for Dashe residents to cross to Xiamen Island. Before the end of 2020, the area around the ferry was full of stalls selling fish and processed local products, serving tourists and residents. However, at the end of 2020, except for those selling local specialties in their own stores, all other vendors were asked to leave. On the west side of the peninsula, the fish trading activity centered on Dragon King's Palace pier has disappeared, and the management of the "Shi Li Chang Di" has been entrusted to outsourcing companies, for unified management [105]. Irregular events and stalls are set up in response to tourists, as well as consumption activities including parking lots, vendor markets, open-air concerts, retail stores, etc. There are often small vendors gathering at the entrance (without paying stall fees) and on the left side of the entrance to the "Shi Li Chang Di" stall spots (vendors here must pay stall fees to the venue manager), mainly selling food and drinks. Bands of all sizes perform on the lawn of Shi Li Chang Di; open-air concerts are mostly held online (Online fans reward tips.) and offline (free), attracting many tourists. Tourists on the lawn can rent canopies from businesses to enjoy the sea view (Supplementary Table S2).

The consumption behavior in the coastal zone is no longer purely physical buying and selling behavior. More consumption comes from experiential services, such as renting canopies and barbecue equipment, and renting charging equipment; another type of cloud consumption is the benefit from Internet-celebrity live broadcasts. Take the bouquet sales on the Jimei Peninsula as an example. The raw materials come from Kunming, Yunnan (the largest fresh-cut flower trading market in Asia), which is 2000 km away. All procurement processes are cloud consumption (Douyin live broadcast room, Huawu app). There are sales around the "Shi Li Chang Di", and tourists come from all over the country.

The transformation of the coastal zone has resulted in the disappearance of space/resources, such as salt farms, breeding ponds, and tidal flats. The coastal-zone space in the production landscape period was an open and wide area, but the characteristic of the consumption landscape is the fragmentation of space due to consumption.

4. Discussion

4.1. The Embeddedness of Heterogeneous Spaces

Before 1920, the elements on the Jimei Peninsula were fishery production landscapes for gathering, fishing, and hunting [106]. In the process of fishery development, various fishing cultures emerged, mostly based on maritime safety and praying for smooth sailing at sea [48]. For example, the Tianfei Palace (Mazu), which protects the living, and the Dragon King Palace, which protects the dead, are major cultural landscape elements, respectively protecting fishers who go to sea on the Jimei Peninsula and those who cannot return.

After 1920, education was embedded into the Jimei Peninsula, creating the intellectual landscape of coastal-zone production technology. From 1949 to 1966, policy factors led to the reclamation of the landscape, with the purpose of expanding the oyster production area. Although this was the most significant land-use change on the peninsula, the landscape essentially remained centered on fishery production. Among the changes, Ao Garden, which was built on reclaimed land in 1960 [107], was integrated into the cultural landscape of the Jimei Peninsula on the former site of Tianfei Palace. It was not until 1984 that the government reclaimed the sea in order to expand the new urban area [44]. Thus, the nature of the area that was originally focused on fishing began to change. Because the purpose of the new space was to accommodate the growing population of the special economic zone, the new urban area included residential areas and hotels (⑨, ⑬); it became a heterogeneous space with different fishery landscapes embedded in the governance landscapes. Although land reclamation is a local or regional change, once it is linked to global issues it may lead to effects in the socio-economic system [108,109].

The spatial location of the new urban area also separates the fishing village from the sea. In 2006, due to the implementation of the CEP, a large number of fishers lost their access to the sea, which also led to a change in the spatial layout of the fishing village. The effect of the tourist city increased the demand for the service industry; due to loss of access to the sea, fishers filled these professions. The old fishers who lost their livelihoods had to rent out their houses to outsiders. The middle-aged people worked as security guards and the women worked as cleaning staff [32]. The fishing village no longer relied on fishing, but on sea-related consumer services instead.

Young tourists from various provinces in China and local fishermen have created a parallel space–society relationship in the horizontal society of the coastal zone, just like the heterogeneity of consumption and production in space [110]. On the other hand, the loss of vertical intersection and transmission between groups is also a problem and a dilemma that must be faced in the future.

There are heterogeneous elements embedded in the Jimei Peninsula, and they have become part of coastal tourism products (Figure 5). Due to the shift caused by policies, the fishing villages and fishing landscapes on the coast have become pure consumption spaces after being separated from the sea.

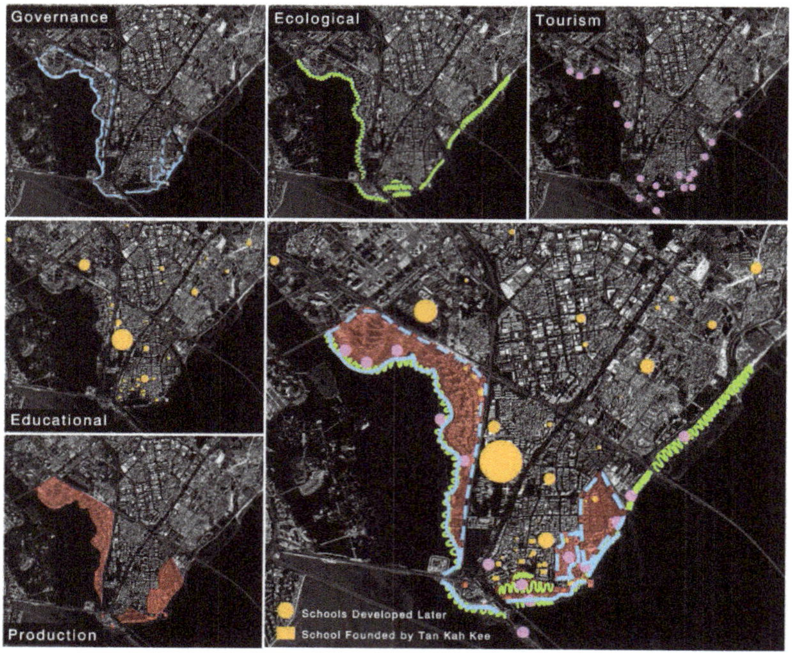

Figure 5. Spatial distribution map of heterogeneous landscapes.

4.2. Loss of Fishery Services

Post-structural geographer Murdoch [111], who focuses on heterogeneous spatial combinations, proposed that one of the core features of post-structural geography is that space and place are intersected by processes and operations originating from internal or external factors. We may infer that the current Jimei Peninsula is also a post-structural space. This is because, from a geographical perspective, the changes in land use have caused the fragmentation of spaces, whether it is internally generated (such as Jimei School Village) or due to outside factors (such as the Shi li Chang Di). Post-structural space may be attractive in the context of globalization, but from the perspective of coastal zone protection, the results are not optimistic.

After 2014, the attention on coastal zones shifted to humanistic aspects [112]; for example, correlations between land use and socioeconomic changes along the Atlantic coast have been used to assess ecosystem services. However, the results showed that land use changes and income and population growth negatively affect ecosystem services [113]. This study, using the Jimei Peninsula in the Pacific as an example, also showed the continued decline of cultural ecosystems after land-use changes. Fishing culture is the main axis of the Jimei Peninsula's cultural ecosystem services. The highly productive tidal flats provide fishing villages with a stable food supply. The gods located on the east and west coasts provide spiritual comfort for fishers who go to sea and their families at home. The knowledge transfer of navigation technology also contributes to the conservation of marine fishery resources. Of course, the proper provision of tourism and entertainment activities can also effectively maintain the function of ecosystem services. However, if external operations are too harmful, they may be counterproductive.

The tidal flats formed by the coastal currents in Zhejiang and Fujian are the basis of the Jimei Peninsula ecosystem services. However, the policy-led spatial expansion was all in response to managing the population growth. Although the spatial pressure was alleviated, it still brought socioeconomic risks [108]. In addition to reducing the area of the tidal flats, land reclamation also affected the productivity of the coastal zone. More importantly, the method of land reclamation affected the ecosystem service functions provided by the coastal zone, which is mainly reflected in the blocking of the connection of the fishing villages to the sea and the irreversibility of destroying natural nurseries and salt fields. The economy of the coastal zone and the consumption power brought by tourism have replaced the productivity of the tidal flats.

Hein, van Koppen [114] proposed that when fishery spaces are transformed into leisure and entertainment spaces the scale of the ecosystem service space also changes accordingly, diverging from the region to the country. However, on the Jimei Peninsula, a by-product of the transformation from a production to a service industry is the disruption to the fishery knowledge chain that had been passed down from generation to generation within the fishing village. That is, the fishery knowledge that relies on observing tides, wind directions, climate, astrology, and ocean currents, and acting according to the solar signs, will be completely lost. The embeddedness of heterogeneous landscapes on the Jimei Peninsula is the main cause of the spatial fragmentation of the coastal zone, and the ecosystem services of the fishing culture are also lost with the changes in land uses.

However, changes in land use, increases in demand [113], and lengthening of time scales [115], will lead to damage to ecosystem services. At present, the Jimei Peninsula seems not to have considered the long-term impact of cultural and ecological loss on the next generation. Even if the scale is different, there can be eco-economic practices, such as on the UK's coast [116], the Swedish coastal zone [117], Satoumi in Japan [118], and so on. Multi-generational participation in sea activities can stabilize social relations and promote the dialogue of experience and cultural inheritance [119]. Wilson, Acheson [120] pointed out that after the environmental integration on a complex scale, those with environmental knowledge can develop a strong social structure, and the traditional fishery economic action is easier to continue. This study holds the view that the cooperation between coastal zone and local residents is the best way to maintain cultural ecosystem services.

4.3. Commodity Chains, Circuits, and Networks

The productivity of the coastal zone involves shellfish harvesting on tidal flats; the shellfish are sold as seafood for consumption. However, the commodity chain is limited by time and transportation costs; thus, the producers are the fishers, and the consumers are local residents.

The COVID-19 pandemic has affect coastal planning and management [121] all over the world. The Jimei Peninsula was extremely depressed during the epidemic, and the residents near the coast now have increased demands for healthcare after the pandemic [122]. However, the coastal zone took advantage of the government's strategic tourism promotion to strengthen its tourism image after the pandemic. The heterogeneous landscape embedded in the coastal zone brought in many outsiders. The scenery of the coastal zone is a commodity; the outsourced management company used it and organized activities, especially the purchase of "invisible goods" in the cloud, such as live-streaming rewards, live-streaming goods, etc. However, every "moment" of consumption between consumers, sellers, products, and origins forms a product circuit with no starting point and no end point. These moments are mediated through different apps to create a product circuit in the coastal zone. Everyone can follow the products and the different tourist experiences through links. In this way, certain moments of prioritization in the flow of goods are eliminated [36]; there is no end-consumer or production end, because every moment of the product is connected to people's activities in the coastal zone [123]. In recent years, part of Fujian Provence coastal zone has been actively transformed into a tourism complex [29,66,124]. The consumption circuit with the coast as the background is spreading in China. For example, the fishing village of Xunpu, which is two hours' drive from Xiamen, connects each consumption moment with a trip shoot (a trip with a photographer), which collages the image of "a girl who is a daily fisher" [125], and transforms the fishing village into a space dominated by consumption.

Bouquets are a luxury product. Behind the product is a product network constructed by retailers [36]. In the Jimei Peninsula coastal zone there are a large number of bouquet vendors, and business is booming. It is also the only product that can be seen carried by consumers, and is widely disseminated as images by social apps. Its luxury characteristics are very different from the original landscape characteristics of the coastal zone, such as production, culture, navigation, and ecology. The consumption landscape represented by bouquet retail is a blooming flower grafted onto the transformation of the coastal-zone-governance landscape. The spatial embeddedness of luxury goods symbolizes the complete qualitative change in the coastal zone towards a consumption landscape.

This is the cumulative effect of the CEP and tourist cities. In the long run, the consumption pattern formed after the transformation of the coastal zone will prove to be more influential on urban development.

4.4. The Changing Nature of the Coastal Zone

In the 19th century, the coastal zone provided happiness and health to humans, with its value surpassing its economic output [125]. Therefore, since the 19th century, the coastal zone has seemingly become synonymous with holidays and leisure.

Since 1994, to adapt to the needs of the governance of the western sea area and the development of tourism, the Jimei Peninsula has used the CEP to promote the peninsula's function as more consumer-oriented, thereby realizing the transformation of the urban economy. The urgency of coastal-space governance has attracted governance, power, and mobile capital, as well as flowing capital; this move is no different from the city of Paris' use of urban renewal to increase the flow of people and money in the consumer space [126]. The coast of contemporary China is similar to the Paris Avenue, becoming an important public display center and promoting the circulation of goods, money, and people. The transformation of governance caused the coastal landscape to become like merchandise in a department store window.

This study found that the governance landscape in the coastal zone is the key to transforming consumption practices. Luxury consumption appeared through the promotion of social apps, connecting all moments of cloud consumption and creating another invisible environment. In addition, in the process of landscape construction, the control of coastal tourism landscapes still needs to be discussed, because all forms of production and consumption are remotely governed, incorporated, and disciplined based on performance, and extending the possibility of consumption politics, which is another possible direction for discussion.

5. Conclusions

This study identified the heterogeneous elements and their effects on the functional changes in the Jimei Peninsula's coastal zone through time and space and explored the changing patterns of the coastal zone under rapid economic development. The results showed that the Jimei coastal zone has transformed its functions following the needs of urban development, and the embeddedness of heterogeneous spaces has fragmented the coastal landscape. Tourism-oriented governance of the landscape is a radical approach that breaks with the past. The transformation of coastal-zone functions has led to the loss of traditional fishery knowledge and skills, along with the disintegration of its related cultural ecosystem services. Coastal -one consumption has developed from a short and narrow commodity chain into a commodity network shaped by moments. This invisible network is the nature of the coastal-zone-consumption landscape. The transformation model of fishing villages in the southeastern coastal areas of China can use the case of the Jimei Peninsula as a reference.

The Jimei Peninsula belongs to the Xiamen Special Economic Zone, and its coastal-zone governance and economic transformation have important significance for other coastal areas. The results of this study can encourage local governments to respect ecology and culture while carrying out economic construction and give priority to protecting the coastal ecosystem to develop a healthy blue economy. Although we collected as many documents as we could and used triangulation to confirm the correctness of the data, this study was still limited by a lack of cultural and historical materials and the loss of first-hand knowledge due to aging. For example, the spatial structure of the Jimei Peninsula still needs more documents for more precise spatial localization. In addition, whether the transformation of the Dashe settlement into a cultural and creative settlement in recent years will promote the revival of fishery literature and art is a topic worthy of continued attention in the future.

Supplementary Materials: The following supporting information can be downloaded at: https://www.mdpi.com/article/10.3390/w16060913/s1, Table S1: Interview summary; Table S2: Changes in Coastal Consumption Patterns.

Author Contributions: Conceptualization, S.-C.T.; data curation, H.W.; formal analysis, H.W.; funding acquisition, S.-C.T. and X.-F.Z.; investigation, S.-C.T. and H.W.; methodology, S.-C.T.; project administration, S.-C.T.; resources, S.-C.T.; software, H.W.; supervision, S.-C.T. and S.-H.L.; validation, S.-C.T., S.-H.L. and H.W.; visualization, H.W.; writing—original draft, S.-C.T. and H.W.; writing—review and editing, S.-C.T. All authors have read and agreed to the published version of the manuscript.

Funding: This research was funded by a scientific research start-up fund of Jimei University, China, grant number Q202214 and Xiamen Academician Workstation (Artistic Digital Research on the Maritime Silk Road Civilization), grant number YSZ202301.

Institutional Review Board Statement: This study was approved by the Science and Technology Ethics Committee at Jimei University (JMU202307039). No harm was caused to the participants during the study.

Informed Consent Statement: Informed consent was obtained from all subjects involved in the study.

Data Availability Statement: The data presented in this study are available on request from the corresponding author.

Acknowledgments: The study was made possible thanks to all the interviewees in the coastal zone. Thanks to three anonymous reviewers for their constructive suggestions; all deficiencies in the article are the responsibility of the authors.

Conflicts of Interest: The authors declare no conflict of interest.

References

1. Fagan, B. *Fishing: How the Sea Fed Civilization*; Gūsa Press, Walkers Cultural Enterprise, Ltd.: New Taipei City, Taiwan, 2022.
2. Oppenheimer, M. Climate change and environmental pollution: Physical and biological interactions. *Clim. Chang.* **1989**, *15*, 255–270. [CrossRef]
3. Holligan, P.; Reiners, W. Predicting the responses of the coastal zone to global change. In *Advances in Ecological Research*; Elsevier: Amsterdam, The Netherlands, 1992; Volume 22, pp. 211–255.
4. Li, Z.; Luan, W.; Zhang, Z.; Su, M. Research on the Interactive Relationship of Spatial Expansion between Estuarine and Coastal Port Cities. *Land* **2023**, *12*, 371. [CrossRef]
5. Yu, H. China's coastal ocean uses: Conflicts and impacts. *Ocean. Coast. Manag.* **1994**, *25*, 161–178. [CrossRef]
6. Wang, Y. Coastal management in China. *Ocean. Manag. Glob. Chang.* **2003**, *469*.
7. SOA. *Report of Marine Environmental Quality in China*; State Oceanic Administration: Beijing, China, 2006.
8. Mingjiang, Z.; Mingyuan, Z.; Jing, Z. Status of harmful algal blooms and related research activities in China. *Chin. Bull. Life Sci.* **2001**, *13*, 54–59.3.
9. Suo, A.; Guan, D.; Sun, Y.; Lin, Y.; Zhang, M. Advances in coastal landscape ecology and its role in the construction of marine ecological civilization. *Acta Ecol. Sin.* **2016**, *36*, 3167–3175.
10. Hou, X.; Liu, J.; Song, Y.; Li, X. Environmental-ecological effect of development and utilization of China's coastline and policy recommendations. *Bull. Chin. Acad. Sci.* **2016**, *31*, 1143–1150.
11. Du, P.; Hou, X.; Xu, H. Dynamic Expansion of Urban Land in China's Coastal Zone since 2000. *Remote Sens.* **2022**, *14*, 916. [CrossRef]
12. Zong, S.; Hu, Y.; Zhang, Y.; Wang, W. Identification of land use conflicts in China's coastal zones: From the perspective of ecological security. *Ocean. Coast. Manag.* **2021**, *213*, 105841. [CrossRef]
13. Zou, S.; Xie, S.; Yang, Y. *Encyclopedia Geography of China—Xiamen Peninsula*; World Book Publishing Guangdong Co., Ltd.: Guangzhou, China, 2016; pp. 4–5.

14. Li, Y.; Sun, Y.; Li, J. Heterogeneous effects of climate change and human activities on annual landscape change in coastal cities of mainland China. *Ecol. Indic.* **2021**, *125*, 107561. [CrossRef]
15. Bao, J.; Wu, D.-T. *Space, Scale and System: A Geographical Study of China's Land and Sea Coordinated Development Strategy*; Southeast University Press: Nanjing, China, 2016.
16. Lin, X.-H.; Peng, X.; Li, X.-J. Research on the characteristics of my country's coastal zone economic development under the new situation. *Mar. Econ.* **2019**, *9*, 12–19.
17. Huang, L.-H.; Jiang, Y.; Lin, C.; Li, T.-W.; Chen, F.; Wang, W.-Y. Research on the coupling coordination relationship between Xiamen Port development and coastal eco-nvironment evolution. *Environ. Pollut. Prev.* **2020**, *42*, 890–893, 900.
18. Wang, L.-B. Investigation of the Social and Cultural Changes of Sea Island Fishing Villages—A Case Study on Dayian Island Fishing Village. In *The Sixth Postgraduate Symposium of the School of Ethnology and Sociology, Minzu University of China*; Minzu University of China: Beijing, China, 2012; pp. 89–96.
19. Dai, Z.; Zhang, H.; Zhou, Q.; Tian, Y.; Chen, T.; Tu, C.; Fu, C.; Luo, Y. Occurrence of microplastics in the water column and sediment in an inland sea affected by intensive anthropogenic activities. *Environ. Pollut.* **2018**, *242*, 1557–1565. [CrossRef] [PubMed]
20. Wang, T.; Hu, M.; Song, L.; Yu, J.; Liu, R.; Wang, S.; Wang, Z.; Sokolova, I.M.; Huang, W.; Wang, Y. Coastal zone use influences the spatial distribution of microplastics in Hangzhou Bay, China. *Environ. Pollut.* **2020**, *266*, 115137. [CrossRef] [PubMed]
21. Xue, X.-Z.; Hong, H.-S.; Charles, A.T. Cumulative environmental impacts and integrated coastal management: The case of Xiamen, China. *J. Environ. Manag.* **2004**, *71*, 271–283. [CrossRef]
22. Xiamen Ocean and Fisheries Bureau. *Flying on the Waves: Oral Records of Key Figures in the Twenty Years of Xiamen Coastal Zone Integrated Management from 1996 to 2016*; Xiamen University Press: Xiamen, China, 2018.
23. Nield, R. *China's Foreign Places: The Foreign Presence in China in the Treaty Port Era, 1840–1943*; Hong Kong University Press: Hong Kong, China, 2015.
24. Bracken, G. Treaty Ports in China: Their Genesis, Development, and Influence. *J. Urban Hist.* **2019**, *45*, 168–176. [CrossRef]
25. *Compiled by the Compilation Committee of "China's Island Annals" Annals of China's Islands, Southern Coast of Fujian, Fujian Volume 3*; Ocean Press: Beijing, China, 2014; p. 788.
26. Compiled by Xiamen Ocean and Fisheries Bureau. *Proceedings of the Xiamen Marine Economic Development Strategy and Marine Environmental Protection Seminar*; Ocean Press: Beijing, China, 2006; p. 298.
27. Central People's Government of the People's Republic of China. *Reply of the State Council on the Overall Urban Planning of Xiamen City*; State Council, Ed.; State Council: Beijing, China, 2000.
28. Wen, Z. China's domestic tourism: Impetus, development and trends. *Tour. Manag.* **1997**, *18*, 565–571. [CrossRef]
29. Gu, M.; Wong, P.P. Coastal zone management focusing on coastal tourism in a transitional period of China. *Ocean. Coast. Manag.* **2008**, *51*, 1–24. [CrossRef]
30. Mansvelt, J. *Geographies of Consumption*; SAGE Publications Ltd.: London, UK, 2005.
31. Zhang, Y.-Y.; Zou, Z.; Tsai, S.-C. From Fishing Village to Jimei School Village: Spatial Evolution of Human Ecology. *Int. J. Environ. Sustain. Prot.* **2022**, *2*, 33–43. [CrossRef]
32. Zou, Z.; Zhang, Y.-Y.; Lee, S.-H.; Tsai, S.-C. The Transformation of Coastal Governance, from Human Ecology to Local State, in the Jimei Peninsula, Xiamen, China. *Water* **2023**, *15*, 2659. [CrossRef]
33. Liu, Y. *2021 China Consumer Report: Changes in Lifestyle Consumption*; Mazars: Hong Kong, China, 2021.
34. Miller, D. *Material Culture and Mass Consumerism*; John Wiley & Sons: Hoboken, NJ, USA, 1997.
35. Hughes, A. Retailers, knowledges and changing commodity networks: The case of the cut flower trade. *Geoforum* **2000**, *31*, 175–190. [CrossRef]
36. Tsai, S. *Theory on Adaptive Mode and Creative Destruction under the Shift-Reset: A Case Study of Water Distribution in Pingtung Plain*; National Taiwan Normal University: Taipei, Taiwan, 2020.
37. Tsai, S.; Zou, Z.; Chu, T. (Eds.) On the Post-Structure Geography Perspective of Regional Integration: A Case Study of Water Distribution in Pingtung Plain, Taiwan. In Proceedings of the 2021 IEEE 3rd International Conference on Architecture, Construction, Environment and Hydraulics (ICACEH), Miaoli County, Taiwan, 24–26 December 2021; IEEE Xplore: New York, NY, USA, 2021.
38. Deleuze, G.; Guattari, F. *Capitalisme et Schizophrenie 2. Mille Plateaux*; Shanghai Bookstore Press: Shanghai, China, 2010.
39. Sun, J.-Q. *Xiamen Jimei District Chronicle*; Zhonghua Bookstore: Beijing, China, 2013.
40. *China Geography Encyclopedia Series Editorial Committee China Encyclopedia Geography-Xiamen Peninsula 2016*; World Book Publishing Company, Nanfang Daily Publishing House: Guangzhou, China, 2016.
41. Wang, C.; Guo, X.-F.; Fang, J.; Li, Q.-S. Characteristics of seasonal spatial expansion of Fujian and Zhejiang Coastal Current and their bay effects. *J. Appl. Oceanogr.* **2018**, *37*, 1–8.
42. Xiamen Jimei District Culture and Tourism Bureau. *Jimei Humanistic Customs: Southern Fujian Cultural Style*; Xiamen University Publishing House: Xiamen, China, 2019; p. 312.
43. Chief Editor of Jimei School Secretariat. *Jimei Weekly*; National Library Document Microform Reproduction Center: Xiamen, China, 2006.
44. Compiled by Xiamen Jimei District Chronicles and Local Chronicles Compilation Committee. *Local Chronicles of the People's Republic of China, Fujian Chronicles, Xiamen Jimei District Chronicles*; Zhonghua Book Company: Beijing, China, 2013; p. 867.
45. Compiled by Xiamen Local Compilation Committee. *Xiamen City Annals Volume 3*; Fangzhi Publishing House: Beijing, China, 2004; p. 2353.
46. Mengsou, Z. *Brief History of Jimei*; Unpublished: Xiamen, China, 2002.

47. Editor-in-Chief Jiang Fuyuan; Editorial Committee of "Xiamen Traffic Chronicle". *Xiamen Traffic Chronicle*; People's Communications Press: Beijing, China, 1989; p. 359.
48. Zhifeng, L. *Survey on Cultural Heritage Resources of Traditional Agricultural, Forestry and Fishery Production Customs in Fujian and Taiwan*; Xiamen University Publishing House: Xiamen, China, 2014; p. 315.
49. Zhuan, M. A Brief History of Jimei. Unpublished: Xiamen, China, 2000.
50. Jiaqing, S. *Records of Jimei District, Xiamen City*; Records of Jimei District: Xiamen, China, 2013.
51. Jimei District of Xiamen City People's Government Physical Geography Xiamen: Jimei District of Xiamen City People's Government. 2023. Available online: http://www.jimei.gov.cn/F566/qqgk/zrdl/ (accessed on 21 December 2023).
52. Jitang, W. *Time Adds Value in Jimei—Old Photos*; Xiamen University: Xiamen, China, 2017; p. 452.
53. Juexiang, C. *Jimei History*; Qiaoguang Printing Company Limited: Yueqing, China, 1963; p. 168.
54. Fusheng, C. *Xiamen Fishermen Customs*; Lujiang Publishing House: Xiamen, China, 2013.
55. Jimei District People's Government Biography of Tan Kah Kee Xiamen: Jimei District People's Government. 2023. Available online: http://www.jimei.gov.cn/F566/F567/rwls/rw/201602/t20160205_313315.htm (accessed on 22 December 2023).
56. Risheng, Z. *Jimei School's 80-Year History*; the Compilation Team of "Jimei School's 80-year History"; Lujiang Publishing House: Xiamen, China, 1993; p. 363.
57. Xiamen Jimei District Culture and Tourism Bureau. *Jimei Humanities and Strong Culture in the Village*; Xiamen University Publishing House: Xiamen, China, 2019; p. 179.
58. Gaoshu, Z. *Jimei*; Central Documentation Publishing House: Beijing, China, 2005; p. 600.
59. Zheng Daxian, T.X. *Research on Ecological Function Zoning in Fujian Province*; China Environmental Science Press: Beijing, China, 2007; p. 332.
60. Shaojian, H. *Changes in Xiamen's Sea Areas: Huge Changes in Our Home*; Xiamen Evening News: Xiamen, China, 2014.
61. Luoping, Z. *Digital Modeling and Environmental Research on the Bay in Fujian Province Xiamen Bay*; Ocean Publishing Company: Beijing, China, 2009; p. 215.
62. Islam, K.S. Looking at the Prospects of ICZM Implementation in Bangladesh from the Successful ICZM Model of Developing Countries (Xiamen Model, China). Master's Thesis, Xiamen University, Xiamen, China, 2009.
63. Mingding, Z. A Study on the Population Ecology of the Chinese White Dolphin in Xiamen Bay. Master's Thesis, The Third Ocean Research Institute of the Ministry of Natural Resources, Xiamen, China, 2021.
64. Weiqi, C.; Yan, L.; Huasheng, H.; Xiaofeng, H. Evaluation of tourism and entertainment value on the east coast of Xiamen Island. *J. Xiamen Univ. Nat. Sci. Ed.* **2001**, *4*, 914–921.
65. Fujian Provincial Ocean and Fisheries Bureau. Xiamen: Vigorously Crack Down on Illegal Fishing and Seize 27 Illegal Boats. 2019. Available online: http://hyyyj.fujian.gov.cn/xxgk/hydt/jcdt/201911/t20191115_5098410.htm (accessed on 28 May 2023).
66. Mengshi, H. Research on the Regulation of Urban Construction Land in Harbin from the Perspective of Stock Planning. Ph.D. Thesis, Harbin Institute of Technology, Harbin, China, 2020.
67. Qinhua, F. Research on Coastal Zone Strategic Environmental Assessment Based on Ecosystem Management Theory. Master's Thesis, Xiamen University, Xiamen, China, 2006.
68. Wang, H.; Zou, Z.; Tsai, S.-C. Exploring Environmental Restoration and Psychological Healing from Perspective of Resilience: A Case Study of Xinglin Bay Landscape Belt in Xiamen, China. *Int. J. Environ. Sustain. Prot.* **2022**, *2*, 44–54. [CrossRef]
69. Huasheng, H.X.X. *A Review of Ten Years of Comprehensive Coastal Zone Management in Xiamen*; Xiamen University Publishing House: Xiamen, China, 2006; p. 126.
70. Net, X. This Coastline of Xiamen is Going to Be "Beautiful"! Construction of the Sea Walk/Camping Beach Xiaman: Xiaman House. 2021. Available online: http://news.xmhouse.com/bd/202106/t20210609_724576.htm (accessed on 28 February 2024).
71. Headline, T. Jimei Romantic Coastline 2022. Available online: https://www.toutiao.com/article/7061823966238966272/?source=seo_tt_juhe (accessed on 8 January 2024).
72. Television XRa. 3.1km! Jimei Maritime Ecological Landscape Corridor is about to open!*Sina* 23 May 2023.
73. People Net. Let Traditional Folk Culture Help Light Up the Night Economy Xiamen's Thousand-Year-Old Jimei Temple Is Transformed into a Creative Market: People Net. 2023. Available online: http://fj.people.com.cn/BIG5/n2/2023/0828/c181466-40548421.html (accessed on 21 December 2023).
74. Bai, Q.; Chen, J.; Chen, Z.; Dong, G.; Dong, J.; Dong, W.; Fu, V.W.K.; Han, Y.; Lu, G.; China Coastal Waterbird Census Group; et al. Identification of coastal wetlands of international importance for waterbirds: A review of China Coastal Waterbird Surveys 2005–2013. *Avian Res.* **2015**, *6*, 12. [CrossRef]
75. Kan, Z.; Chen, B.; Yu, W.; Chen, G.; Ma, Z.; Hu, W.; Liao, J.; Du, H. Forecasting land-cover change effects on waterbirds in Xiamen Bay, China: Determining prospective species winners and losers. *Mar. Environ. Res.* **2023**, *188*, 106003. [CrossRef] [PubMed]
76. You, T.-F.; Chen, X.-Y.; Lin, J.-X.; Ye, Q.-T. Investigation on nekton resources of spring in the west sea areas of Xiamen. *J. Fish. Res.* **2016**, *38*, 386–393.
77. Ma, C.; Liu, Y.; Zhuang, Z.-D.; Xu, C.-Y.; Shen, C.-C.; Tsai, J.-D. Analysis on the resource status and change reason of Branchiostoma balcheri in Xiamen Amphioxus Natural Reserve. *J. Fish. Res.* **2022**, *44*, 44–51. [CrossRef]
78. Editor-in-Chief such as Zeng Chengkui and others. *China Ocean Record*; Editorial Committee of "China Ocean Record", Ed.; Elephant Publishing House: Zhengzhou, China, 2003; p. 1320.

79. Bennett, N.J. In Political Seas: Engaging with Political Ecology in the Ocean and Coastal Environment. *Coast. Manag.* **2019**, *47*, 67–8[illegible] [CrossRef]
80. Mokhtar, M.B.; Ghani Aziz, S.A.B.A. Integrated coastal zone management using the ecosystems approach, some perspectives i[illegible] Malaysia. *Ocean. Coast. Manag.* **2003**, *46*, 407–419. [CrossRef]
81. Gamarra, N.; Costa, A.; Ferreira, M.; Diele-Viegas, L.; Santos, A.; Ladle, R.; Malhado, A.; Campos-Silva, J. The contribution o[illegible] fishing to human well-being in Brazilian coastal communities. *Mar. Policy* **2023**, *150*, 105521. [CrossRef]
82. Huang, L.-M.; Wang, J.-Q.; Shih, Y.-J.; Li, J.; Chu, T.-J. Revealing the Effectiveness of Fisheries Policy: A Biological Observation o[illegible] Species Johnius belengerii in Xiamen Bay. *J. Mar. Sci. Eng.* **2022**, *10*, 732. [CrossRef]
83. Yongxun, W.; Yafei, W.; Jingwen, Z.; Qiang, W. Coastal zone use transformation and its ecological and environmenta[illegible] effects—Taking Fujian coastal zone as an example. *J. Environ. Sci.* **2021**, *41*, 3927–3937. [CrossRef]
84. Fujian Provincial Bureau of Statistics. GDP of Xiamen City 2022. Available online: https://gdp.gotohui.com/show-18172[illegible] (accessed on 19 February 2024).
85. Shanshan, Y.; Liming, L. Xiamen's GDP Growth rate LAST Year Ranked First among 15 Similar Cities. 2023. [CrossRef]
86. Rongchang, L. Research on Regional Tourism Industry Agglomeration Measurement and Economic Effects. Master's Thesis Xiamen University, Xiamen, China, 2020.
87. Jimei District Statistics Bureau. *Jimei District 2022 National Economic and Social Development Statistical Bulletin*; Jimei Distric[illegible] Statistics Bureau: Xiamen, China, 2023.
88. Xiamen Jimei District Statistics Bureau. *Statistical Yearbook of Jimei District, Xiamen City*; Liren Color Printing Company Limited o[illegible] Fuzhou: Fuzhou, China, 2022.
89. Xiamen Jimei District Statistics Bureau. *Statistical Yearbook of Jimei District, Xiamen City*; Liren Color Printing Company Limited of Fuzhou: Fuzhou, China, 2023.
90. Xiamen Jimei District Statistics Bureau; Xiamen Jimei District Development and Reform Bureau. *Statistical Yearbook of Jime[illegible] District, Xiamen City*; Huisheng Printing Company Limited of Zhangzhou City: Zhangzhou, China, 2013.
91. Xiamen Jimei District Statistics Bureau. *Statistical Yearbook of Jimei District, Xiamen City*; Huisheng Printing Company Limited of Zhangzhou City: Zhangzhou, China, 2019.
92. Xiamen Jimei District Statistics Bureau. *Statistical Yearbook of Jimei District, Xiamen City*; Huisheng Printing Company Limited of Zhangzhou City: Zhangzhou, China, 2020.
93. Xiamen Jimei District Statistics Bureau. *Statistical Yearbook of Jimei District, Xiamen City*; Huisheng Printing Company Limited of Zhangzhou City: Zhangzhou, China, 2021.
94. Xiamen Jimei District Statistics Bureau; Xiamen Jimei District Development and Reform Bureau. *Statistical Yearbook of Jimei District, Xiamen City*; Huisheng Printing Company Limited of Zhangzhou City: Zhangzhou, China, 2014.
95. Xiamen Jimei District Statistics Bureau; Xiamen Jimei District Development and Reform Bureau. *Statistical Yearbook of Jimei District, Xiamen City*; Huisheng Printing Company Limited of Zhangzhou City: Zhangzhou, China, 2015.
96. Xiamen Jimei District Statistics Bureau; Xiamen Jimei District Development and Reform Bureau. *Statistical Yearbook of Jimei District, Xiamen City*; Huisheng Printing Company Limited of Zhangzhou City: Zhangzhou, China, 2017.
97. Xiamen Jimei District Statistics Bureau; Xiamen Jimei District Development and Reform Bureau. *Statistical Yearbook of Jimei District, Xiamen City*; Huisheng Printing Company Limited of Zhangzhou City: Zhangzhou, China, 2018.
98. Xiamen Jimei District Statistics Bureau; Xiamen Jimei District Development and Reform Bureau. *Statistical Yearbook of Jimei District, Xiamen City*; Huisheng Printing Company Limited of Zhangzhou City: Zhangzhou, China, 2016.
99. Wenjing, S.; Tong, X.C.X.; Wanhong, C.; Buren, H.; Zifeng, Z. *History of the Development of Fujian Merchants, Xiamen Volume*; Xiamen University Publishing House: Xiamen, China, 2016; p. 345.
100. Compiled by Xiamen Jimei District Caring for the Next Generation Working Committee. *Jimei's Yesterday, Today and Tomorrow*; Compiled by Xiamen Jimei District Caring for the Next Generation Working Committee: Xiamen, China, 2009.
101. Chinese People's Political Consultative Conference. *Jimei Literature and History Materials Volume 18*; Chinese People's Political Consultative Conference: Xiamen, China, 2015; p. 193.
102. Xiamen Municipal Party Committee Party History Research Office; Jimei School Committee; Alumni Association Editor of Jimei. *Tan Kah Kee and Jimei School Village*; Printing Factory of Jimei: Xiamen, China, 1994; p. 29.
103. Su, H. From Chaos to Governance, from Governance to Beauty, and from Beauty to Diversion. Why Is the "Shi Li Chang Di" in Jimei, Xiamen, so Prominent?: People Net. 2023. Available online: http://fj.people.com.cn/n2/2023/0705/c181466-40482227.html (accessed on 18 January 2024).
104. Editorial Board of "Fujian Fishery History" FFS (Ed.) *Fujian Fishery History*; Fujian Science and Technology Press: Fuzhou, China, 1988; p. 466.
105. Maoyi, L.C.S.; Guoliang, C. *Jimei School Village*; Cultural Relics Publishing House: Beijing, China, 1984; p. 30.
106. Xu, L.; Ding, S.; Nitivattananon, V.; Tang, J. Long-Term Dynamic of Land Reclamation and Its Impact on Coastal Flooding: A Case Study in Xiamen, China. *Land* **2021**, *10*, 866. [CrossRef]
107. Cai, R.; Liu, K.; Tan, H.; Yan, X. Climate change and China's coastal zones and seas: Impacts, risks, and adaptation. *Chin. J. Popul. Resour. Environ.* **2021**, *19*, 304–310. [CrossRef]
108. Crang, P. Displacement, Consumption, and Identity. *Environ. Plan. A Econ. Space* **1996**, *28*, 47–67. [CrossRef]
109. Murdoch, J. *Post-Structuralist Geography-Aguide to Lelational Space*; SAGE Publications Ltd.: London, UK, 2006.

110. Ramesh, R.; Chen, Z.; Cummins, V.; Day, J.; D'elia, C.; Dennison, B.; Forbes, D.; Glaeser, B.; Glaser, M.; Glavovic, B.; et al. Land–Ocean Interactions in the Coastal Zone: Past, present & future. *Anthropocene* **2015**, *12*, 85–98. [CrossRef]
111. Magalhaes Filho, L.; Roebeling, P.; Villasante, S.; Bastos, M.I. Ecosystem services values and changes across the Atlantic coastal zone: Considerations and implications. *Mar. Policy* **2022**, *145*, 105265. [CrossRef]
112. Hein, L.; van Koppen, K.; de Groot, R.S.; van Ierland, E.C. Spatial scales, stakeholders and the valuation of ecosystem services. *Ecol. Econ.* **2006**, *57*, 209–228. [CrossRef]
113. Fu, B.; Zhang, L. Land-use change and ecosystem services: Concepts, methods and progress. *Prog. Geogr.* **2014**, *33*, 441–446. [CrossRef]
114. Turner, R.K.; Lorenzoni, I.; Beaumont, N.; Bateman, I.J.; Langford, I.H.; McDonald, A.L. Coastal management for sustainable development: Analysing environmental and socio-economic changes on the UK coast. *Geogr. J.* **1998**, *164*, 269–281. [CrossRef]
115. Söderqvist, T.; Eggert, H.; Olsson, B.; Soutukorva, Å. Economic valuation for sustainable development in the Swedish coastal zone. *AMBIO J. Hum. Environ.* **2005**, *34*, 169–175. [CrossRef]
116. Uehara, T.; Mineo, K. Regional sustainability assessment framework for integrated coastal zone management: Satoumi, ecosystem services approach, and inclusive wealth. *Ecol. Indic.* **2017**, *73*, 716–725. [CrossRef]
117. Watanabe, T. *Create the Caspian Sea with the Participation of Generations*; FujiGreenLetter: Japan, 2020; pp. 20–22.
118. Wilson, J.A.; Acheson, J.M.; Johnson, T.R. The cost of useful knowledge and collective action in three fisheries. *Ecol. Econ.* **2013**, *96*, 165–172. [CrossRef]
119. Milanes, C.B.; Montero, O.P.; Cabrera, J.A.; Cuker, B. Recommendations for coastal planning and beach management in Caribbean insular states during and after the COVID-19 pandemic. *Ocean. Coast. Manag.* **2021**, *208*, 105575. [CrossRef] [PubMed]
120. Tsai, S.-C.; Wang, H.; Lee, S.-H.; Zou, Z. Cognition and Interaction: From the Perspective of Daily Therapeutic Landscape of the Coastal Zone. *Behav. Sci.* **2023**, *13*, 794. [CrossRef]
121. Du Gay, P. *Production of Culture/Cultures of Production*; Sage: London, UK, 1997.
122. Ma, Y.; Zheng, W. Research on the Development of East China Sea Coastal Tourism Complex in Quanzhou Central City. *J. Mudanjiang Univ.* **2013**, *22*, 106–108. [CrossRef]
123. Chan, V.; Yan, L.; Huasheng, H.; Xiaofeng, H. Evaluation of Tourism and Entertainment Value on the East Coast of Xiamen Island. *J. Xiamen Univ. Nat. Sci. Ed.* **2001**, *4*, 914–921.
124. Wang, D.; Zhu, L. "Being a fisher girl for a day": How to construct collective imagination by short video punching—A field trip to a topic in online celebrity. *Media Obs.* **2024**, *2*, 45–54. [CrossRef]
125. Tuan, Y.-F. *Topophilia: A Study of Environmental Perception, Attitudes, and Values*; The Commercial Press: Beijing, China, 2018.
126. Harvey, D. *Paris, Capital of Modernity*; Socio Publishing: Taipei, Taiwan, 2007.

water

MDPI

Article

The Spatial and Governance Dilemma of Small and Medium-Sized Italian Ports (SMPs): Maritime Spatial Planning (MSP) as a Potential Response

Fabio Carella [1,*], Paolo De Martino [1,2], Folco Soffietti [1], Vittore Negretto [1] and Francesco Musco [1]

[1] Department of Architecture and Arts, Università Iuav di Venezia, 30135 Venezia, Italy; pdemartino@iuav.it (P.D.M.); fsoffietti@iuav.it (F.S.); vnegretto@iuav.it (V.N.)
[2] Department of Architecture, Delft University of Technology, 2628 CD Delft, The Netherlands
* Correspondence: fcarella@iuav.it

Abstract: The Italian coast has about 700 ports, which are different in typology, dimension, role, and ownership. Historically, this has led to the significant fragmentation of governance and space and a lack of cooperation that ports and cities still experience today. Among all ports, small and medium-sized ports (SMPs), such as marinas, small touristic harbors, and moorings, are the most affected. Unlike the main ports, where spatial and strategic regulation planning fall under the port authority's responsibilities, SMPs are a combination of public and private management and are, therefore, excluded from national and regional planning and larger strategies. Improving SMPs' cooperation at the regional level can drive more effective sustainable management among related activities (tourism and the fishing sector) and reduce pressures on the land–sea interaction (LSI). In filling the gaps, this article challenges the existing legal framework, planning tools, approaches, and initiatives and may pave the way to establishing a better-integrated national governance for SMPs. In conclusion, this paper identifies two main opportunities that can support the steady establishment of governance and the systematic harmonized development of these SMPs. The first one is offered by maritime spatial planning (MSP) as a strategic and legal tool whereby SMPs are recognized and, if financially supported, could find incentives and measures for their development. The second one is through European projects, programs, and initiatives such as Framesport as drivers in establishing a common ground among public and private interests and as a cooperation engine at a local scale.

Keywords: maritime spatial planning; governance; small and medium-sized ports; land–sea interaction; common-ground strategy

1. The Premise: Small Ports' Challenges

Worldwide, ports are under pressure due to challenges related to infrastructure, global trade, production capacity, financing, regulation compliance, safety and security, sustainability, digitalization, and a lack of community support [1–4]. Ports are fascinating transitional landscapes, gateways through which people and cultures move [5–7]. The largest part of the world economy travels through ports, making them key players in the global production and supply chains that we today depend on. They connect people and provide opportunities for them to thrive [8]. On the one hand, a port, by definition, is a contact area with liminal space that provides a safe environment for ships. On the other, it is a space of conflict, especially when it comes to designing the relationship between coastal and sea uses [9–11]. They are threshold spaces where solid and liquid mix, where large-scale economies meet urban palimpsest, and where the city takes on a landscape connotation [12,13]. However, ports are also controversial by nature, as they are the epicenters of contemporary challenges, from extreme weather conditions to demographic and social pressures [5].

These global challenges do not only concern large ports. SMPs are also affected by global events. Massive tourism is putting pressure on ports and cities, triggering new strategies on both local and regional scales that aim to generate new services and attractions for people and companies [14].

These maritime challenges, however, can be used as opportunities as long as the ports and communities around them remain open to change, stemming from the green and digital transformation taking place in the maritime and shipping industry. In large ports, it seems difficult to achieve a sustainable transition due to the strong path dependencies between energy and global economies [15–17]. Meanwhile, SMPs are interesting laboratories that promote economic, social, cultural, and environmental innovations with the potential to lead a sustainable transition [18,19].

All scales are interrelated, and strategies at the regional and local levels must also take into consideration the broader macro-objectives and strategies established by the European Union, such as the European Green Deal, the Circular Economy, and the recent Repower EU plan, to rapidly reduce the impacts on the coast and in the marine environment [20–22].

As pointed out by several scholars, while larger ports can profit from a plurality of initiatives related to sustainability, smaller ports' systematic sustainability management is still lacking or very rare [4,18]. Looking at the economic dimension of ports, the current fragmented situation linked to the volatile economic demands can offer new opportunities for small and medium-sized ports [18]. Although these are often very responsive in dealing with supply chain dynamics and related logistics systems, SMPs' lack of governance makes their responsiveness not as effective [18].

Therefore, this contribution argues that while major ports are hard to change due to their dependencies on global-scale phenomena, small ports can be catalysts for more sustainable development at local-scale interventions, but there is a need to establish a new governance model [19].

Among the various critical issues affecting ports, the governance modality is particularly relevant in small Italian ports. Unlike the main ports, there is a lower capacity for integrated management and a lack of coordination. Moreover, these issues reflect the territorial planning dimension, especially in the urban regulatory plans and policies ranging between different administrative levels [19]. The port authorities are responsible for the main ports' regulation and spatial and strategic planning and are consequently also in charge of Maritime State Property, the drafting of port plans, and the implementation of inter-port coordination strategies. On the other hand, minor ports have very different characteristics [23]. A mixture of public and private management often coexists in adjacent spaces, generating a set of conflicts and criticalities for which management responsibility is not clear. Furthermore, the definition of a strategic coordination framework can also interfere with local governance issues typical of many small ports. While major ports are on public property and entrusted to public actors' management (or in consortia with the public), the smaller realities, although also on public property, are entrusted (or rather granted) to private actors. The substantial difference between these management models is that the first seeks growth and collective economic stability, while the second, instead, is based on the principle of entrepreneurship and business and is, therefore, oriented to the maximum profit achievement at the expense—often—of the collective interest [11,18,19].

In framing the meaning of SMPs, this article investigates the following questions: where and how are SMPs considered or included in the legal framework at the European and national levels? What role could minor ports play in response to spatial and institutional fragmentation? Are there any tools or approaches that may facilitate the establishment of a common governance practice?

The thesis that this article outlines, and begins to explore, lies in investigating the opportunities offered by MSP (maritime spatial planning) and European projects as possible integrative and multi-disciplinary tools for designing land–sea interactions in SMPs and promoting cooperation.

2. The Regulatory Framework: Gaps and Opportunities

2.1. European Level

The European Commission regulation, specifically the one establishing a framework on market access to port services and the financial transparency of ports, shows a clear effort directed mainly toward the Trans-European Transport Network (TEN-T) ports. These ports are "by definition essential for the international and intra-European trade exchanges, and therefore for the European internal market, and/or the cohesion within the EU" [24].

This sectoral approach responds efficiently to economic needs and aspects related to the movement of goods and people on a large scale. Yet, it completely ignores the heterogeneity of the European port territory, which is made up of approximately 70,000 km of coasts characterized by a wide diversity in the type and organization of ports. However, this richness and diversity of the European port city territories do not seem to be recognized as an added value by the EU, which, therefore, does not impose a defined regulatory framework on SMPs since they "do not have a significant role for the European transport system" [24].

Addressing the size of a port is always a bit of a problematic question. The European Port Governance report [14] already raised some epistemological questions, such as, "is the size determined by the surface of the port area, the volume of goods handled, the number of passengers that pass through the port, the financial turnover, the staff employed or a combination of these and other factors?" Specifically, the European Port Governance report frames SMPs "as the ones with an annual volume of goods handled in all the ports managed by the port authority as less than or equals 10 million tonnes" [14].

The fact that there is no common definition of what SMPs mean is highlighted by several studies, including a recent article by Gerlitz that questions the role of medium-sized ports as drivers of regional innovation and development in Research and Innovation Strategies for Smart Specialisation (RIS3) in the Baltic Sea Area [18].

In line with the European 2020 strategy and in response to the gaps in policies and definitions, RIS3 is configured as a regional-scale strategy for SMPs' regional development and as an advancement of the policy. It represents a tool that aims at improving the efficiency of funds' distribution between European regions and contributing to the European objectives on smart growth, the UN's sustainable development goals, and the recently launched European Green Deal [20]. RIS3 recognizes that small (and medium-sized) ports can play a leading role in identifying functions such as (i) enhancing blue economy competitiveness, (ii) contributing to regionalization processes, and/or (iii) facilitating the setup of multiport gateways. However, based on the analysis of 37 regions, Meyer's article highlights a very low recognition of SMPs as drivers of regional innovation in the Baltic Sea Area under the RIS3 policy, and, thus, it remains a rather sectoral reading of small ports. It provides a very technical and functional approach to SMPs, instead of addressing challenging topics like sustainability, smart specialization, and blue growth as driving forces for socio-cultural development [18].

Ports are, in fact, functional machines that respond, among other objectives, to creating network connections. This is clearly pointed out by the European regulations on mobility and the Trans-European Transport Network [25]. According to this regulation, the Trans-European Transport Network should "allow the seamless, safe and sustainable mobility of persons and goods, ensuring accessibility and connectivity for all regions of the Union, and contributing to further economic growth and competitiveness in a global perspective. Those specific objectives should be achieved by establishing interconnections and interoperability between national transport networks in a resource-efficient and sustainable way. For example, rail interoperability could be enhanced by innovative solutions". It continues, "the core network has been identified on the basis of an objective planning methodology. That methodology has identified the most important urban nodes, ports and airports, as well as border crossing points. Wherever possible, those nodes are connected with multimodal links as long as they are economically viable, environmentally sustainable and feasible until 2030. The methodology has ensured the interconnection of all Member States and the integration of the main islands into the core network". This regulation mainly focuses on large connections and major ports while ignoring "small" or "minor" ports, two words that are missing from the regulation [25].

Within the Sustainable and Smart Mobility Strategy document [26], mobility is recognized as one of the most relevant sectors at the European level, whether it is mobility related to tourism or the handling of goods and industrial production. Free movement within European borders has offered many opportunities for economic but also social and cultural growth, but at the expense of the environment and the loss of biodiversity. Putting mobility on track for the future, therefore, means working on reducing these impacts in the short and medium term (2030/2035) and on achieving zero-emission goals in the long run (2050) [26].

As far as ports are concerned in "creating zero-emission airports and ports", the document underlines that "inland and sea ports have a great potential to become new clean energy hubs for integrated electricity systems, hydrogen and other low-carbon fuels, and testbeds for waste reuse and the circular economy". In addition, coastal areas and ports should be a priority "in all EU waters ultimately aiming at zero pollution to air and water from shipping for the benefits of sea basins" [26]. However, the document remains vague in defining concrete development actions, and once again, it ignores the theme of SMPs and the role that they could play as engines of sustainable and more resilient spatial and socio-cultural developments.

Many steps are still to be made in achieving climate neutrality in the short and long run. For this reason, in 2019, the European Commission released the European Green Deal, committing to climate neutrality by 2050 [20]. This requires a radical transformation that involves some key themes, including making transport sustainable for all, leading the third industrial revolution, cleaning our energy system, renovating buildings for greener lifestyles, and ultimately working with nature to protect our planet and health [20].

Among the different pillars, two are related to water: "environment and ocean" and "blue economy" [27]. In fact, to fully embed the blue economy into the Green Deal and the recovery strategy, the Commission has adopted a new approach for a sustainable blue economy in the EU, stating it can contribute to climate change mitigation by developing offshore renewable energy, decarbonizing maritime transport, and greening ports. It will make the economy more circular by renewing the standards for fishing gear design, ship recycling, and the decommissioning of offshore platforms. In addition, developing green infrastructure in coastal areas will help preserve biodiversity and landscapes while benefiting tourism and the coastal economy. Even though all of these issues are connected to each other, and despite ports being a transversal theme, the Green Deal mainly considers large ports, and "to ensure a fair contribution from the maritime sector to the effort to decarbonize our economy, the Commission proposes to extend carbon pricing to this sector" [20].

In the last 10 years, given the above-mentioned challenges, there has been a strong pressure on marine planning to try to coordinate a quite complex set of activities at sea while also preserving the landscape, ecosystem, and cultural heritage [28–30]. As such, in 2014, the European Commission implemented legislation on maritime spatial planning (MSP) with the main objective of reducing conflicts among sea uses; creating synergies between different activities; increasing cross-border cooperation between EU countries to develop shared plans on renewable energy, shipping routes, and the protection of the environment by assigning protected areas; calculating impacts on ecosystems; and identifying opportunities for multiple space uses [31].

MSP proposes a significant step forward by looking at the sea not as an empty space but as a possible extension of the city onto the water. MSP is a relatively new planning approach aimed at analyzing and organizing human activities in the sea space to achieve ecological, economic, and social objectives [28–31]. The European Directive 2014/89/EU has also made MSP mandatory in the planning policies of all coastal Member States. The Directive requires the EU Member States to have developed a national maritime spatial plan by 31 March 2021, with a minimum review period of 10 years. The plans are aimed at establishing a reference framework for the planning of maritime spaces (contained within 12 nautical miles from the coast) in order to promote maritime economies' sustainable growth, marine areas' sustainable development, and the sustainable use of marine resources through an ecosystemic approach [31].

In Italy, the MSP Directive was implemented via the Italian Legislative Decree 17 October 2016, n. 201 [32], together with the recommendations adopted by the Decree of the President of the Council of Ministers of 1 December 2017 [33]. However, the Italian maritime plans have not yet been adopted, and they are still in the public consultation phase [11].

Sustainability aspects are central in the European agenda, and this is why, in 2020, the European Commission launched the New European Bauhaus to also better connect the European Green Deal to the lives of people. This calls on all Europeans to imagine collectively building a sustainable and inclusive future by focusing on three main aspects: sustainability, requiring the identification of solutions toward circularity to better face climate change; aesthetics, requiring the identification of solutions that are beautiful to our eyes beyond pure technicality; and inclusion, from valuing diversity to securing accessibility and affordability. In response, Portugal proposed the Bauhaus of the Seas Sails (BOSS) as a continental mobilization around the first and most decisive global natural space: the sea. The manifesto, among its various objectives, focuses on "reconciling with the sea as a territory of trans-geographic continuity through site-specific ecosystems and entanglements of humans and non-humans. This reconciliation opens new possibilities to the strategic needs of the New European Bauhaus" [34].

Even in this case, aspects related to ports and the role that SMPs can play in creating a more integrated and sustainable future are missing. This gap is also considered as an opportunity to further explore.

So far, previous analyses have provided an overview of European policy documents dealing with water and sustainability. Although, on the one hand, they have highlighted enormous opportunities in recognizing the value of the sea and its economies, on the other, they have encountered a significant gap in acknowledging SMPs and their value as drivers of sustainable and resilient development in any of the initiatives mentioned above.

Today, rethinking the coastline means rethinking the future of a port (large or small) in relation to its surroundings. The theme of water is mainly addressed as a space for the movement of goods and people, but it would be interesting to understand its socio-cultural values and how to imagine new forms of livability with/on water, starting from the contributions that SMPs can make.

2.2. National Level

The Italian coast has about 700 ports, which are different in typology, dimension, role, and ownership, making the relationship between small and large ports and territories a controversial one throughout history (Figure 1) [19]. As pointed out by several scholars, the reasons behind the conflict are diverse.

Firstly, there has been an absence of processes for the decentralization of port activities that are no longer compatible with urban dynamics. As a result, except for very few cases, such as Genoa, in many Italian cities, functional ports are within cities and very close to residential areas. Secondly, conflicts also result from the presence of several different non-aligned planning tools (port plans and municipal plans, as well as national and regional planning). For instance, the lack of dialogue among authorities has historically resulted in separate regulations, disputes on competencies, and inefficient plans for the land–sea interaction. Finally, friction can be identified in the presence of different temporalities of ports' and cities' transformation. This refers to the different economic models that ports and cities are subject to, as well as the governance period of port city authorities [12,16,35].

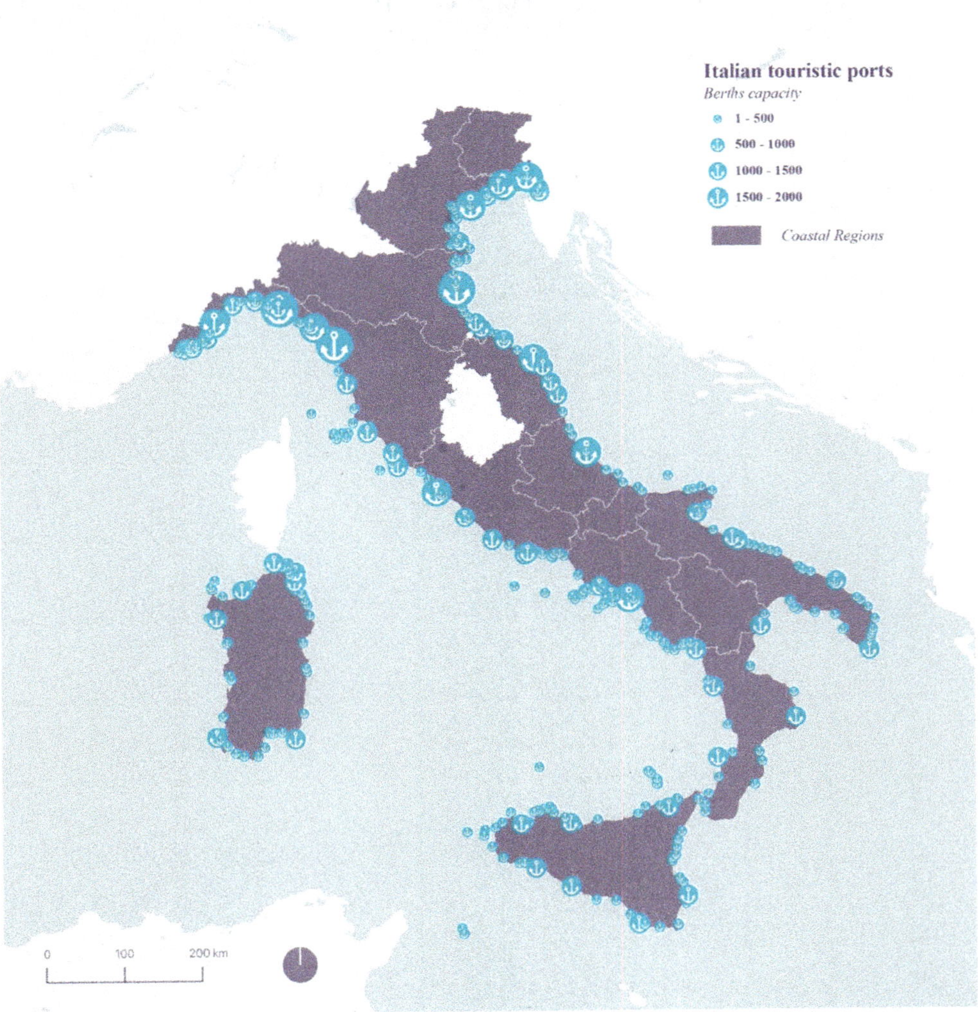

Figure 1. Italian touristic ports. Maps developed by the authors. Data source: SID il portale del mare. URL: https://www.mit.gov.it/documentazione/sid-il-portale-del-mare (accessed on 17 November 2023).

The first Italian port regulation that introduced the governance of port authorities dates back to 1994 with the law n.84 [36]. Prior to this, there was no real legislative discipline except for a pure classification of ports (Royal Decree 2 April 1885, n. 3095) [37]. However, this mainly had to do with economic aspects between the local, national, or regional bodies necessary to finance port development throughout history [16]. The Royal Decree, which was introduced in a period when Italian coast defense military activities played a significant role, divided the port system into two main categories: ports for navigation defensive purposes and ports with commercial functions. The second category was then divided into an additional four classes based on their importance in the national context. The first one included ports heading important lines of communication, dealing with international trade and classified as places of general interest to the state, which financed 80% of these, while the remaining 20% were financed by municipalities. The second class included ports that affected general navigation and safety, serving only or mostly as refuges and military protection. Naples, Genova, and Venice, for example, belonged to this category, and their administration and management were paid for by the state. The third class included ports and docks with trade at a provincial level (50% financed by the state and 50% financed by municipalities). Finally, the fourth class included ports focused on local trade, financed only by municipalities. The classification of ports (at least for the first three classes) was dictated exclusively by quantitative parameters, i.e., by the number of goods handled. The fourth class was for "all the other ports", such as "inlets, gulfs and beaches both on the continent and on the islands". This classification of ports has prevailed for a long time, since it also affects the distribution of expenditures for the construction and maintenance of infrastructures, the drafting of port regulatory plans, and so on [37].

The Royal Decree was the first important attempt to classify ports, introducing a hierarchy among Italian ports that still exists today.

An evolution of the discipline arrived only in 1994. The law introduced a qualitative parameter for the ports belonging to the second category, which were classified based on their vocation: commercial, industrial and petroleum, passenger service, fishing boat, or tourist and recreational. In addition, in this classification, minor ports were considered, for the first time, as basins or small docks, yet not relevant on a national scale [36]. Instead, it reduced minor ports to pure terminological aspects, not even capable of fully grasping the different territorial complexities and the changes that ports on a global scale were facing.

Since the 1960s, with the arrival of containers, ports have become nodes of a global network increasingly connecting the world but quite often detached from local territories and cultures [38–40]. At that time, the concept of the port-city-territory started to appear, which ideally meant that large and minor ports would work in synergy and be governed under the same institutional umbrella. This did not really happen, and SMPs have consequently been left out of national and regional planning.

The recent port Legislative Decree 4 August 2016 n. 169 introduced port systems, merging multiple port authorities into a 16-port-authority system. This process recognized a profound change in the relationship between ports and larger regions and shifted the attention to the main ports, delegating the management of minor ports to sectoral disciplines, local organizations, and systems of concessions [23].

The Plan of Port Systems, which built upon the "Guidelines for the drafting of Port Regulatory Plans" introduced in 2004 by the Superior Council of Public Works, became a technical–functional tool in the resolution of conflicts between ports and cities. According to the guidelines, the port plan was tasked with strategically defining the port's different functional areas. The plan's strategy is divided into two main areas: the first pertains to the "operational" port related to the economy and efficiency of port activities, and the second pertains to the port–city interaction areas, where port flows and urban spaces usually coexist. These land–sea interaction areas are indeed suitable places to experiment with projects that can influence strategic plans, like reconciling the development needs of the ports with the objectives of urban and environmental quality [41].

The Legislative Decree n. 232/2017 introduced a strategic planning document aimed at defining the development objectives and the systemic planning contents of the port authorities and identifying and outlining the areas intended for strictly port and retro-port functions, the port–city interaction areas, the last-mile road and rail infrastructural connections with the individual ports of the system, and the crossing axis between ports and cities [42].

Except for a few examples, there are not many plans in Italy that have put these indications into practice, and there is, therefore, still no real systemic planning that connects ports (major and small–medium-sized) to each other and to the larger territory. It is necessary to innovate the discipline and the planning framework by questioning the current models and investigating whether minor ports can play a role in generating new maritime mindsets [38,43,44].

3. Small and Medium-Sized Ports (SMPs) "within" the Italian MSP Plan

This section of the article briefly analyzes the Italian maritime spatial draft plan with reference to the methodological structure and the contents of the fourth phase of the plan entitled "Planning: vocations, specific objectives, specific measures (by areas and sectors)". It focuses on how SMPs are addressed and highlights cooperation and development opportunities, arguing that these could trigger new cognitive processes, particularly with the vocations of coastal territories.

Recently, and especially in the last 20 years, there has been increasing concern in marine planning for better coordinating a quite complex set of activities at sea that also have to preserve the landscape, ecosystem, and cultural heritage [11,31,44]. Today, thanks to maritime spatial planning (MSP), this complexity, which historically has been conceived and planned as a dividing line between land and sea, is becoming a multifunctional place, a transition space, and, consequently, an area in which to integrate land and water planning. Although the challenges that water cities are facing today are evident, and although flows of people and goods move across territories globally, we can argue that land and water are still planned in a disconnected manner. This is due to a land-based approach that has always seen the sea as a blank canvas [11,28,31,44].

Nowadays, Europe's sea is filled with continuous, simultaneous activities, primarily set by well-established sectoral planning systems, such as the maritime transport of commercial goods and related ports, fisheries, and tourism. In addition to these, it is significant to also consider the prospects offered by new technologies, such as offshore wind, submarine cables and pipelines, offshore hydrocarbon exploration, research and cultivation, and aquaculture and marine biotechnology [44,45].

In response to the imbalance between natural and anthropogenic aspects, there emerges maritime spatial planning and, consequently, the blue economy, focused on economic activities based on and actively beneficial to the ocean, promoting sustainable development as its basis [44–47].

To provide a more coherent approach to maritime issues, the Integrated Maritime Policy of the European Union (IMP) was adopted, calling for increased coordination between different policy areas under a comprehensive policy "umbrella". The MSP pillar of the EU's IMP is regulated by the European Directive 2014/89/EU, which made MSP mandatory in all coastal Member States' (MSs') planning policies and requires each identified MS to have developed a national maritime spatial plan by 31 March 2021 [31].

By asserting that MSP plans "need to take into consideration Land-Sea interactions", the MSP Directive paves the way for the possible application of MSP in supporting the strategic development and evolution of SMPs [28].

Over the last ten years, the Italian government has been actively working to develop and shape maritime policy and tools in support of the blue economy and marine conservation. The MSP process in Italy, currently in the final stages (public consultation phase), started in 2016 with the transposition of the Directive 2014/89/EU within the Italian legislation through the Legislative Decree 201/2016 and the guidelines of the DPCM (Decree of the President of the Council of Ministers) of 1 December 2017, officially activating the implementation of the three Italian Maritime Spatial Plans (PSGM), one for each maritime area: Adriatic, Tyrrhenian-Western Mediterranean, and Ionian Central-Mediterranean [31–33].

The maritime spatial plan does not alter the existing planning tools. On the contrary, it is configured as a superordinate planning level and strategic level aimed at ensuring an alignment of the various existing sectoral plans (territorial coordination plans, landscape plans, etc.). For this reason, the maritime spatial plan represents a strategic planning tool, providing guidelines and criteria for new possible directions for the use of the sea.

Due to its juridical nature, maritime spatial plans are conceived as programs capable of directly influencing not only those areas concerning marine waters but also those concerning terrestrial activities that can have effects on marine waters [11].

The Legislative Decree 201/2016 also defined a new model of multilevel governance, constituting new entities and assigning new roles to the various levels of ordinances. In particular, Art. 8 of the Decree designated the Ministry of Infrastructure and Transport (MIT) as the Competent Authority (CA) for MSP, while Art. 6 established an Interministerial Coordination Table (TIC), chaired by a representative of the Presidency of the Council of Ministers (Department for European Policies) and participated in by almost all other ministries. Finally, Art. 7 established the Technical Committee (TC) with the function of preparing, for each maritime area, maritime spatial management plans.

The TC is a multilevel body composed of representatives from ministerial entities: three referents from the MIT; two representatives for each Ministry involved in marine and maritime issues (Environment and Energy Security (MASE), Agriculture, Food Sovereignty and Forestry (MASAF), and Culture (MIC)); and one representative from each of the coastal regions (fifteen in total). The TC is responsible for ensuring the consistent implementation of the MSP guidelines (according to the 2017 DPCM) and the plan processes' coordination, while the Scientific Team (ST) is responsible for the operational and scientific–technical support aspects. The ST consists of 25 researchers from different scientific backgrounds and from three different institutions: CORILA, CNR-ISMAR, and IUAV University of Venice [32,33].

To facilitate harmonization between the three management plans in the respective maritime areas and according to the operational guidelines represented by Legislative Decree 201/2016 and DCPM 01/12/2017, six methodological steps have been defined to allow their implementation:

1. Describe the initial status and current and expected trends by assembling a collection of data for each sector and using them to gain a cognitive framework to support the analysis and planning process.
2. Perform an analysis of the interaction (conflicts and synergies) between uses and impacts on environmental components aimed at defining the relationships among activities and uses and supporting the following phases.
3. Set a vision and strategic objectives for the individual sector based on existing strategies, plans, and standards through the collected information.
4. As a result of the three first phases, create a plan, including vocations, specific objectives, and measures (by areas and sectors). Each sub-area has defined a vision with a 10-year horizon and defined specific objectives based on the strategic ones identified in phase three. Current uses and activities recognized in phase one and their relationships in phase two have allowed the fragmentation of the sub-area into planning units, which are regulated through measures of the activities.
5. Evaluate its effectiveness by implementing a monitoring program in order to achieve the goals defined in phases three and four. It also supports an adaptive approach, which allows the plan to be tailored to different contexts and needs that may change and emerge over time.
6. Perform activities for the consolidation, implementation, and updating of the plan, progressively developed through the activities related to the monitoring system in phase five, during and after the approval process of the plans, to feed their consolidation and support the implementation process.

The SMPs' analysis particularly focuses on steps three and four of the Adriatic MSP plan process, which are the core of the plan and the identification of strategic (SO) and specific objectives and measures, in addition to spatial choices and geometries (zoning). Indeed, the elements (objectives, zoning, and measures) included in these two phases are relevant to promoting, supporting, and developing new governance practices [33].

SMPs are considered in two specific sectors of the MSP plan: implicitly in "Maritime Transport and Ports (MT)" and more directly in "Coastal and Maritime Tourism (T)".

Within the Maritime Transport and Ports strategic objectives, the plan intends to increase the competitiveness of Italian ports and foster the sharing of best practices aimed at energy efficiency and environmental sustainability. The plan directly mentions the roles of port authorities and related major commercial and passenger ports, and more implicitly, it lays the foundation for medium and small ports (SO_MT | 04).

On the other hand, the plan also aims to promote integration and dialogue among existing planning systems, seeking, in particular, to bring to the same table the authorities responsible for strategic port, land, and sea plans to ensure an exchange of information and avoid potential planning conflicts (SO_MT | 05). The establishment of these tables would also facilitate dialogue between public and private actors and consequently strengthen cooperation between major and minor ports [48].

Regarding the "maritime and coastal tourism" sector, the SMP theme related to land–sea interactions becomes a priority for the plan. In fact, strategic objective T | 02 recognizes the persuading role marinas play in favoring actions aimed at transforming these ports into nodes of connection for various types of transport [48].

As a strategy for their sub-area, several Adriatic Italian coastal regions identified specific objectives for recreational boating, including the role of SMPs in diversifying touristic opportunities while ensuring accessibility to waterways and environmental sustainability (e.g., Friuli-Venezia Giulia, Emilia Romagna, and Marche). Moreover, the Veneto Region highlighted the importance of developing slow and experiential tourism on the coastal strip in synergy with the inland area, ending littoral and pleasure boating, encouraging the redevelopment of SMPs, integrating land and sea planning systems, and protecting the landscape characteristics of the coastal system and architectural features of seaside towns [48].

To achieve the objectives, each marine plan includes a set of national measures common to the entire Italian marine space that are valid for all three maritime areas (Adriatic included). For some sub-areas falling within the territorial waters co-planned by coastal regions and ministries, more detailed measures have also been defined.

To help increase the competitiveness of Italian ports and the sharing of "best practices", as reported by the strategic objective MT | 04, the plan proposes a measure to bring the performance and functionality of Italian ports up to the standards required for obtaining the different existing certifications, such as European Clean Ports, Environmental Management System (EMS), PERS (Port Environmental Review System), and Environmental Port Index. Certifications might assume a relevant function for SMPs, both to overcome the fragmentation of public–private management in which many SMPs find themselves and to promote a new governance system by harmonizing and setting a common ground among different SMPs [48].

As for the tourism sector and, in particular, for the strategic objective of fostering coherent planning actions on land and at sea, including for tourism purposes, the plan intends to design and develop recreational boating monitoring activities. Plus, in order to acquire adequate knowledge of traffic flows and define management measures for the sustainable development of the sector, the measure also seeks to consolidate existing initiatives through collaboration between regions and local operators/entities. This measure, if implemented, further reinforces the need to bring SMPs' public and private owners closer together to improve data flows and obtain scientific evidence to regulate and facilitate a more cohesive and harmonized development [48].

4. MSP as a Tool to Renew Small Ports: The Interreg Italy–Croatia Framesport Experience

The need to rethink SMPs was identified within EU programs and addressed via strategic EU projects. These programs addressed ports from a regional perspective that allowed local knowledge and needs to be taken into consideration. Meanwhile, a general approach to developing SMPs focused on fostering their connectivity and greening, as indicated in the European Green Deal. This was the case for a set of EU projects taking place in the Adriatic Sea, namely, ECOMAP—Eco-sustainable management of marine and tourist ports; SUSPORT; and FRAMESPORT. All of these projects were developed through the Interreg program Italy–Croatia, which looked to strengthen collaboration among neighboring countries and promote common transboundary actions [19].

This paper focuses on the Framesport project because it offered the opportunity to test out a strategy definition method for SMPs by supporting their recognition (e.g., through MSP) and identifying them as promoters of sustainable development.

Ports and cities in the Adriatic Sea are separated from a spatial, cultural, and institutional perspective—a separation that has direct implications on the quality of the land–sea interaction. Historically, this has generated a chaotic mixture of spaces that today are characterized by an uncertain planning regime. Fragments of an industrial past, and quite often obsolete infrastructures at the edge of the port and city, are some of the tangible results of this uneven growth between the city and its port. As a consequence of this fragmentation, four main degrees of separation can be identified:

1. Spatial separation between the nautical-tourist routes and inland territories;
2. Environmental separation of ports that need to make better use of existing resources and energy sources;
3. Technological separation of ports that need to identify solutions to be safer and better connected;
4. Fractures between global and local economies that require new competitive strategies.

The activities carried out within the Framesport project dealt with the territorial complexities linked to the small Italian and Croatian ports. It also aimed at defining a methodology for the construction of a solution-and-scenario abacus capable of facing the plurality of economic, social, and environmental challenges that afflict coastal territories today. Adriatic SMPs, such as marinas and touristic harbors, are currently experiencing rather challenging spatial, social, and economic situations due to a surplus of supply compared to demand levels and the increased users' average age. Boosting competitiveness is a priority for these realities in both the Italian and Croatian contexts, which would highly benefit from new business models, measures, and actions aimed at recovering their overall efficiency and attractiveness [19].

This experimentation was carried out by IUAV and CORILA as part of the Framesport project to set the objectives of SMPs' reconceptualization as catalysts of new social, cultural, spatial, and environmental values, which often escape the gaze of regional and national planning. Particular attention was dedicated to the collection of data and the mapping of these realities as a fundamental process for the construction of future scenarios.

Framesport is aimed at supporting homogeneous and integrated improvements in Adriatic small ports' sustainability, competitiveness, and attractiveness through the following measures:

- The delivery of a strategic framework orienting their future development in the long run, also by improving their connection with the neighboring territories and populations, as well as enhancing and diversifying the overall touristic opportunities.
- The realization of an ICT platform as a virtual space available for users and stakeholders, containing results from the implementation of pilot actions and the best practices, suggestions, and proposals for Adriatic SMPs' development and management.
- An increase in competencies for the harmonized planning and management of SMPs, contributing to elevating their role as drivers of the sustainable growth of coastal areas.

This strategy is the result of a complex analysis and consultation work with stakeholders achieved through meetings and semi-structured surveys. It will support decision makers' and port management's decisions in the coming years to ensure that small Adriatic ports are greener, connected, competitive, and safer. The state-of-the-art study informed the definition of the different key aspects composing the strategy. The state of the art is composed of EU directives, regional strategies, and national laws; data on ports describing their current state; an analysis of demand and supply; the results of Framesport pilots; and inputs from local stakeholders via dedicated surveys and workshops [19].

The methodology for the common-strategy definition began with the identification of objectives (already included in the Framesport focus) by considering European directives and strategies and national laws and strategies on tourism and sustainability. The combination of analysis and educated guesses from collected data verified via stakeholder engagement with a set of dedicated events and a questionnaire informed the definition of a vision and strategic actions. Moreover, the results from Framesport tools and the analysis of demand and supply reinforced the selection of actions and aggregation in dedicated areas of interventions linked to the tools' macro-topics.

The main areas of intervention were identified as follows:

1. Governance and planning (land/sea), where the discussion pointed out the fragmentation of tools and a lack of strategic vision.
2. Tourism, where the urgency to diversify routes and better connect the coast to the inland territory, among other topics.
3. Maritime culture and involvement, where different actors discussed the uses related to marinas and leisure boating and how these will change in the future.
4. Landscape and heritage protection and the challenges related to climate change. This condition of great uncertainty due to climate extremes highlighted the need to identify clearer actions capable of protecting heritage from changing water conditions.
5. A green transition. This needs to take place to allow SMPs to grow sustainably within the territory. SMPs are also facing problems due to obsolete infrastructures. They are often insufficient and, therefore, need to build a strategic and interconnected vision to reconnect ports and regional territories.
6. Climate change and risk management. These are also important topics that need to be addressed by the strategy by, for example, introducing monitoring systems that deal with changing risk mitigation.
7. Coastal and water management, as well as ecosystems and coasts' physical assets, to ensure high-quality standards, including in connection with river routes and land.
8. Maritime transport: the need for a structural rethinking of land–sea interactions. Today, the line between ports and cities quite often appears as a fragmented and chaotic space. In fact, there is a territorial fracture that manifests itself in abandoned or underused spaces that can play a significant role in terms of economic and new ecological connections between SMPs and the larger regions [19].

The actions resulting from stakeholders' engagement and analysis of needs were divided according to this categorization and assigned to one of the specific objectives (connected, green, competitive, and safe) and one macro-topic (sustainable growth, business development, and system management). The strategy identified the stakeholders to be involved in each action (government and management, private sector and industry, and end-user research) and its level of priority (Figure 2).

The proposed strategy aims at systematizing the plurality of challenges at stake and, in line with the feedback collected through interviews with key stakeholders, seeks to conceptualize SMPs from a sustainable perspective (environmental, economic, and social). The strategy looks at the spatial and governance redesign of land–sea interaction with a specific focus on the sectors and identified actions.

All of these different dimensions of separation challenge the current understanding SMPs, requiring decision makers to move away from a planning approach that has conceived them as isolated and punctual dots. On the contrary, current and future challenges are urging decision makers to design them as greener, safer, and more competitive. While big ports are hard to change due to the scale of relations and to a complex governance structure and, therefore, institutional rigidity, SMPs have a different scale and can play a key role in testing key changes and eventually connecting them to the larger context. Thus, working on SMPs is essential to adapting to climate, environmental, and social transitions [19].

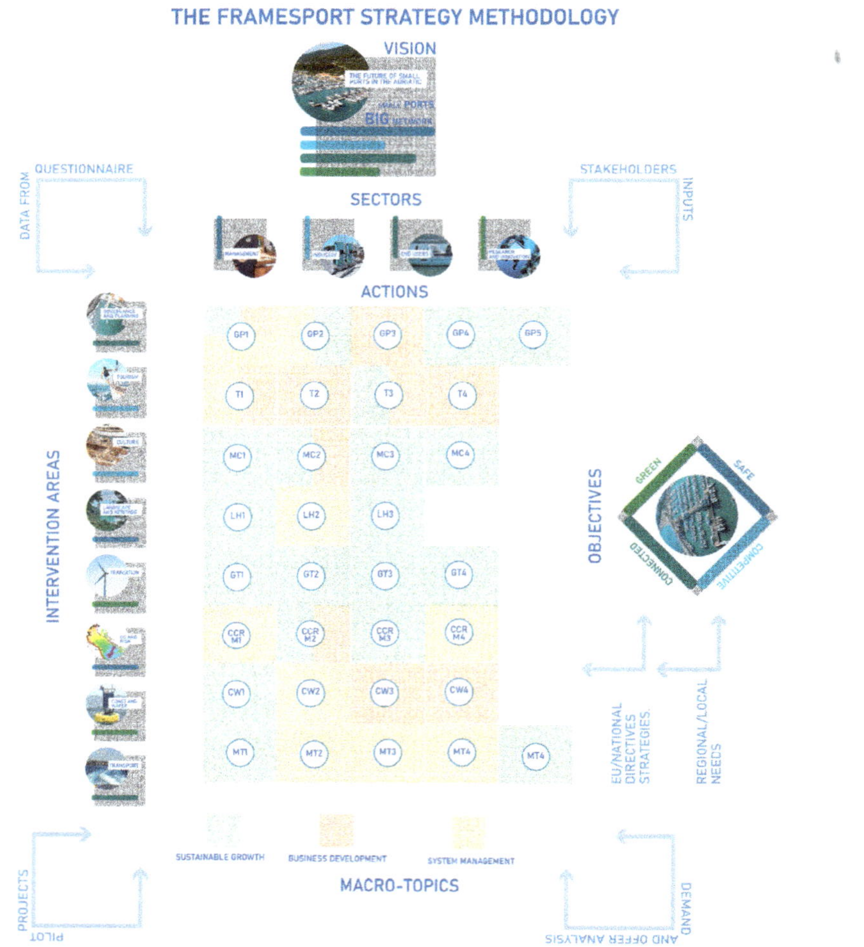

Figure 2. Framesport strategy. The scheme shows the interconnections between data, objectives, intervention areas, sectors, and macro-topics.

5. Conclusions

Port cities are under pressure due to a plurality of spatial, social, environmental, and economic challenges. Climate change and sea-level rise, specifically, are pushing a multitude of stakeholders to identify solutions that can improve ports and cities' relationship in the short, medium, and long terms and that also respond to spatial and institutional fragmentation. With all current public discussion and research focused on big ports, SMPs often escape the gaze of regional and national planning, with tangible consequences for spatial and governance divisions. This article argues that major ports are hard to change due to strong economic, regulatory, and cultural dependencies (e.g., energy dependence), and transition will likely require longer timeframes. Therefore, juxtaposing a parallel focus on smaller ones, such as minor harbors, marinas, and touristic harbors, according to their fragmented nature (private and public), might trigger new opportunities and sustainable practices: from the energy sector to new touristic experimentations, such as fishing tourism. In response, the need for new approaches and models (governance and methodologies) becomes clear. SMPs could be fully recognized as exemplary solutions to design a broader, trans-disciplinary, and more coordinated vision, resulting in a holistic transition approach applicable to all port sizes and also improving the connection between large and small ones.

European and national policy documents reveal a lack of specific regulations on SMPs. Despite recognizing ports as relevant engines for urban development at multiple scales, these documents do not pay enough attention to aspects of the management and development of SMPs—not even from a terminological point of view.

Designing maritime spaces and the interaction between land and water requires in-depth knowledge of the territory and its vocations. Plus, there is a need to consider activities, functions, and land uses that start on land and continue at sea, and vice versa, such as port-related logistics, the movement of goods and people, the energy sector, tourism, and fishing. This intricate interweaving of activities and flows, belonging to different logics and dynamics by their nature, has traditionally been planned through an approach that ignored their interactions.

This growing attention to the sea's history and its spaces is reflected in current political, economic, and spatial planning discourses around the sea, as demonstrated by a significant amount of research. Directive No. 2014/89/EU and Legislative Decree No. 201 of 17 October 2016 provide a framework for maritime spatial planning (MSP) in Italy. Even if it does not directly work on ports, the MSP plans provide a strategic orientation on a national scale concerning maritime areas and address the coastline sea uses with a specific focus on land and sea interactions. MSP represents the first attempt where SMPs are explicitly considered and included, not only in the specific objectives but also in the national measures aiming to promote and set a common strategy. The MSP plan paves the way to overcome public and private barriers by proposing measures to improve efficiency and striving for European certificates that would unlock economic incentives.

Despite the great opportunities offered by the plan and the process, there are still doubts regarding the economic availability that would enable some measures to be actually implemented.

Following existing research on MSP, this article claims that small and medium ports, through MSP, might have a relevant role in better managing passenger and good flows, promoting sustainable mobility, and valorizing tangible and intangible heritage. Dealing with these assets will drive SMPs toward a coordinated strategy and a new governance model.

These themes and challenges were also the starting point for the Framesport project, which represented an interesting laboratory that looked at the SMPs in the Adriatic between Italy and Croatia. As its main output, the Framesport project defined a strategic and common framework that might be the answer to the pressing spatial and institutional fragmentation that characterizes the small Adriatic ports and beyond. This means providing a multi-scale and multi-sectoral categorization capable of bringing together challenges due to climate change with the issues of heritage protection, energy, culture, infrastructure, and logistics, responding to sustainability requirements in a more systematic way. The strategy is a document that discusses an uncertain future, yet one filled with new opportunities for SMPs. Particular attention was given to the interaction between land and water as a transversal principle and as a space of tangible and intangible opportunities for an enlarged reconnection between ports and inland regions.

The need for a regional approach, one that can harmonize and connect ports in a given area that may deal with similar challenges (e.g., subsidence in the Adriatic case) and have similar opportunities (e.g., desalination), has emerged. A connection between ports at the regional level would be relevant for SMPs in sharing costs and efforts to maximize resource uses and limit impacts on connected ecosystems. To what extent and, especially, how MSP may be a game changer for SMP management in the Italian context is still to be fully understood, mainly depending on the plans' implementation at a national level and their possible downscaling to further local tests.

Author Contributions: F.C.: conceptualization, writing, investigation, and visualization; P.D.M.: conceptualization, writing, and investigation; F.S.: writing, investigation, and visualization; V.N.: investigation; F.M.: investigation and supervision. All authors have read and agreed to the published version of the manuscript.

Funding: This research was funded by the INTERREG V-A IT-HR CBC Programme—"Strategic", Subsidy Contract—Project: "Framework initiative fostering the sustainable development of Adriatic small ports, FRAMESPORT" (Project ID: 10253074) and by EASME through Grant Agreement number 887390—MSPMED—EMFF-MSP-2019, corresponding to the call for proposal Call EMFF-MSP-2019 (Maritime Spatial Planning), Topic: EMFF-MSP-2019, Type of action: EMFF-AG for Projects on Maritime Spatial Planning (MSP). The APC was funded by Università Iuav di Venezia within the activities of the Iuav Planning and Climate Change Lab.

Data Availability Statement: For additional information regarding the Framesport project, please visit the following website: https://framesport.eu. For data on Italian MSP, please visit https://www.sid.mit.gov.it/mappa, and for more information regarding the MS Italian draft plans, please visit the following webpage: https://www.mit.gov.it/documentazione/pianificazione-dello-spazio-marittimo.

Conflicts of Interest: The authors declare no conflicts of interest.

References

1. Santos, C.F.; Agardy, T.; Andrade, F.; Crowder, L.B.; Ehler, C.N.; Orbach, M.K. Major challenges in developing marine spatial planning. *Mar. Policy* **2021**, *132*, 103248. [CrossRef]
2. Lam, J.S.L.; Yap, W. A Stakeholder Perspective of Port City Sustainable Development. *Sustainability* **2019**, *11*, 447. [CrossRef]
3. Monios, J.; Wilmsmeier, G. Between path dependency and contingency: New challenges for the geography of port system evolution. *Elsevier J. Transp. Geogr.* **2016**, *51*, 247–251. [CrossRef]
4. Nebot, N.; Rosa-Jiménez, C.; Ninot, R.P.; Perea-Medina, B. Challenges for the future of ports. What can be learnt from the Spanish Mediterranean ports? *Ocean. Coast. Manag.* **2017**, *137*, 165–174. [CrossRef]
5. Hein, C. *Port Cities: Dynamic Landscape and Global Networks*; Routledge: New York, NY, USA, 2011.
6. Hein, C. Port Cities: Nodes in the Global Petroleumscape between Sea and Land. *Technosphere Mag.* **2016**. Available online: https://www.anthropocene-curriculum.org/contribution/port-cities-nodes-in-the-global-petroleumscape-between-sea-and-land (accessed on 17 November 2023).
7. Russo, M.; Attademo, A.; Formato, E. *Transitional Landscapes*; Quodlibet: Macerata, Italy, 2023.
8. ARUP, Port Resilience Framework for Action. 2022. Available online: https://www.arup.com/perspectives/publications/research/section/port-resilience-framework-for-action (accessed on 17 November 2023).
9. Abspoel, L.; Mayer, I.; Keijser, X.; Warmelink, H.; Fairgrieve, R.; Ripken, M.; Kidd, S. Communicating Maritime Spatial Planning: The MSP Challenge approach. *Mar. Policy* **2021**, *132*, 103486. [CrossRef]
10. Furlan, E.; Slanzi, D.; Torresan, S.; Critto, A.; Marcomini, A. Multi-scenario analysis in the Adriatic Sea: A GIS-based Bayesian network to support maritime spatial planning. *Sci. Total Environ.* **2020**, *703*, 134972. [CrossRef] [PubMed]
11. Ramieri, E.; Bocci, M.; Brigolin, D.; Campostrini, P.; Carella, F.; Fadini, A.; Farella, G.; Gissi, E.; Madeddu, F.; Menegon, S.; et al. Designing and implementing a multi-scalar approach to Maritime Spatial Planning: The case study of Italy. *Mar. Policy* **2024**, *159*, 105911. [CrossRef]
12. Moretti, B. Beyond the Port City: The Condition of Portuality and the Threshold Concept. JOVIS: Berlin, Germany, 2020.
13. Russo, M. Harbourscape: Between Specialization and Public Space. In *The Fluid City Paradigm. Waterfront Regeneration as an Urban Renewal Strategy*; Carta, M., Ronsivalle, D., Eds.; Springer International Publishing: Cham, Switzerland, 2016.
14. European Sea Ports Organisation. *European Port Governance. Report of An Enquiry Into The Current Governance Of European Seaports*; European Sea Ports Organisation: Brussels, Belgium, 2010.
15. David, P.A. Path Dependence—A Foundational Concept for Historical Social Science. *J. Hist. Econ. Econom. History* **2007**, *1*, 91–114. [CrossRef]
16. De Martino, P. The Central Tyrrhenian Sea Port Authority. A critical juncture for the Campania port system? *Portus Plus Online Mag. Rete* **2020**, *9*, 1–18.
17. Arrow, K.J. Path dependence and competitive equilibrium. In *History Matters: Essays on Economic Growth, Technology, and Demographic Change*; Stanford University Press: Redwood City, CA, USA, 2004; pp. 23–35.
18. Gerlitz, L.; Meyer, C. Small and Medium-Sized Ports in the TEN-T Network and Nexus of Europe's Twin Transition: The Way towards Sustainable and Digital Port Service Ecosystems. *Sustainability* **2021**, *13*, 4386. [CrossRef]
19. EU. *Common Methodology for Strategy Definition: Sharing a Common Strategy Structure*; 2023; p. 11. Available online: https://programming14-20.italy-croatia.eu/documents/2144190/0/D.3.3.1_Common+methodology+for+strategy+definition.pdf/fd226365-debe-581a-91e9-9f96be053009?t=1686491120077 (accessed on 17 November 2023).

20. EC. *Delivering the European Green Deal*; 2019; Available online: https://ec.europa.eu/info/strategy/priorities-2019-2024/european-green-deal/delivering-european-green-deal_en (accessed on 17 November 2023).
21. EC. *Communication from the Commission to the European Parliament, the Council, the European Economic and Social Committee and the Committee of the Regions. A New Circular Economy Action Plan for a Cleaner and More Competitive Europe*; European Commission: Brussels, Belgium, 2020.
22. EC. *Communication from the Commission to the European Parliament, the European Council, the Council, the European Economic and Social Committee and the Committee of the Regions Repowereu Plan*; European Commission: Brussels, Belgium, 2022. Available online: https://eur-lex.europa.eu/resource.html?uri=cellar:fc930f14-d7ae-11ec-a95f-01aa75ed71a1.0001.02/doc_1&format=pdf (accessed on 17 November 2023).
23. MIT. Strategic Plan for Ports and Logistic. 2014. Available online: https://www.confetra.com/wp-content/uploads/PNL.pdf (accessed on 17 November 2023).
24. EU. *Directive (EU) 2021/1187 of the European Parliament and of the council of 7 July 2021 on Streamlining Measures for Advancing the Realisation of the Trans-European Transport Network (TEN-T)*; European Commission: Brussels, Belgium, 2021; Available online: https://eur-lex.europa.eu/legal-content/EN/TXT/PDF/?uri=CELEX:32021L1187 (accessed on 17 November 2023).
25. EU. *Regulation (EU) no 1315/2013 of the European Parliament and of the Council of 11 December 2013 on Union GUIDELINES for the Development of the Trans-European Transport Network and Repealing Decision No 661/2010/EU (Text with EEA relevance)*; European Commission: Brussels, Belgium, 2013; Available online: https://publications.europa.eu/resource/cellar/f277232a-699e-11e3-8e4e-01aa75ed71a1.0006.01/DOC_1 (accessed on 17 November 2023).
26. EU. *Communication from the Commission to the European Parliament, the Council, the European Economic and Social Committee and the Committee of the Regions. Sustainable and Smart Mobility Strategy—Putting European Transport on Track for the Future*; European Commission: Brussels, Belgium, 2020. Available online: https://eur-lex.europa.eu/resource.html?uri=cellar:5e601657-3b06-11eb-b27b-01aa75ed71a1.0001.02/doc_1&format=pdf (accessed on 17 November 2023).
27. EC. *European Green Deal: Developing a Sustainable Blue Economy in the European Union*; European Commission: Brussels, Belgium, 2021. Available online: https://ec.europa.eu/commission/presscorner/detail/en/ip_21_2341 (accessed on 17 November 2023).
28. Maragno, D.; Dall'omo, C.F.; Pozzer, G.; Bassan, N.; Musco, F. Land–Sea Interaction: Integrating Climate Adaptation Planning and Maritime Spatial Planning in the North Adriatic Basin. *Sustainability* **2020**, *12*, 5319. [CrossRef]
29. Zaucha, J.; Gee, K. *Maritime Spatial Planning. Past, Present, Future*; Palgrave Macmillan Cham: Berlin, Germany, 2019; pp. 121–149. [CrossRef]
30. Howells, M.; Ramírez-Monsalve, P. Maritime Spatial Planning on Land? Planning for Land-Sea Interaction Conflicts in the Danish Context. *Plan. Pract. Res.* **2022**, *37*, 152–172. [CrossRef]
31. European Union. *Directive 2014/89/EU of the European Parliament and of the Council of 23 July 2014 Establishing a Framework for Maritime Spatial Planning*; J. Eur. Union L257; European Union: Brussels, Belgium, 2014; pp. 135–145.
32. GU. DECRETO LEGISLATIVO 17 ottobre 2016, n. 201 Attuazione della direttiva 2014/89/UE che istituisce un quadro per la pianificazione dello spazio marittimo. (16G00215) (GU Serie Generale n. 260 del 07-11-2016). 2016. Available online: https://www.gazzettaufficiale.it/eli/id/2016/11/07/16G00215/sg (accessed on 17 November 2023).
33. GU. DECRETO DEL PRESIDENTE DEL CONSIGLIO DEI MINISTRI 1 Dicembre 2017 Approvazione delle Linee Guida Contenenti dli Indirizzi e i Criteri per la Predisposizione dei Piani di Gestione dello Spazio Marittimo. (18A00392) (GU Serie Generale n. 19 del 24-01-2018). 2017. Available online: https://www.gazzettaufficiale.it/eli/id/2018/01/24/18A00392/sg (accessed on 17 November 2023).
34. EC. Bauhaus of the Seas 2020. Available online: https://bauhaus-seas.eu/ (accessed on 17 November 2023).
35. Pavia, R.; Di Venosa, M. *Waterfront. From Conflict to Integration*; LISt Lab Laboratorio. Internazionale Editoriale: Trento, Italy, 2012.
36. Assoporti. Confronto L. 84-94 Testo Vigente e Testo con Integr. D. Lg.vo Riforma AP 7.9.2016. 2016. Available online: https://www.assoporti.it/media/1574/confronto-l-84-94-testo-vigente-e-testo-con-integr-d-lgvo-riforma-ap-792016.pdf (accessed on 17 November 2023).
37. GU. Royal Decree 1885 n. 3095. 1885. Available online: https://www.gazzettaufficiale.it/eli/gu/1885/05/27/123/sg/pdf (accessed on 17 November 2023).
38. Hoyle, B. Global and local change on the port-city waterfront. *Geogr. Rev.* **2000**, *90*, 395–417. [CrossRef]
39. Hoyle, B.; Pinder, D. (Eds.) *European Port Cities in Transition*; Belhaven Press, British Association for the Advancement of Science, Annual Meeting, University of Southampton: London, UK, 1992.
40. Hoyle, B.S. *Port Cities in Context: The Impact of Waterfront Regeneration*; Transport Geography Study Group, Institute of British Geographers: London, UK, 1994.
41. Ministero Infrastrutture e Trasporti. Linee Guida per la Redazione dei Piani Regolatori Portuali (art. 5 Legge n. 84/1994). 2004. Available online: https://docs.dicatechpoliba.it/filemanager/189/info%20corso%20A.A.%202014_2015/supporti%20didattici/slides%20del%20corso/linee%20guida/lineeguidaprp_40.pdf (accessed on 17 November 2023).
42. GU. Decreto Legislativo 13 Dicembre 2017, n. 232 Disposizioni Integrative e Correttive al Decreto Legislativo 4 Agosto 2016, n. 169, Concernente le Autorita' Portuali. (18g00024) (GU Serie Generale n. 33 del 09-02-2018). 2017. Available online: https://www.gazzettaufficiale.it/eli/id/2018/02/09/18G00024/sg (accessed on 17 November 2023).
43. Hein, C.; Luning, S.; van de Laar, P. Port City Cultures, Values, and Maritime Mindsets: Defining What Makes Port Cities Special. *Eur. J. Creat. Pract. Cities Landsc.* **2021**, *4*, 7–20.

44. Couling, N.; Hein, C. *The Urbanisation of the Sea. From Concepts and Analysis to Design*; nai010 Publisher: Rotterdam, The Netherlands, 2020.
45. Jouffray, J.-B.; Blasiak, R.; Norström, A.V.; Österblom, H.; Nyström, M. The Blue Acceleration: The Trajectory of Human Expansion into the Ocean. *One Earth* **2020**, *2*, 43–54. [CrossRef]
46. European Union. *Blue Growth*; European Union: Brussels, Belgium, 2012; Available online: https://ec.europa.eu/maritimeaffairs/policy/blue_growth_en (accessed on 17 November 2023).
47. Rodríguez-Rodríguez, D.; Malak, D.; Soukissian, T.; Sánchez-Espinosa, A. Achieving Blue Growth through maritime spatial planning: Offshore wind energy optimization and biodiversity conservation in Spain. *Mar. Policy* **2016**, *73*, 8–14. [CrossRef]
48. MIT. Pianificazione dello Spazio Marittimo. 2022. Available online: https://www.mit.gov.it/documentazione/pianificazione-dello-spazio-marittimo (accessed on 17 November 2023).

water

MDPI

Review

Solutions Based on Nature to Face Water Stress: Lessons from the Past and Present

Daniel O. Suman [1,*], Manuela Morais [2] and Carlos Hiroo Saito [3]

[1] Department of Environmental Science and Policy, Rosenstiel School of Marine, Atmospheric, and Earth Science, University of Miami, 4600 Rickenbacker Causeway, Miami, FL 33149, USA
[2] Laboratory of Water, School of Sciences and Technologies, University of Évora, Parque Industrial e Tecnológico, R. da Barba Rala No. 1, 7005-345 Évora, Portugal; mmorais@uevora.pt
[3] Center for Sustainable Development, Department of Ecology, Institute of Biological Sciences, University of Brasilia, Campus Universitário Asa Norte, Brasilia 70910-900, DF, Brazil; carlos.h.saito@hotmail.com
* Correspondence: dsuman@earth.miami.edu; Tel.: +1-305-742-8762

Abstract: Nature-based solutions (NbS) to water scarcity, environmental degradation, climate change, and biodiversity losses are enjoying increasing implementation throughout the world. This manuscript reviews three case studies from Brazil, Panama, and Portugal that illustrate NbS and searches for commonalities that may assist their usefulness in new sites. The Tijuca Forest in Rio de Janeiro is a remarkable story of centuries of forest management and restoration that initially aimed at providing water security for the capital of the country during the XIX Century while it was still a monarchy. Today, it is recognized as a UNESCO World Heritage Site. The Panama Canal Watershed produces water for canal operations, electricity generation, and drinking water for half the country's population. Traditional water mills and weirs near streams in the Alentejo Region, Portugal, have largely been abandoned due to the damming of the Guadiana River. Yet today, weirs are increasingly recognized for their important contribution to water provisioning in this dry region. All have a primary goal related to water provisioning, yet their ecosystem benefits are multiple. The cases offer important lessons for adaptation to climate change, cultural benefits from traditional human activities, and concerns about social equity.

Keywords: ecosystem restoration; watershed management; nature-based solutions; reforestation; weirs; governance; Tijuca Forest; Rio de Janeiro; Panama Canal; Alentejo Region; Portugal

1. Introduction

In recent decades freshwater scarcity has been increasingly recognized as a global systemic risk and elevated to the center of global concern [1]. The fact that the theme of World Water Day 2007 was "Coping with Water Scarcity" illustrates this [2].

Although water scarcity often has its roots in water shortage, there are several ways of defining water scarcity, so it can be considered a relative concept. Scarcity can occur at any level of supply or demand because of the lack of correspondence between availability and demand, or it can be related to access or the expected water quality related to the increasing degradation of groundwater and surface water [3]. Temporal or geographical imbalances between availability and demand are largely responsible for water scarcity [4,5] because these can generate conflicts between water users. The imbalance between the forcing drivers relative to demand (high population growth and economic development) on the one hand and the forcing drivers relative to availability (droughts and wide climate variability) on the other, can make this water scarcity dramatically real and acute. Thus, spatial and temporal considerations are important determinants of water scarcity, and therefore, to water security [6]. A feature of any "desirable" water-secure future must be the existence of healthy ecosystems, which require water to maintain the life of these ecosystems [7].

Usually, coupled water scarcity and water availability are analyzed by considering surface water because surface fresh water is the most readily available resource to meet water demands and, thus, is considered the first indicative of blue-water availability [8]. Nevertheless, considering a systemic view of water (its whole cycle and flows), it is necessary to consider the interdependence between blue and green water and how the entire landscape can help to produce and maintain water [9]. Water is essential for maintaining the ecological integrity of aquatic and terrestrial ecosystems, as well as for providing the quality that allows for the subsistence and well-being of the human beings who depend on these ecosystems [10]. However, the use of water resources has already exceeded the alarm level in many regions of the world [11], resulting in the decline of the water table [12,13], degradation of aquatic ecosystems [14], and losses of wetlands [15–17]. Furthermore, the impacts of climate change, which result in changes in abiotic factors, such as precipitation and temperature levels, affect species and biological populations and how they interact with other organisms and their habitats, which alters the structure and function of ecosystems and the goods and services that natural systems provide to society [18]. In addition to insufficient water quantity, water pollution also aggravates local water-scarcity risks. These phenomena can be exacerbated when they occur in regions with few water resources or where water resources are poorly managed, resulting in imbalances between the demand for water and the supply capacity of the natural system [19].

The ecological concepts of sources, sinks, and recycling can help to provide a useful understanding of water basins, water cycles, and flows [20]. When considering the systemic view, applied not only for the elements in a structural perspective of landscapes [21] but for the entire process and feedback controls, we can see not only that healthy ecosystems need water but also that healthy ecosystems can produce water [22]. In other words, water should absolutely be considered connected to the landscape where it runs. The way to recover water should rely on nature-based solutions [23–25].

This paper reflects on the role of ecosystems in water production and conservation and how this systemic approach to water conservation can be improved by analyzing past experiences as supportive lessons, similar to the way ancient urban facilities could inspire us to harvest water in cities in water-scarce regions [26]. Examples from different areas of the world illustrate the long existence of what are presently named "nature-based solutions" (NbS).

2. The Nature-Based Solutions and Water Security

Nature-based solutions (NbS) are defined as "actions to protect, sustainably manage and restore natural or modified ecosystems that address societal challenges effectively and adaptively, simultaneously providing human well-being and biodiversity benefits" [27]. The NbS concept aims at highlighting positive outcomes for society (so-called 'solution') as a gift offered or promoted by nature if we consider and respect its flows and chains. It may present some overlap with other concepts used to propose ecosystem management for societal benefits, such as ecological engineering, green/blue infrastructure, ecosystem approach, ecosystem-based adaptation/mitigation, ecosystem services approach, or natural capital [28]. NbS can be an important strategy to increase water security supported by a systemic approach that considers water security as "the capacity of a population to safeguard sustainable access to adequate quantities of acceptable quality water for sustaining livelihoods, human well-being, and socio-economic development, ensuring protection against waterborne pollution and water-related disasters, and preserving ecosystems in a climate of peace and political stability" [29]. Scott et al. [30] (p. 281) present a different definition that is more ecosystem-based and adds a perspective of resilience: "Water security constitutes the sustainable availability of adequate quantities and qualities of water for resilient societies and ecosystems in the face of uncertain global change". Stable and resilient ecosystems can reduce uncertainty, enhance overall water security by improving both water availability and quality, prevent water-related disasters, and achieve increased water-body resilience to climate change [31]. Ecosystem restoration in Iceland presented positive results on runoff dynamics and flood prevention, addressing a decrease in high discharge peaks and a reduction of erosion rates in the rivers due to the lower soil erodibility [23]. In China, a mangrove landscape restoration and protection project in the Fengxinglong Ecological Park at the junction of the Sanya and Linchun Rivers provided a reduction of shoreline erosion while water was being purified [32]. The contributions of mangrove forests and salt marshes to a reduction of flooding in deltas throughout the world are well documented [33]. Additionally, ecosystem restoration of mangroves provides greater carbon sequestration benefits than afforestation for mitigating global climate change [34].

Vegetation–water interdependencies should be considered because vegetation influences water recycling from local to continental scales. Within this coupled system, tropical trees extract deep soil water and pump it into the atmosphere, circulating water through evapotranspiration to precipitation processes [35]. Vegetation should be regarded as a water 'recycler' [31], illustrating the interdependence between green and blue water [9].

NbS can offer the main or only viable solution in each particular context, facing challenges that gray infrastructure cannot address (due to aging or inadequacies with respect to climate change). For example, the third Rhone River Correction Project in the canton of Valais, Switzerland, aims to return space to the floodplain that had been eliminated. After recognizing that the past interventions in that river were incapable of flood and disaster prevention, authorities decided to widen the river corridor, creating additional space for the river to improve the landscape and biological diversity and restore the ecological functions of the river and the natural dynamics of the ecosystem as much as possible [36]. By embedding an ecosystem services approach, NbS can represent innovative solutions within the principles and practice of integrated water resources management (IWRM), leading to water resilience and improving people's livelihoods.

The following sections describe three cases of NbS implementation from Brazil, Panama, and Portugal. Subsequently, we comment on what these experiences can teach us today, as we aim for a sustainable future where ecosystem services can be fully provided under an ecocentric framework.

3. Some Nature-Based Solution Case Studies

The three case studies whose sites are shown in Figure 1 represent different continents (Europe, North America, and South America), latitudes, and climatic regimes. Together, they illustrate that NbSs have applicability in diverse ecosystems and climatic regions and suggest lessons that can be appropriate to other interventions throughout the world.

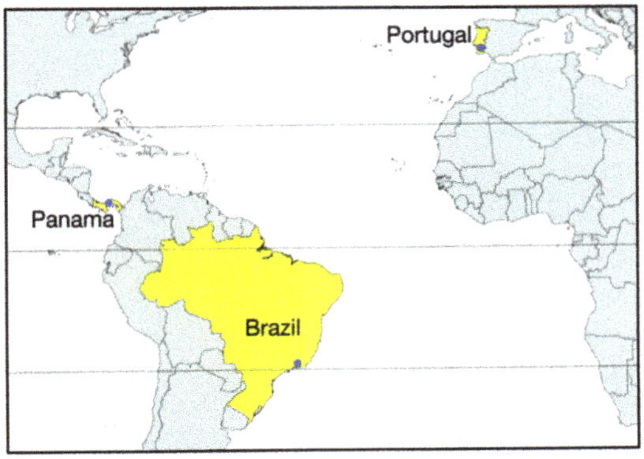

Figure 1. Map of the three case studies discussed in this manuscript.

3.1. The Forest Landscape Restoration in Tijuca Tropical Rain Forest, Brazil

"Rio de Janeiro's Carioca Landscapes between the Mountain and the Sea (Brazil)" was inscribed on UNESCO's World Heritage List in 2012 (Figure 2). According to UNESCO's Decision 36 COM 8B.42, an outstanding contributing factor was that "*the mountains and open green areas of the Tijuca National Park, together with Corcovado and the hills around the Guanabara Bay still retain a similar combination of forest and open observation points as at the time of colonization and allow access to vistas of the city from many high vantage points that demonstrate very clearly the extraordinary fusion between culture and nature in the way the city has developed*". The Tijuca Forest, the core landscape of this set that impressed UNESCO, remains less than twenty kilometers from the city center, occupying a large area of land on the southern slopes of the Tijuca and Papagaio peaks, encompassing the Alto da Boa Vista and the Pedra do Conde [37].

Figure 2. Rio de Janeiro's Carioca Landscapes between the mountain and the sea (Photo, Keyko Carolina Saito).

Presently, the Tijuca Forest belongs to a protected area, the Tijuca National Park (22°55′–23°00′ S, 43°11′–43°19′ W), with a total of 39.51 km^2 of Atlantic Forest biome containing rich biodiversity, including several endangered fauna species, among others, the margay (*Leopardus wiedii*), the brown howler monkey (*Alouatta guariba*), and the endemic frog *Thoropa lutzi* [38–40]. Additionally, this protected area has a total of 254 taxa recorded among ferns and lycophytes, with seven species identified as endangered, including *Anemia blechnoides*, *Lytoneuron tijucanum*, *Pteris congesta*, *Anemia gardneri*, *Asplenium cariocanum*, *Doryopteris rediviva*, and *Lytoneuron quinquelobatum* [41]. The mountainous and rugged terrain includes Pico da Tijuca, Corcovado, and Pedra da Gávea, the last, considered to be the largest seaside monolith in the world. The Park presents a unique natural beauty, contrasting the green of the forest with the ancient rocky surfaces and the sea. The rocks that form the mountains of Tijuca National Park are aged around 1.7 billion years. The park's climate, due to the geographic orientation of the Tijuca Massif, presents abundant rainfall, with no dry period in winter. The annual precipitation typically varies from 2000 to 2500 mm, although 3300 mm can be reached in exceptionally rainy years [38].

Sites located up to 500 m high have a humid tropical climate (Aw type in the Köppen classification, with mean monthly temperatures varying from 25 °C in February to 19 °C in June, and an annual mean temperature of about 22 °C [38]. Meanwhile, above this altitude, temperate conditions occur. The hydrography of Tijuca National Park is quite extensive, with several small streams (Maracanã, Carioca, Cachoeira, and das Almas, among others, with very modest water flow [42]). It is estimated that 20% of the total average annual rainfall is absorbed by the preserved tropical rainforest, with the help of the leaf litter and its associated biogenic activity on the top of the soil, which provides high infiltration rates and storage of rainwater in the soil (Figure 3). Due to these roles of the forest cover, only 30% of the precipitation converges into the river channels during the rainy periods, resulting in modest flow, but with a perennial supply of basic river discharges [43]. Several waterfalls are also present, the most famous being the Cascatinha Taunay.

Figure 3. Atlantic Forest biome, Brazil, with its structure of intercepting and infiltrating rainfall (Photo, Carlos Hiroo Saito).

Today's impressive landscape resulted from an enormous effort to restore the tropical rainforest. In fact, the Tijuca Forest is a replanted tropical forest, resulting from pioneering reforestation experiences with native species and forest and watershed management started in the 19th century. Initially, it aimed to provide water security for Rio de Janeiro, the capital of colonial Brazil, which continued as the capital of the Brazilian Empire after the independence proclamation in 1822 and as the capital of the Republic of Brazil from 1989 to the mid-20th century.

The forest remained intact throughout almost the entire 18th century, as land squatters that supplied the city with food were restricted to the edge of the urban area on land where the new city would later emerge. The location preferences of peasants probably were guided by the abundance of water for food production.

In the second half of the 18th century, coffee plantations emerged in Rio de Janeiro and quickly spread around the hills surrounding the city, due to favorable local climate conditions and soil. However, by the beginning of the 19th century, the abandoned coffee farms and eroded slopes resulted in severe environmental degradation. From 1817 to 1818, Dom João VI, King of the United Kingdom of Portugal, Brazil, and the Algarves, and still living in Brazil, issued orders to (1) stop the forest devastation in the springs near the city; (2) plant trees next to the springs of some rivers; and (3) carry out assessments allowing the government to expropriate areas with springs and place them under protection. These measures were not effective and had only a modest implementation.

It is important to note the main source of freshwater in the region was the small rivers and streams that flowed from the hills surrounding Guanabara Bay because there were no large rivers nearby. Additionally, the wells in the lower regions around the bay were almost always subject to the intrusion of brackish water. The degraded hills, due to coffee plantations, put the provision of drinking water in Rio de Janeiro at risk, aggravated by the severe droughts in the empire's capital (now an independent country from Portugal) during the second quarter of the 19th century (particularly 1824, 1829, 1833, and 1844). The Brazilian Emperor at that time (Dom Pedro II), was cognizant of the relationship between land use and water availability, as well as the importance of vegetation to produce a continuous supply of water [44].

It took more than ten years for the government to implement appropriate land purchases, and several more to initiate forest management practices. From 1862 to 1874, Major Manuel Gomes Archer, the designated forest manager, implemented an extensive reforestation process, working personally with a few slaves and following the following criteria of prioritizing some slopes more threatened by erosion and areas for restoring the water flow of rivers and streams; employing seedlings, in opposition to seeds or already grown young trees; and using apparently random combinations of species, based on personal knowledge of the distribution and incidence of the species in the region's original forests [41]. During this 13-year period, over 60,000 trees were planted, with a survival rate of around 80 percent [37]. In his last report, written in 1874, Major Archer concluded that all stream sources with restored vegetation had released more water or remained at a stable level compared to the same period before 1862 [42]. After Major Archer, Baron d'Escragnolle continued reforestation and, additionally, began landscaping work for beautification purposes, aimed at public use and contemplation. Under the coordination of the renowned French landscaper Auguste Glaziou, the forest received corners, leisure areas, fountains, and lakes.

In the 19th century, capturing water from the Tijuca National Park became a strategic activity that was essential for supplying the city's water supply. At the end of that century, a complete network of small dams was established, capturing water from several rivers in the park. After many years of relative abandonment, the 1940s saw a revitalization of the Tijuca Forest, under the leadership of Raymundo Ottoni de Castro Maya, with the opening of restaurants and the consolidation of internal roads and landscape projects by Roberto Burle Marx.

Finally, in 1961, the area of the Tijuca Massif gained the status of a protected area under the name of the Rio de Janeiro National Park, with 33 km^2, changing its name to the present one (Tijuca National Park) on 8 February 1967. On 4 July 2004, a Federal Decree expanded the Park's limits to the present extension of 39.51 km^2, incorporating locations such as Parque Lage, Serra dos Pretos Forros, and Morro da Covanca. Today Tijuca National Park is considered the largest urban park in the world and leads park visits in Brazil (more than 4 million visitors in 2023).

The success of reforestation activities in the past led to the present UNESCO's recognition of landscape value. In addition to the original central target ecosystem service of water security, the entire ecosystem has benefited, supporting today's high biodiversity and offering social opportunities for leisure and climate regulation for residents. This example shows how an NbS can jointly deliver a group of ecosystem services, even when it initially targeted only one.

3.2. *Panama Canal Watershed*

The 82 km long Panama Canal is one of the world's major shipping routes. Approximately 4% of global maritime trade (about 14,000 transits annually) passes through the canal, with primary routes to and from East Asia and the east coast of the USA, as well as Europe to East Asia and the west coast of the Americas. The canal also contributes about USD 2.5 billion annually to the Panamanian treasury or about 3% of the nation's GDP. The direct contributions of the Panama Canal Authority to the National Treasury were USD 2.08 billion in 2021 or about 3.1% of the country's GDP in that year [45,46]. The Panama Canal Authority (ACP) (referred to today simply as "Canal de Panamá") operates the canal, governed by its organic law that guarantees a significant degree of autonomy by the National Constitution and requires that the ACP lead the coordination with other institutions with sectoral authority in the canal area. The canal operates with a series of locks that raise and lower vessels from the ocean to the freshwater Lake Gatun, which was created by damming the Chagres River over a century ago. Ocean-going vessels travel 33 km through Lake Gatun before being lowered to the other ocean through another series of locks. In 1935 an additional dam was constructed upstream on the Chagres River to provide additional water security for canal operations—resulting in the formation of Lake Alajuela (Figure 4).

Figure 4. Map of the Panama Canal Watershed showing the canal and Gatun and Alajuela Lakes and protected areas (in color). (Source: https://pancanal.com/; https://es.m.wikipedia.org/wiki/Archivo:Areas_protegidas_de_la_cuenca_hidrografica_de_panama.jpg#globalusage accessed on 13 August 2024).

After more than a century of operations with the original locks, in 2016, the ACP completed a major project that constructed wider locks handling larger Neopanamax vessels. Unlike the original locks that release some 200 million liters of water to the ocean during each transit, the new locks have the capability of incorporating water reuse chambers resulting in less water per transit. The ACP claims that this water recycling saves the equivalent water used in six daily transits [47]. Nevertheless, since 2019, the recycling chambers have not been used consistently, and as a result, water use for canal operations is higher than ever [48]. Water recycling has important potential because climate-change predictions for Panama suggest that precipitation may decrease by about 10% over current values in the coming decades [49,50]. The canal watershed has experienced significant droughts in recent years (partially explained by strong El Niño events predicted to be more extreme with climate change) that have led the ACP to reduce transits by one-third in 2023–2024 [51]. The 2023 season was the third driest in the history of the canal watershed, with 30% less precipitation than expected, and as a result, the canal's reservoirs only could store about half of the normal water volume [52]. Normally, some 36–39 vessels transit the canal daily. However, in 2023, the ACP was forced to reduce daily transits to 22 [53]. Low water flows during the 5-month dry season (January to May) compounded with strong El Niño events in 1982–1983 and 1997–1998 also resulted in major decreases in canal transits.

Additionally, Lake Gatun and Lake Alajuela are also the source of drinking water for about 2 million residents of Panama City and Colon (about half of the country's population). The lower water levels in the canal reservoirs resulted in decreased supplies of potable water to metropolitan areas in 2023–2024 [54]. Added to the decrease in precipitation in the canal watershed from climate change, the increasing water demands from growing urban populations (the population of metropolitan Panama City increased from 1,216,000 to 2,016,000 persons between 2000 and 2024 [55].), and the land-use changes resulting from the deforestation of the canal's watershed have increased the vulnerability of future operations of the canal.

The Panamanian Government's policy until the 1980s promoted the "conquest of the jungle" and encouraged small farmers to colonize forested lands and convert them to pasture for cattle ranching and the cultivation of crops [56]. Wadsworth [57] compared maps from 1952 and 1978 and concluded that, during this period, about 30% of the watershed had been deforested. Other studies note that, between 1974 and 1991, the canal watershed lost about 43% of its forests to urbanization, cattle ranching, and crop cultivation [58,59]. Constructed and administered by the United States, the Panama Canal and the colonial Canal Zone began a period of gradual transition to Panamanian control with the signing of the Panama Canal Treaties in 1976, resulting in complete Panamanian administration at the close of the century. During this transition period, the binational canal administration increasingly recognized the importance of protecting the watershed of the Chagres River to guarantee the water supply for canal operations. A new environmental ethic based on the importance of canal operations developed during the administrative transition of the canal. Perspectives of scientists, canal administrators, environmentalists, and politicians coalesced around recognition of the importance of the canal's watershed and the green forest infrastructure that provided water for the canal's operations [56]. Additional concerns focused on the increased sedimentation in the two lakes caused by the erosion of deforested lands in the watershed. To this end, Panama created Chagres National Park in 1984 to protect the upper watershed of the Chagres River, and the efforts of today's ACP and the Ministry of the Environment have significantly reduced deforestation within the park boundaries. At the same time, the creation of the national park created adverse impacts on the thousands of small farmers who live inside the boundaries of the protected area and are limited in the land that they can farm using shifting agricultural techniques.

The passage of Law 19 in 1997, which created the Panama Canal Authority, increased efforts to protect the watershed. The ACP adopted its regulation (Agreement 116) formalizing its environmental and watershed management responsibilities [60]. The formal canal watershed, as defined by the ACP, consists of 343,521 ha covering seven political districts, or about 4.5% of the national territory. The average annual precipitation in the canal watershed ranges from 2500 to 4000 mm, with higher values in the primary tropical forest of Chagres National Park [61]. Some 127,326 ha lie on the western side of the canal (37.08% of the watershed). Much of the land on the canal's western flank has been deforested and today is grassland for cattle ranching. Three national parks (Chagres, Soberanía, and Camino de Cruces) account for much of the land on the 216,195 ha on the eastern side of the canal, and indeed, some two-thirds of the forested lands in the watershed [62]. Although urbanization and agricultural expansion have accounted for land conversion on the eastern side of the canal, the land degradation is more extreme on the western portion of the canal's watershed.

Recognizing the importance of wise land use on the watershed, the ACP has embarked on significant efforts during the past 20 years, particularly in the lands on the western bank of the canal which have seen 75% of ACP's investment in reforestation and sustainable agricultural activities (Martinez, R., personal communication, 10 June 2024). One of the initial incentives to promote both environmental and social sustainability was a cooperative program of the ACP and ANATI (National Land Titling Authority) to survey land holdings and grant land titles to small farmers. Formal titles created legal security for small farmers and set the foundation for further support from the ACP. By 2023, more than 10,000 land titles had been registered, largely in the Capira District in the western portion of the canal's watershed [51]. The ACP has created six watershed committees and 26 local committees of locally elected members to facilitate ACP watershed management efforts and convey community needs and concerns to the ACP [51].

ACP's efforts have been channeled through its Program of Environmental Economic Incentives (PIEA), which began in 2010 with the aim of fostering sustainable land uses and forest protection, thus providing water resources for canal operations and drinking water for half of Panama's population. Interventions in rural areas, described below, improve environmental conditions and the livelihoods of rural residents, income for families, and employment. PIEA activities have been concentrated in the watersheds of the Cirí–Trinidad Rivers in the extreme western area of the canal watershed.

Reforestation for conservation and commercial purposes is one major PIEA project. Establishing vegetative protective cover helps increase dry-season water flows into Lake Gatun and avoids sedimentation of the lake reservoirs and associated dredging costs [63,64]. By the end of 2021, over 3000 ha had been reforested, with the planting of 790,000 trees. Over 1200 small- and medium-sized agricultural producers have benefited in Capira, Colón, and La Chorrera Districts. These benefits include shade for crops, wind breaks, soil improvement, water conservation (quantity and quality), increased water infiltration and reduction of erosion and sedimentation, carbon fixation, and support for biodiversity. The ACP also utilizes payments for environmental services (PES) as incentives to maintain the existing forested lands in the canal watershed. The ACP approved regulations formalizing PES application in 2018 [65]. By 2023, over 400 private owners of some 5150 ha of existing forested lands have received financial benefits (about USD 100/ha/yr) under this program [51,66].

PIEA has generated significant support for agroforestry projects—including planting coffee, cacao, and fruit trees. The ACP supported the creation in 2012 of the Asociación de Productores de Café de la Subcuenca de Ríos Cirí Grande y Trinidad (ACACPA) to support sustainable and environmentally friendly production and commercialization of coffee [67]. Shade-grown coffee farms have replaced the deforested lands used for cattle ranching (Figures 5 and 6). ACP incentives include support for plantation establishment and payment for farmers' work for the initial year; provision of materials, tools, and fertilizer; and technical support for an additional two years as the coffee farms become sustainable. ACACPA's 90 members successfully planted some 4000 ha of shade-grown, organic robusta coffee by the end of 2021. Panama's Ministry of Commerce and Industry recognized the trademark of "Cuencafé" in 2018, and today, it is sold as a whole bean or ground.

Figure 5. Shade-grown coffee berries in the Panama Canal Watershed (Photo, Jorge Muñoz).

Additional PIEA efforts focus on improved agricultural practices to reduce soil erosion (and sedimentation of the canal's reservoirs), increase yields, and improve farmers' livelihoods. Interventions include the promotion of improved grasses for pasture, planting of riparian buffers, living fences, planting of trees in pastures, cattle crossings of streams to avoid erosion, crop rotation, and organic fertilizers.

Figure 6. Shade-grown coffee bushes in the Panama Canal Watershed (Photo, Jorge Muñoz).

The ACP recognizes the close relationship between wise land use in the canal's watershed to guarantee sustainable flows of water for canal operations and provision of potable water for half of Panama's population while simultaneously supporting the livelihoods of thousands of small farmers who reside in the canal watershed. Will the ACP's impressive efforts be sufficient to guarantee sufficient water for these multiple needs in the future? Additional areas of action must include water-conservation measures by all users, a revision of water pricing, repair and update of potable water infrastructure to minimize water losses, continual use of the water-recirculating chambers in the Neopanamax locks, and increased reforestation of the canal watershed. The current discussion surrounds the possible extension of the canal's legal watershed to the west with the diversion of water from the Río Indio or to the east with a transfer of water from the reservoir of the Bayano River [68,69]. Water for the canal is clearly a primary national objective for Panama. The future portends growing conflicts among watershed stakeholders over the use of increasingly scarce water [19]. We believe that reforestation efforts in the existing canal watershed should be prioritized and enhanced before diverting water from adjacent watersheds, which would lead to significant adverse social and environmental impacts.

3.3. *The Ancient Weirs in Riverbeds—Alentejo Region, Southern Portugal*

The Alentejo is Portugal's largest region (Figure 7). However, despite occupying around one-third of the national territory (31,551 km^2), it is characterized by an aging and poorly qualified population, with a low capacity to attract young people. Its gross domestic product (GDP) per capita is below the national average. This region has the highest rate of desertification and the lowest population density in Portugal (5% of the national population) [70].

Historically, the main sector of economic activity in Alentejo has been extensive agriculture and livestock farming—well adapted to climatic and environmental conditions [70,71]. The rural population is traditionally linked to the valorization of the natural heritage in a territory with high biodiversity and many endemic species of conservation interest. However, today, the structure of this territory is potentially threatened by increasing changes in land use, with the progressive replacement of traditional extensive agriculture by intensive irrigated agriculture [72,73]. This production model emerged on a large scale after the 2002 construction of the Alqueva Dam, mainly due to a long-standing public investment in the Alqueva irrigation system [48]. The project, with a long history (the first Alentejo Irrigation Plan dates from 1957), was designed to supply water to Portugal's driest region, which is most prone to water scarcity and physical and human desertification processes [74,75].

Figure 7. Location in red of the Alentejo Region in Portugal (38°34′0.12″ N; 7°54′0″ W) (https://en.wikipedia.org/wiki/Alentejo_Region accessed on 13 August 2024).

The Alentejo's typical Mediterranean climate is characterized by warm and dry summers and cool and wet winters. The mean annual air temperature is around 18 °C, and the mean annual precipitation is around 900 mm with a strong north–south gradient, reaching values below 500 mm in the southeastern part of the region [76]. Thus, the river flow regime is intermittent, with torrential flows in autumn/winter (November, December, and January) and interruptions in summer (June, July, August, September, and October), with long dry reaches, even on the main international Guadiana River (regulated since 2002 due to the construction of the Alqueva Dam system) [77].

In the past, the rural population, uneducated and impoverished, adapted the agricultural production model to the cycles of nature, taking advantage of natural resources and finding natural solutions [72]. An example of these practices are the water mills located on the banks of the Guadiana River and its tributaries, used to produce flour from different cereals (Figure 8). The existence of water mills dates to at least the 13th century [78]. However, in the 19th century, the number of mills reached its peak, due to the increasing population, and continued into the 20th century as a natural mode of production for large plantations [79,80].

Figure 8. Water mills located on the banks of the Guadiana River, Southern Portugal (Municipality of Mérola). The weir is located in front of the mill. (Photo, Luís Pavão).

In 2002, during the construction of the Alqueva Dam, around a hundred mills of different shapes and sizes were identified, built in such a way that their architectural structure was not damaged when the flow rose (i.e., during winter after flood events) and submerged the structure for periods of up to four months [80]. In very rainy years, when the water level remained very high, preventing the mill from operating on the main river (i.e., the Guadiana River), millers would search for similar structures on adjacent tributaries with less water flow, where they would remain until the water level dropped in the Guadiana.

Guita [80] identified 80 mills in an area of 1279 km^2 on the Guadiana and its tributaries, the smallest with drainage basins of 15 to 20 km^2. In the municipality of Mértola, he noted that one mill existed for every 16 km^2 of surface area. Mill activity remained active until the 1960s when more than 30 mills were in operation. The five remaining mills operating in the 1980s closed before 1990.

These structures were kept in operation by weirs built in the riverbed with diversions that channeled water where it was needed to grind grains. On the Guadiana River and its tributaries, the dams were made of stones, fitted perpendicularly to the flow of water, with the thickest section facing the outside of the construction, forming an arch closed at the top by slabs [80] (Figure 9). These technologically simple but very resistant structures, designed by master craftsmen, demonstrate knowledge of hydraulic engineering that is remarkable for its simplicity, having withstood the passage of centuries. Today, after more than 100 years in some cases, the weirs are completely integrated into the landscape and form part of the cultural heritage. From an ecological perspective, they constitute important water-retention zones, creating lentic habitats that are maintained all year round, even during dry summer periods. Consequently, they are important refuge areas for aquatic and terrestrial species, promoting biodiversity, ecological integrity, and the conservation of ecosystems. On the other hand, these weirs, due to their heritage value, and also because they retain water in natural pools, are important recreational areas for local populations and are visited and preserved by local people.

Figure 9. General view of a water mill and weir on a temporary stream (Guadiana River basin, Portugal) (Photo, Manuela Morais).

The mills, often built at pristine river sites, had the disadvantage of being relatively isolated and distant from the population centers. Also, for this reason (i.e., the distance to the supply centers), great care was taken to maintain the weirs that supplied the water retained in the natural pools and additionally provided easily accessible food (fish caught in natural pools).

Abandonment of the mills and weirs was due to several factors, including the substantial reduction of bread baked in farmhouses, because of the mechanization of agriculture [79]. Also decisive in this process were the migratory flows of the rural population towards the industrialized centers of Portugal and Europe in the 1960s. Additional factors were (1) the transformation of people's eating habits, in which bread lost its value; (2) the lack of able men to work as millers; (3) the colonial war between Portugal and its African countries in the 1960s, which lasted until April 1974, with young people being recruited all over the country; and (4) the competition from milling factories and the increased production and widespread consumption of industrialized bread [71,79]. Generally, it was precisely these reasons that led to the deactivation of traditional milling systems in Portugal and other European countries.

The abandonment of mills, particularly the lack of maintenance of many weirs, led to the desiccation of the natural pools and, consequently, the loss of biodiversity and ecosystem degradation, increasingly being aggravated by the extreme drought phenomena and the extension of the dry summer period resulting from climate change in the region.

4. Discussion

The three case studies briefly described above illustrate how ecosystem balance is delicate and how human interventions in supporting natural processes can effectively enhance ecosystem services. Although the cases vary in time, space, and scale, we can distill several important lessons from their experiences, which can offer useful information in other restoration scenarios.

The cases are illustrative of compliance with global agreements and policy goals. All are fully aligned with the 2030 Agenda for Sustainable Development (AfSD), with its Sustainable Development Goals (SDGs) and its Target 6.6 ("*By 2020, protect and restore water-related ecosystems, including mountains, forests, wetlands, rivers, aquifers and lakes*") to support the achievement of the goal to ensure availability of water for all (SDG 6). Additionally, they exemplify the recent United Nations General Assembly recognition of the importance of the ecosystem approach for the integrated management of land, water, and living resources to face water scarcity when it approved the United Nations Decade on Ecosystem Restoration (2021–2030). The UNGA's A/RES/73/284 Resolution asserts that member states should "foster political will, the mobilization of resources, capacity-building, scientific research and cooperation and momentum for ecosystem restoration at the global, regional, national and local levels, as appropriate" and "to mainstream ecosystem restoration into policies and plans to address current national development priorities and challenges due to the degradation of marine and terrestrial ecosystems, biodiversity loss and climate change vulnerability, thereby creating opportunities for ecosystems to increase their adaptive capacity and opportunities to maintain and improve livelihoods for all".

The Brazil and Panama cases highlight the important benefits, particularly of water provisioning, resulting from the restoration and protection of tropical rainforests. Deforestation of the Tijuca Rainforest adjacent to Rio de Janeiro about two centuries ago caused severe soil erosion and diminished the water supply for this important city in the Portuguese colony. The Brazilian Emperor (Dom Pedro II) understood the relationship between a sustainable water supply and wise forest management, protection of vegetation near springs, and reforestation. As a result of these farsighted decisions, the Tijuca Rainforest (now a remnant of the endangered Mata Atlântica) has gained national protection as Tijuca National Park and international recognition as a UNESCO World Heritage Site of cultural importance. Several centuries later, in Panama in the 1970s, administrators of the Panama Canal became aware that the increasing deforestation in the canal watershed threatened the operations of the canal, particularly during the 5-month dry season. To protect the forest and its contribution to freshwater, the country designated several national parks on the eastern bank of the canal that strictly limited further forest clearing. During the past decade, canal authorities have dedicated significant efforts to support forest protection on the western side of the canal in areas with many small farms. These efforts include reforestation projects, payments to farmers to protect existing forests, and support for the cultivation of shade-grown coffee on lands that were previously pasture. The Panama experience is recent compared to the Brazilian case, yet the results to date are positive.

The interdependence between ecosystem restoration and water, particularly in tropical forest ecosystems, is clear. Forested headwater regions and also their protected springs are considered critical in sustaining water provision because they control flows within a catchment basin/watershed [81,82]. The role of forests in water provision and regulation was recognized and attributed to several categories of ecosystem services, including provisioning services (freshwater provision) and regulating services (flood regulation and water purification, among others) [83], as can be seen in Figure 10. These functions are a consequence of interactions between ecological processes and ecosystem structures [84]. According to Bonell [85], forest soils have many macropores associated with the high density of roots and soil fauna activity that facilitate infiltration and the vertical bypassing of the unsaturated matrix to reach the saturated zone more quickly. These characteristics help forests both to regulate and provide water for the catchment basin.

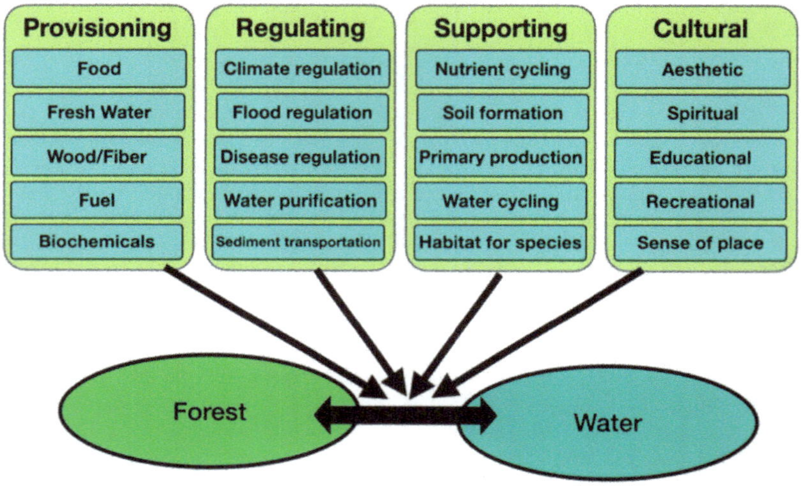

Figure 10. Ecosystem services and benefits that forests provide. Figure based on [83].

The role of forests in water provision and regulation can be seen in a systemic view by its action on the hydrological cycle. It can pump water into the atmosphere by evapotranspiration [35] or promote infiltration, storing water during the wet season and then slowly releasing it during the dry season. In this way, the Tijuca Forest in Brazil keeps rivers and streams with a modest but everlasting flow, and in Panama, forests also provide water for economic purposes, particularly during the dry season. Similarly, the ancient weirs in the riverbeds of Portugal's Alentejo Region address water scarcity through water storage. NbSs recognize interconnections between surface and groundwater and between green and blue water in a systemic way. Restored forests can retain more water, and this change in the hydrological distribution increases the availability of water for more plant growth. Meanwhile, more water will infiltrate and, therefore, will be available to return to surface water by groundwater surface discharge at springs. Ecological restoration per NbSs slows down water cycling; decreases erosion and flooding; reduces water flushed away; and improves the timing of water provisioning, keeping water in the local system. It is the opposite of the intensification of the water cycle [86].

All three cases illustrate that, while there may have been a sole principal objective of NbS activity, multiple benefits commonly result. The pioneering restoration experiences with native species in the Tijuca Forest aimed to solve the problem of the water supply to Rio de Janeiro. However, in the 20th century, restaurants opened, and landscaping projects were implemented in the protected forest. Today Tijuca National Park is an important site for tourism and public recreation. As a UNESCO World Heritage Site and Biosphere Reserve, the site is recognized for its spectacular cultural landscape and high biodiversity in the endangered Atlantic Forest. As an urban forest, Tijuca also serves as an important climate regulator and helps reduce air pollution. Likewise, in recent times, the protection of the forests surrounding the Panama Canal had the primary purpose of providing a sustainable water supply for canal operations. Nevertheless, today, the forests are also important for providing potable water for half of Panama's population and also generating electricity. The national parks in the watershed are also biodiversity hotspots, important links in the Mesoamerican Biological Corridor, and sites for ecotourism. Other efforts in the watershed assist small farmers to adopt sustainable agricultural techniques and improve their livelihoods. Thus, the ecosystem and social benefits of forest protection are multiple. Similarly, in Portugal's Alentejo Region, weirs on streams that served as hydraulic structures to support water mills that ground cereal grains in the past today help to moderate stream flow and retain water, particularly during dry seasons; attract wildlife; and are sites where people can recreate. All these examples illustrate that, while NbSs may initially focus on one ecosystem service, it can benefit a broad array of ecosystem services that also benefit local communities. A key feature of NbS is the capability to deliver groups of ecosystem services together—even if only one was originally targeted by the intervention [21]. Moreover, ecosystem services (as a concept per se detached and distinguished from economic fields such as valuation or privatization) link societal and environmental systems [87].

NbS are well-suited to address different variables of climate change (increased temperatures, changes in annual precipitation, extended dry days, and heavier rainfall). Climate scenarios for the Panama Canal Watershed and the Alentejo Region in Portugal suggest futures of diminished annual rainfall, and all three sites are projected to experience extended periods of consecutive dry days [88,89]. Dereczynski et al. [89] predict longer dry seasons, shorter wet seasons, and increased rainfall during heavy rain events in the Tijuca Forest. Similar scenarios are predicted for Panama and Central America [90]. Forest protection in the Tijuca Forest and Panama helps to maintain water flows in streams during dry periods and mitigate rapid runoff carrying high sediment loads during heavy rain events. Additionally, reforestation and the protection of existing forests serve as carbon sinks and help mitigate climate change.

The IPCC estimates [88] that a decrease in water availability is expected in the Alentejo Region in Portugal, with huge implications for the hydrological regime of intermittent rivers. An extension of the dry season is expected, which will cause fragmentation of aquatic habitats and loss of biodiversity. Consequently, these pressures threaten the sustainability of ecosystems, the provisioning of ecosystem services, and ultimately, human well-being. More broadly for the Mediterranean region, historical records show that the frequency and intensity of droughts have increased since 1950 and that the length of the dry period without flow is extending over time [91]. Temporary rivers with intermittent hydrological characteristics are, therefore, expected to cover more than 50 percent of the global hydrographic network [77,92]. Currently, it will be difficult to restore the ancient mills to produce cereal flour as was conducted in the past. However, the weirs and their supporting hydraulic structures that retain water are gaining increased importance by promoting water retention during the dry season. During the summer, these weirs, perfectly integrated into the riverbeds, are sites where aquatic and terrestrial species converge (Figure 11). Aquatic species recolonize the system from these sites after the flow restarts with the first autumn rains. Terrestrial species seek refuge in these cooler places that provide water and food. On the other hand, these hydraulic structures, many of them integrated into the Guadiana Valley National Park (https://natural.pt/protected-areas/parque-natural-vale-guadiana?locale=en accessed on 13 August 2024), are protected as places of leisure for local populations. These restoration actions with natural materials (stones and sand), which rescue ancestral knowledge rooted in regional culture, contribute to compliance with environmental policies, and, above all, constitute local NbSs that promote biodiversity and natural heritage by increasing (i) the water-retention capacity; (ii) the availability of habitats for aquatic and terrestrial species; (iii) the water quality and ecosystem integrity; (iv) the groundwater recharge; and (iv) ecosystem services also related to the promotion of the values of built heritage. In Southern Portugal, within the scope of different national and European policies, of which the Water Framework Directive stands out requiring the achievement of Good Status for all water bodies (Directive 2000/60/CE), actions that include the rehabilitation/reinforcement of ancient infrastructures (mills and dams) are a priority [93].

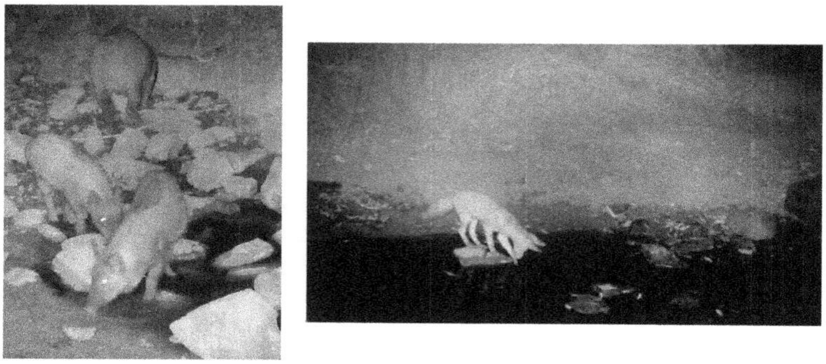

(**a**) (**b**)

Figure 11. A family of wild boars (**a**) and a fox (**b**) drinking water in a weir photographed with a night camera (Guadiana River watershed, Southern Portugal). (Photo, André Oliveira).

NbSs often tend to emphasize ecosystem benefits above benefits to communities. Both are integral components of the coupled human–natural system. The UN Resolution on the United Nations Decade on Ecosystem Restoration (2021–2030) warns that ecosystem restoration "needs to be carried out in ways that balance social, economic and environmental objectives and with the engagement of relevant stakeholders, including indigenous peoples and local communities". While the Tijuca Forest provides a sustainable water supply for nearby urban residents and areas for recreation and cultural appreciation, the case suggests that social impacts may not always have been so positive. Due to its historical context as a Portuguese colony, and prior to the abolition of slavery, the Emperor determined priorities, targeting the well-being of the Empire's capital (Rio de Janeiro), and using slaves for reforestation. Were communities displaced by restoration efforts? In Panama, the designation of Chagres National Park in 1984 strictly limited the ability of small farmers residing within the park's boundaries to clear additional land for agricultural purposes. While authorities have offered opportunities for additional livelihood activities, questions remain regarding social equity. On the other hand, canal authorities are actively engaged in promoting sustainable agricultural techniques that reduce sedimentation and rapid runoff from streams in the watershed.

Finally, an important factor in the successful implementation of NbSs is good governance, institutional coordination, meaningful public involvement, and strong political will.

In Panama, the ACP possesses clear competence to coordinate interventions of other state agencies and non-governmental organizations in the canal's watershed through clear statutory and constitutional mandates. The ACP encourages citizen involvement through watershed management councils that bring community concerns to the ACP and also permit authorities to explain their concerns and interventions. This is an example of nested governance [94].

The Tijuca Forest case in Brazil offers a different historical context. Abandoned coffee cultivation and the risk of strong water scarcity due to droughts led to degraded slopes in the capital of the Brazilian Empire. As restoration was a direct order of the Emperor and numerous slaves were used as workforce, the case is not completely compatible with today's realities. Nevertheless, the Tijuca Forest highlights some lessons about governance.

This large ecological restoration effort required 13 years and, considering the present labor laws, today could take much longer to meet the goal. On the other hand, modern machines and reforestation tools could hasten positive outcomes. Both a strong commitment from the government and compliance with restoration requirements are essential. Moreover, considering the long timeframe required to obtain results and climate-change scenarios, restoration efforts demand strong political will that persists despite short-term political changes.

Finally, after successful ecological restoration, the result must be protected. The creation of a protected area (Tijuca National Park plus recognition of World Heritage status by UNESCO) combines environmental protection with recreational uses. Similarly, a large portion of the Panama Canal Watershed enjoys national park designation. Protected status accompanied by an environmental education program with historical and cultural components could help people, politicians, and decision makers appreciate the restored landscapes.

Restoration efforts in Brazil and Panama offer clear and strong economic and social benefits. The benefits of ecosystem restoration in Alentejo are not so immediate, however. Most of the Portuguese water mills and their weirs have been abandoned, even though they constitute an important cultural heritage and are integrated into the rivers' ecological functioning. In a few rare cases, parish councils consider them to be recreational areas for local people and have undertaken restoration efforts. New incentives for restoration may provide opportunities for Alentejo. The European Union's new economic growth strategy has placed climate change at the center of political decisions through the launch of the European Green Deal (EGD) [95]. In line with this strategy, the EU launched the Water4All partnership in the Horizon Europe program as an integrated approach to water security in Europe because 'water is central to all components of the EGD'. The Water4All partnership was created to combat issues of fragmentation in research and innovation on water-related topics and to facilitate change in water security and management in Europe (in the sense of the more holistic approach referred to by Scott et al. [96]. In this context, initiatives and calls for projects have been promoted to implement more sustainable solutions together with stakeholders. The European Union aims to encourage cooperation and provide technical assistance, particularly to European Neighborhood Policy South countries, to develop effective green growth and climate-governance mechanisms, especially given the region's particular vulnerability to climate change [97]. In this context, NbSs gain increasing relevance as actions that address societal challenges through protection, sustainable management, and ecosystem restoration, benefiting both biodiversity and human well-being. The restoration of weirs, which are an integral part of the functioning of rivers and landscapes in Southern Portugal, could be framed within this more holistic approach. They can function as living examples of NbSs rescued from the past, which, if well managed, would contribute to water retention, the promotion of biodiversity, and the fight against climate change in a region where water will be a scarce resource.

5. Conclusions

Growing worldwide recognition of the importance of water and concern about its limited availability to meet human and ecological needs considering global climate change are leading to new strategies and solutions, including NbSs. Guidance for the transition to a socio-environmental approach to ecosystem management exists at the global level through the 17 UN Sustainable Development Goals [98], at the European level through the European Green Deal [95], and through policies and programs in many nations. We must invest in a new productive rationality based on the limits of the laws of nature, ecological potential, social justice, the production of social meanings, and human creativity that integrates "knowledge" without exclusion.

Nature-based solutions present a credible means to address key societal issues, such as climate change, biodiversity loss, and environmental justice. Policy dialogues and global awareness-raising initiatives need to be actively and consistently implemented to promote participation, develop a broad knowledge base, and support the adoption of nature-based solutions on a large scale. For example, the promotion of interdisciplinary or even transdisciplinary research into approaches to water scarcity, as well as the integration of stakeholders, offers the possibility of creating clear frameworks and understanding ecosystem functions and the coupled human–natural system. In fact, the factors affecting the demand for water, such as changes in land use, perceptions of water scarcity, and attitudes towards water use, are framed in social science understandings of how these factors can be influenced by government policies and social norms [99]. Furthermore, a new opportunity is emerging to make the social sciences more effective in improving water management and understanding the drivers of water scarcity [100,101]. This approach promotes public and private investment in nature-based solutions and is an essential component of the transformational change needed to make our societies and economies more sustainable and just.

Author Contributions: Conceptualization, D.O.S., M.M. and C.H.S.; investigation, D.O.S., M.M. and C.H.S.; writing—original draft preparation, D.O.S., M.M. and C.H.S.; writing—review and editing, D.O.S., M.M. and C.H.S.; funding acquisition, D.O.S., M.M. and C.H.S. All authors have read and agreed to the published version of the manuscript.

Funding: The European Union's ERASMUS AMIGO 2019-1-KA107-060632 funded the authors' travel between Portugal (University of Évora), Brazil (University of Brasilia), and the USA (University of Miami).

Data Availability Statement: No new data were created or analyzed in this study. Data sharing is not applicable to this article.

Conflicts of Interest: The authors declare no conflicts of interest. The funders had no role in the design of the study; in the collection, analyses, or interpretation of data; in the writing of the manuscript; or in the decision to publish the results.

References

1. UN-Water. *Coping with Water Scarcity: A Strategic Issue and Priority for System-Wide Action*; UN-Water: Geneva, Switzerland, 2006.
2. UN-Water. *Coping with Water Scarcity: Challenge of the Twenty-First Century*; UN-Water: Geneva, Switzerland, 2007.
3. Steduto, P.; Faurès, J.-M.; Hoogeveen, J.; Winpenny, J.; Burke, J. *Coping with Water Scarcity—An Action Framework for Agriculture and Food Security*; Food and Agriculture Organization of the United Nations: Rome, Italy, 2012.
4. Mekonnen, M.M.; Hoekstra, A.Y. Four billion people facing severe water scarcity. *Sci. Adv.* **2016**, *2*, e1500323. [CrossRef]
5. van Beek, L.P.H.; Wada, Y.; Bierkens, M.F.P. Global monthly water stress: 1. Water balance and water availability. *Water Resour. Res.* **2011**, *47*, W07517. [CrossRef]
6. Mishra, B.K.; Kumar, P.; Saraswat, C.; Chakraborty, S.; Gautam, A. Water security in a changing environment: Concept, challenges and solutions. *Water* **2021**, *13*, 490. [CrossRef]
7. Cosgrove, W.J.; Loucks, D.P. Water management: Current and future challenges and research directions. *Water Resour. Res.* **2015**, *51*, 4823–4839. [CrossRef]
8. Falkenmark, M. Meeting water requirements of an expanding world population. *Philos. Trans. B R. Soc.* **1997**, *352*, 929–936. [CrossRef]
9. Falkenmark, M. Society's interaction with the water cycle: A conceptual framework for a more holistic approach. *Hydrol. Sci. J.* **1997**, *42*, 451–466. [CrossRef]
10. Vanham, D.; Alfieri, L.; Flörke, M.; Grimaldi, S.; Lorini, V.; de Roo, A.; Feyen, L. The number of people exposed to water stress in relation to how much water is reserved for the environment: A global modelling study. *Lancet Planet. Health* **2021**, *5*, e766–e774. [CrossRef] [PubMed]
11. Yang, D.; Yang, Y.; Xia, J. Hydrological cycle and water resources in a changing world: A review. *Geogr. Sustain.* **2021**, *2*, 115–122. [CrossRef]
12. Bierkens, M.F.; Wada, Y. Non-renewable groundwater use and groundwater depletion: A review. *Environ. Res. Lett.* **2019**, *14*, 063002. [CrossRef]
13. Camacho, C.; Negro, J.; Elmberg, J.; Fox, A.D.; Nagy, S.; Pain, D.J.; Green, A.J. Groundwater extraction poses extreme threat to Doñana World Heritage Site. *Nat. Ecol. Evol.* **2022**, *6*, 654–655. [CrossRef]
14. Palmer, M.; Ruhi, A. Linkages between flow regime, biota, and ecosystem processes: Implications for river restoration. *Science* **2019**, *365*, eaaw2087. [CrossRef] [PubMed]
15. Davidson, C.N. How Much Wetland Has the World Lost? Long Term and Recent Trends in Global Wetland Area. *Mar. Freshw. Res.* **2014**, *65*, 934–941. [CrossRef]

16. Verhoeven, J.T.A. Wetlands in Europe: Perspectives for restoration of a lost paradise. *Ecol. Eng.* **2014**, *66*, 6–9. [CrossRef]
17. Spencer, T.; Schuerch, M.; Nicholls, R.J.; Hinkel, J.; Lincke, D.; Vafeidis, A.T.; Reef, R.; McFadden, L.; Brown, S. Global coastal wetland change under sea-level rise and related stresses: The DIVA Wetland Change Model. *Glob. Planet. Chang.* **2016**, *136*, 15–30. [CrossRef]
18. Weiskopf, S.R.; Rubenstein, M.A.; Crozier, L.G.; Gaichas, S.; Griffis, R.; Halofsky, J.E.; Hyde, K.J.W.; Morelli, T.L.; Morisette, J.T.; Muñoz, R.C.; et al. Climate change effects on biodiversity, ecosystems, ecosystem services, and natural resource management in the United States. *Sci. Total Environ.* **2020**, *733*, 137782. [CrossRef] [PubMed]
19. Wheeler, K.G.; Hussein, H. Water research and nationalism in the post-truth era. *Water Int.* **2021**, *46*, 1216–1223. [CrossRef]
20. Seckler, D. *The New Era of Water Resources Management: From "Dry" to "Wet" Water Savings*; International Irrigation Management Institute (IIMI): Colombo, Sri Lanka, 1996. Available online: https://www.iwmi.cgiar.org/publications/iwmi-research-reports/iwmi-research-report-1/ (accessed on 13 August 2024).
21. Saito, C.H. O estruturalismo na ecologia da paisagem. *Braz. J. Ecol.* **1998**, *2*, 47–56.
22. Beschta, R.L.; Ripple, W.J. The role of large predators in maintaining riparian plant communities and river morphology. *Geomorphology* **2012**, *157–158*, 88–98. [CrossRef]
23. Keesstra, S.; Nunes, J.; Novara, A.; Finger, D.; Avelar, D.; Kalantari, Z.; Cerdà, A. The superior effect of nature based solutions in land management for enhancing ecosystem services. *Sci. Total Environ.* **2018**, *610–611*, 997–1009. [CrossRef]
24. WWAP (United Nations World Water Assessment Programme)/UN-Water. *The United Nations World Water Development Report 2018: Nature-Based Solutions for Water*; UNESCO: Paris, France, 2018; ISBN 978-92-3-100264-9.
25. Michels-Brito, A.; Ferreira, J.C.R.; Saito, C.H. The source-to-sea landscape: A hybrid integrative territory management approach. *Sci. Total Environ.* **2024**, *931*, 172961. [CrossRef]
26. Saito, C.H.; Morais, M.M. Learning from the past: What cultural heritage can teach us about water storage and management. In *The Palgrave Handbook of Global Sustainability*; Brinkmann, R., Ed.; Palgrave Macmillan: Cham, Switzerland, 2022; pp. 437–457. [CrossRef]
27. Cohen-Shacham, E.; Walters, G.; Janzen, C.; Maginnis, S. *Nature-Based Solutions to Address Global Societal Challenges*; IUCN: Gland, Switzerland, 2016; ISBN 978-2-8317-1812-5. [CrossRef]
28. Nesshöver, C.; Assmuth, T.; Irvine, K.N.; Rusch, G.M.; Waylen, K.A.; Delbaere, B.; Haase, D.; Jones-Walters, L.; Keune, H.; Kovacs, E.; et al. The science, policy and practice of nature-based solutions: An interdisciplinary perspective. *Sci. Total Environ.* **2017**, *579*, 1215–1227. [CrossRef]
29. UN-Water. *Water Security and the Global Water Agenda: A UN-Water Analytical Brief*; UN University: Hamilton, ON, Canada, 2013; ISBN 978-92-808-6038-2.
30. Scott, C.A.; Meza, F.J.; Varady, R.G.; Tiessen, H.; McEvoy, J.; Garfin, G.M.; Wilder, M.; Farfán, L.M.; Pablos, N.P.; Montaña, E. Water security and adaptive management in the arid Americas. *Ann. Assoc. Am. Geogr.* **2013**, *103*, 280–289. [CrossRef]
31. WWAP (United Nations World Water Assessment Programme)/UN-Water. *The United Nations World Water Development Report 4: Managing Water under Uncertainty and Risk*; UNESCO: Paris, France, 2012; ISBN 978-92-3-104235-5/978-92-3-001045-4.
32. Zheng, N.; Ma, H.; Peng, C.; Liu, Y.; Grotewal, R.; Klaus, U. Resilient ecology and landscape systems of the Fengxinglong Ecological Park, Sanya. *Landsc. Archit. Front.* **2018**, *6*, 32–41. [CrossRef]
33. Van Coppenolle, R.; Schwarz, C.; Temmerman, S. Contribution of mangroves and salt marshes to nature-based mitigation of coastal flood risks in major deltas of the world. *Estuaries Coasts* **2018**, *41*, 1699–1711. [CrossRef]
34. Song, S.; Ding, Y.; Li, W.; Meng, Y.; Zhou, J.; Gou, R.; Zhang, C.; Ye, S.; Saintilan, N.; Krauss, K.W.; et al. Mangrove reforestation provides greater blue carbon benefit than afforestation for mitigating global climate change. *Nat. Commun.* **2023**, *14*, 756. [CrossRef] [PubMed]
35. Aragão, L.E.O.C. The rainforest's water pump. *Nature* **2012**, *489*, 217–218. [CrossRef] [PubMed]
36. Bender, G. Correcting the Rhone and the Valaisans: Three centuries of works and debate. *Rev. Géographie Alp.* **2004**, *92*, 62–72. [CrossRef]
37. Maya, R.O.C. *A Floresta da Tijuca*; Bloch Edições: Rio de Janeiro, Brazil, 1967. Available online: http://objdigital.bn.br/objdigital2/acervo_digital/div_obrasgerais/drg605826/drg605826.pdf (accessed on 27 June 2024).
38. Dorigo, T.A.; Siqueira, C.C.; Oliveira, J.C.F.; Fusinatto, L.A.; Santos-Pereira, M.; Almeida-Santos, M.; Maia-Carneiro, T.; Reis, C.N.C.; Rocha, C.F.D. Amphibians and reptiles from the Parque Nacional da Tijuca, Brazil, one of the world's largest urban forests. *Biota Neotrop.* **2021**, *21*, e20200978. [CrossRef]
39. Freitas, S.R.; Neves, C.L.; Chernicharo, P. Tijuca National Park: Two pioneering restorationist initiatives in Atlantic forest in southeastern Brazil. *Braz. J. Biol.* **2006**, *66*, 975–982. [CrossRef]
40. Oklander, L.I.; Rheingantz, M.; Rossato, R.S.; Peker, S.; Hirano, Z.M.B.; Monticelli, C.; Dada, A.N.; Di Nucci, D.L.; Oliveira, D.; de Melo, F.R.; et al. Restoration of Alouatta guariba populations: Building a binational management strategy for the conservation of the endangered brown howler monkey of the Atlantic Forest. *Front. Conserv. Sci.* **2024**, *5*, 1401749. [CrossRef]
41. Mynssen, C.M.; Bicalho, M.B.; Sylvestre, L.S.; Rocha, T.; Siqueira, M.F. Ferns and Lycophytes as new challenges: Richness of ferns and lycophytes from Tijuca National Park, an urban forest. *Rodriguésia* **2023**, *74*, e00782023. [CrossRef]
42. Drummond, J. The garden in the machine: An environmental history of the Tijuca Forest (Rio de Janeiro, Brazil), 1862–1889. *J. Environ. Hist.* **1996**, *1*, 83–104. [CrossRef]

43. Coelho Netto, A.L. A interface florestal-urbana e os desastres naturais relacionados à água no maciço da tijuca: Desafios ao planejamento urbano numa perspectiva sócio-ambiental. *Rev. Do Dep. De Geogr.* **2005**, *16*, 46–60. [CrossRef]
44. Martins, M.F.V. A floresta e as águas do Rio: A Inspeção Geral de Obras Públicas e as intervenções urbanas para abastecimento e reflorestamento na primeira metade do século XIX. *Intellèctus* **2015**, *14*, 21–47. [CrossRef]
45. Panama Canal Authority's Direct Contributions to the National Treasury from FY 2016 to FY 2021, by Type. Available online: https://www.statista.com/statistics/1011908/panama-canal-authority-contributions-national-treasury-type/ (accessed on 27 July 2024).
46. Panama: Gross Domestic Product (GDP in Current Prices from 1989 to 2029). Available online: https://www.statista.com/statistics/454680/gross-domestic-product-gdp-in-panama/ (accessed on 27 July 2024).
47. From Cross-Fillings to Long-Term Solutions: How the Panama Canal Is Addressing the Issue of Water Head On. Available online: https://pancanal.com/en/how-the-panama-canal-is-addressing-the-issue-of-water-head-on/ (accessed on 30 July 2024).
48. Calvo Gobbetti, L.; Ríos Córdoba, K. Impacto de la ampliación del Canal de Panamá en las confiabilidades hidrica y de calado. *Rev. I+D Tecnológico* **2024**, *20*, 49–60. [CrossRef]
49. CEPAL (Comisión Económica para America Latina y el Caribe); Fondo Nórdico de Desarrollo (FND); Banco Interamericano de Desarrollo (BID); Secretaria Nacional de Energia de Panamá; y Ministerio de Energia y Minas de la Republica Dominicana. Impactos Potenciales del Cambio Climático en el Ámbito Hidroelectrico en Panamá y la Republica Dominicana. LC/MEX/TS.1217/28. Ciudad de Mexico, Mexico, 2017. Available online: https://repositorio.cepal.org/server/api/core/bitstreams/afca41c0-8cdf-47ab-af33-855265f2d0cd/content (accessed on 13 August 2024).
50. Hidalgo, H.G.; Amador, J.A.; Alfaro, E.J.; Quesada, B. Hydrological climate change projections for Central America. *J. Hydrol.* **2013**, *495*, 4–112. [CrossRef]
51. Canal de Panamá. Informe 2023. Available online: https://pancanal.com/wp-content/uploads/2021/09/Informe-2023 EspFINAL.pdf (accessed on 26 June 2024).
52. Canal de Panamá. Agua y el canal. Available online: https://pancanal.com/agua/ (accessed on 26 June 2024).
53. Maritime Executive. Panama Canal to Add Back Daily Transits as Rainy Season Approaches (16 April 2024). Available online: https://maritime-executive.com/article/panama-canal-to-add-back-daily-transits-as-rainy-season-approaches (accessed on 26 June 2024).
54. Sandoval, Y.; Mojica, Y. Las Bombas del Idaan No Funcionan al 100% Cuando Baja el Nivel del Lago Alajuela; Acuden a Carros Cisterna. La Prensa (Panama). 18 May 2024. Available online: https://www.prensa.com/economia/las-bombas-del-idaan-no-funcionan-al-100-cuando-baja-el-nivel-del-lago-alajuela-acuden-a-carros-cisternas/ (accessed on 26 June 2024).
55. Panama City, Panama Metro Area Population 1950–2024. Available online: https://www.macrotrends.net/global-metrics/cities/22063/panama-city/population#google_vignette (accessed on 30 July 2024).
56. Carse, A. *Beyond the Big Ditch: Politics, Ecology, and Infrastructure at the Panama Canal*; MIT Press: Cambridge, MA, USA, 2014.
57. Wadsworth, F. Deforestation: Death to the Panama Canal. In *US Strategy Conference on Tropical Deforestation*; USAID: Washington, DC, USA, 1978; pp. 22–24.
58. Buckingham, K.; Hanson, C. *The Restoration Diagnostic—Case Example: Panama Canal Watershed*; World Resources Institute: Washington, DC, USA, 2015.
59. Dale, V.H.; Brown, S.; Calderón, M.O.; Montoya, A.S.; Martinez, R.E. Projected land-use change for the Eastern Panama Canal Watershed and its potential impact. In *The Rio Chagres, Panama: A Multidisciplinary Profile of a Tropical Eatershed*; Harmon, R.S., Ed.; Springer: Dordrecht, The Netherlands; Berlin/Heidelberg, Germany, 2010; pp. 337–345.
60. Canal de Panamá. Acuerdo No. 116—Por el cual se aprueba el Reglamento sobre Ambiente, Cuenca Hidrográfica y Comisión Interinstitucional de la Cuenca Hidrográfica del Canal de Panamá (27 July 2006). Available online: https://pancanal.com/wp-content/uploads/2021/09/acuerdo116-1.pdf (accessed on 26 June 2024).
61. Instituto Geográfico Nacional "Tommy Guardia". *Atlas Nacional de Panamá*; Instituto Geográfico Nacional "Tommy Guardia": Panama, Panama, 1988.
62. Simonit, S.; Perrings, C. Bundling ecosystem services in the Panama Canal watershed. *Proc. Natl. Acad. Sci. USA* **2013**, *110*, 9326–9331. [CrossRef] [PubMed]
63. Ogden, F.L.; Crouch, T.D.; Stallard, R.F.; Hall, J.S. Effect of land cover and use on dry season river runoff, runoff efficiency, and peak storm runoff in the seasonal tropics of Central Panama. *Water Resour. Res.* **2013**, *49*, 8443–8462. [CrossRef]
64. Zafirah, N.; Nurin, N.A.; Samsurijan, M.S.; Zuknik, M.H.; Rafatullah, M.; Syaki, M.I. Sustainable ecosystem services framework for tropical catchment management: A review. *Sustainability* **2017**, *9*, 546. [CrossRef]
65. Canal de Panamá. 2610- EAC- 115 Norma Ambiental para Pagos por Servicios Ambientales (PSA) en la Cuenca Hidrográfica del Canal de Panamá (1 March 2018). Available online: https://pancanal.com/wp-content/uploads/2022/03/EAC-115-pagos-por-servicios-ambientales.pdf (accessed on 26 June 2024).
66. Martínez, R. Programa de Incentivos Económicos Ambientales (PIEA). Presentation to University of Miami Students, Watershed Management Section, Canal de Panamá, Corozal, Panama. 11 March 2024.
67. Moreno, O.; Aguilar, J. Towards a knowledge management system for the strengthening of coffee production: A case study in the Panama Canal Basin, Panamá Oeste province. *Green Technol. Sustain.* **2024**, *2*, 100056. [CrossRef]
68. Eco TV (Panama). Bayano es 3 veces más caro que un Embalse en Río Indio, Explicó Canal de Panamá (18 January 2024). Available online: https://www.youtube.com/watch?v=0ArOyH9DIXc (accessed on 24 June 2024).

69. Voz de América. Proyecto en río Indio Buscaría Solucionar crisis del Canal de Panamá, pero se Necesitan Cambios en la Legislación. (8 Feburary 2024). Available online: https://www.vozdeamerica.com/a/rio-indio-solucion-crisis-hidrica-canal-panama/7477229.html (accessed on 24 June 2024).
70. CCDR, *Relatório Portugal 2020 na Região Alentejo 2017*. Orgão de Acompanhamento das Dinâmicas Regionais, Comissão de Coordenação e Desenvolvimento Regional do Alentejo. 2018. Available online: https://www.ccdr-a.gov.pt/wp-content/uploads/2018/07/Relatorio_Portugal2020_2017_F.pdf (accessed on 27 June 2024).
71. Cutileiro, J. *Ricos e Pobres no Alentejo. (Uma Sociedade Rural Portuguesa)*; Livros Horizonte: Lisboa, Portugal, 2004.
72. Pinto-Correia, T.; Godinho, S. Changing agriculture—Changing landscapes: What is going on in the high Valued montado. In *Agriculture in Mediterranean Europe: Between Old and New Paradigms (Research in Rural Sociology and Development*; Ortiz-Miranda, D., Moragues-Faus, A., Arnalte-Alegre, E., Eds.; Emerald Group Publishing Limited: Leeds, UK, 2013; Volume 19, pp. 75–90. [CrossRef]
73. Silveira, A.; Ferrão, J.; Muñoz-Rojas, J.; Pinto-Correia, T.; Guimarães, M.H.; Schmidt, L. The sustainability of agricultural intensification in the early 21st century: Insights from the olive oil production in Alentejo (Southern Portugal). In *Changing Societies: Legacies and Challenges. The Diverse Worlds of Sustainability*; Delicado, A., Domingos, N., de Sousa, L., Eds.; Imprensa de Ciências Sociais: Lisbon, Portugal, 2018; Volume 3, pp. 247–275. [CrossRef]
74. Veiga, B.; Duarte, L.; Vasconcelos, L. A Barragem do Alqueva para quem? Por uma contextualização pluridimensional do desenvolvimento no Alentejo—Portugal. In *IV Encontro Nacional da ANPPAS, Anais do IV Encontro Nacional da ANPPAS*; Associação Nacional de Pós-Graduação e Pesquisa em Ambiente e Sociedade: Brasilia, Brazil, 2008.
75. Sanches, R.; Pedro, J.O. *Empreendimento de Fins Múltiplos de Alqueva*; Empresa de Desenvolvimento de Infraestruturas do Alqueva (EDIA): Beja, Portugal, 2006.
76. Carvalho, A.; Schmidt, L.; Santos, F.D.; Delicado, A. Climate change research and policy in Portugal. *Rev. Clim. Chang.* **2014**, *5*, 199–217. [CrossRef]
77. Skoulikidis, N.T.; Sabater, S.; Datry, T.; Morais, M.M.; Buffagni, A.; Dörflinger, G.; Zogaris, S.; del Mar Sánchez-Montoya, M.; Bonada, N.; Kalogianni, E.; et al. Non-perennial Mediterranean rivers in Europe: Status, pressures, and challenges for research and management. *Sci. Total Environ.* **2017**, *577*, 1–18. [CrossRef] [PubMed]
78. Costa, M.R.; Costa, A.M.; Teixeira, E.R.; Ribeiro, F.V.; Santos, M.; Malobbia, S.; Matias, V. *Património Rural Construído do Baixo Guadiana*; Odiana: Algarve, Portugal, 2004; pp. 1–216. Available online: http://hdl.handle.net/10400.1/1823 (accessed on 13 August 2024).
79. Silva, R. Moinhos e moleiros no Alentejo oriental: Uma perspectiva etnográfica. *Etnográfica* **2004**, *8*, 221–242. [CrossRef]
80. Guita, R. Por baixo de cada moinho do Guadiana está um açude no Guadiana. In Proceedings of the Actes II Jornades de Molinologia = Actas II Jornadas de Molinología, Terrasa, Barcelona and Cérvoles, Lleida, Spain, 30 September–3 October 1998; pp. 351–363.
81. Baker, T.; Kiptala, J.; Olaka, L.; Oates, N.; Hussain, A.; McCartney, M. *Baseline Review and Ecosystem Services Assessment of the Tana River Basin, Kenya*; International Water Management Institute (IWMI): Colombo, Sri Lanka, 2015; Volume 165.
82. Dib, V.; Brancalion, P.H.; Chan Chou, S.; Cooper, M.; Ellison, D.; Farjalla, V.F.; Filoso, S.; Meli, P.; Pires, A.P.; Rodriguez, D.A.; et al. Shedding light on the complex relationship between forest restoration and water services. *Restor. Ecol.* **2023**, *31*, e13890. [CrossRef]
83. Millennium Ecosystem Assessment. *Ecosystems and Human Well-being: Synthesis*; Island Press: Washington, DC, USA, 2005.
84. de Groot, R.S.; Wilson, M.; Boumans, R.M.J. A typology for the classification, description, and valuation of ecosystem functions, goods and services. *Ecol. Econ.* **2002**, *41*, 393–408. [CrossRef]
85. Bonell, M. Progress in the understanding of runoff generation dynamics in forests. *J. Hydrol.* **1993**, *150*, 217–275. [CrossRef]
86. Huntington, T.G. Evidence for intensification of the global water cycle: Review and synthesis. *J. Hydrol.* **2006**, *319*, 83–95. [CrossRef]
87. Engel, S.; Schaefer, M. Ecosystem services—A useful concept for addressing water challenges? *Curr. Opin. Environ. Sustain.* **2013**, *5*, 696–707. [CrossRef]
88. IPCC (Intergovernmental Panel on Climate Change). *Climate Change 2022: Impacts, Adaptation and Vulnerability. Contribution of Working Group II to the Sixth Assessment Report of the Intergovernmental Panel on Climate Change*; Cambridge University Press: Cambridge, UK; New York, NY, USA, 2022. [CrossRef]
89. Dereczynski, C.; Silva, W.L.; Marengo, J. Detection and projections of climate change in Rio de Janeiro, Brazil. *Am. J. Clim. Chang.* **2013**, *2*, 25–33. [CrossRef]
90. Imbach, P.; Chou, S.C.; Lyra, A.; Rodrigues, D.; Rodriguez, D.; Latinovic, D.; Siqueira, G.; Silva, A.; Garofolo, L.; Georgiou, S. Future climate change scenarios in Central America at high spatial resolution. *PLoS ONE* **2013**, *13*, e0193570. [CrossRef]
91. Fader, M.; Giupponi, C.; Burak, S.; Dakhlaoui, H.; Koutroulis, A.; Lange, M.A.; Llasat, M.C.; Pulido-Velazquez, D.; Sanz-Cobeña, A. Water. In *Climate and Environmental Change in the Mediterranean Basin—Current Situation and Risks for the Future. First Mediterranean Assessment Report*; Union for the Mediterranean, Plan Bleu, UNEP/MAP: Marseille, France, 2020.
92. Rosado, J.; Morais, M.; Serafim, A.; Pedro, A.; Silva, H.; Potes, M.; Brito, D.; Salgado, R.; Neves, R.; Lillebø, A.; et al. Key factors in the Management and Conservation of temporary Mediterranean streams: A case study of the Pardiela river, southern Portugal (Guadiana catchment). In *River Conservation and Management*; Boon, P.J., Raven, P.J., Eds.; John Wiley & Sons, Ltd.: Chichester, UK, 2012; pp. 273–283.

93. Directive 2000/60/CE. Directive 2000/60/EC of the European Parliament and of the Council of 23 October 2000 Establishing a Framework for Community Action in the Field of Water Policy. Available online: https://eur-lex.europa.eu/eli/dir/2000/60/oj (accessed on 13 August 2024).
94. Ostrom, E. Polycentric systems: Multilevel governance involving a diversity of organizations. In *Global Environmental Commons Analytical and Political Challenges in Building Governance Mechanisms*; Brousseau, E., Dedeurwaerdere, T., Jouvet, P.A., Willinger, M. Eds.; Oxford University Press: Oxford, UK, 2012; pp. 105–125.
95. European Commission. The European Green Deal. European Commission: Brussels, Belgium. 11.12.2019 COM, 640 Final, 2019. Available online: https://eur-lex.europa.eu/legal-content/EN/TXT/?uri=COM:2019:640:FIN (accessed on 13 August 2024).
96. Scott, A.; Holtby, R.; East, H.; Lannin, A. Mainstreaming the environment: Exploring pathways and narratives to improve policy and decision-making. *People Nat.* **2022**, *4*, 201–217. [CrossRef]
97. Sandri, S.; Hussein, H.; Alshyab, N.; Sagatowski, J. The European Green Deal: Challenges and opportunities for the Southern Mediterranean. *Mediterr. Politics* **2023**, 1–12. [CrossRef]
98. UN General Assembly, Transforming Our world: The 2030 Agenda for Sustainable Development, A/RES/70/1, 21 October 2015. Available online: https://www.refworld.org/legal/resolution/unga/2015/en/111816 (accessed on 27 June 2024).
99. Wolters, E.A. Attitude–behavior consistency in household water consumption. *Soc. Sci. J.* **2014**, *51*, 455–463. [CrossRef]
100. Liu, J.; Yang, H.; Gosling, S.N.; Kummu, M.; Flörke, M.; Pfister, S.; Hanasaki, N.; Wada, Y.; Zhang, X.; Zheng, C.; et al. Water scarcity assessments in the past, present, and future. *Earth's Future* **2017**, *5*, 545–559. [CrossRef] [PubMed]
101. Lund, J.R. Integrating social and physical sciences in water management. *Water Resour. Res.* **2015**, *51*, 5905–5918. [CrossRef]

water

MDPI

Review

A Global Review of Progress in Remote Sensing and Monitoring of Marine Pollution

Jingwu Ma [1,2,3], Renfeng Ma [3,4,5,*], Qi Pan [4], Xianjun Liang [6], Jianqing Wang [7] and Xinxin Ni [8]

[1] Key Laboratory of Ocean Space Resource Management Technology of Ministry of Natural Resources, Zhejiang Institute of Marine Sciences, Hangzhou 310012, China; majingwu12@163.com
[2] Land Consolidation and Rehabilitation Center, Wenzhou Bureau of Natural Resources and Planning, Wenzhou 325027, China
[3] Ningbo University Donghai Academy & Zhejiang Ocean Development Think Tank Alliance, Ningbo 315211, China
[4] Department of Geography and Spatial Information Techniques, Ningbo University, Ningbo 315211, China; 2011087072@nbu.edu.cn
[5] Zhejiang Collaborative Innovation Center & Ningbo Universities Collaborative Innovation Center for Land and Marine Spatial Utilization and Governance Research, Ningbo University, Ningbo 315211, China
[6] Ningbo Junzhi Engineering and Environmental Consulting Co., Ltd., Ningbo 315211, China; nbuandrew@163.com
[7] Ningbo Institute of Oceanography, Ningbo 315832, China; wangjq@nbio.org.cn
[8] Department of International Hotel Management at Qiandaohu, Tourism College of Zhejiang China, Hangzhou 311231, China; nixinxin@tourzj.edu.cn
* Correspondence: marenfeng@nbu.edu.cn

Abstract: With the rapid development of urbanization and industrialization, human activities have caused marine pollution in three ways: land source, air source, and sea source, leading to the problem of marine environments. Remote sensing, with its wide coverage and fast and accurate monitoring capability, continues to be an important tool for marine environment monitoring and evaluation research. This paper focuses on the three types of marine pollution, namely marine seawater pollution, marine debris and microplastic pollution, and marine air pollution. We review the application of remote sensing technology methods for monitoring marine pollution and identify the limitations of existing methods. Marine seawater pollution can be effectively monitored by remote sensing technology, especially where traditional monitoring methods are inadequate. For marine debris and microplastic pollution, the monitoring methods are still in the early stages of development and require further research. For marine air pollution, more air pollution parameters are required for accurate monitoring. Future research should focus on developing marine remote sensing with data, technology, and standard sharing for three-dimensional monitoring, combining optical and physical sensors with biosensors, and using multi-source and multi-temporal monitoring data. A marine multi-source monitoring database is necessary to provide an immediately available basis for coastal and marine governance, improve marine spatial planning, and help coastal and marine protection.

Keywords: marine environmental monitoring; governance; marine pollution; coastal zone management

1. Introduction

The ocean is an important environment for humans, and its various types and scales of currents result in the distribution of liquid and three-dimensional resources, which form an important and unique system of marine resources and environment. The ocean is abundant in biological resources, mineral resources, energy, and other resources, and it is attracting increasing attention from the academic circle and industry. The rapid development of industrialization and urbanization has led to a concentration of human activities along coastlines, resulting in the degradation of coastal bays, marine resources, and the environment [1]. Generally, the coastal natural shoreline and coastal mudflat wetlands have been continuously reduced, the area of mangroves and coral reefs has been greatly reduced, and the marine ecological environment has been polluted. The situation is becoming increasingly serious, with increasing eutrophication of seawater and frequent occurrences of marine ecological disasters such as brown tides, green tides, and red tides posing a serious threat to migratory waterfowl and marine biodiversity. Meanwhile, the above environmental pollution, biological extinction, and natural disasters have posed a threat to the sustainable development of the coastal and marine environment. Dissolved organic matter from sewage treatment plants, humus from farmland, anthropogenic shoreline erosion, and the removal of native vegetation can cause significant increases in turbidity in coastal waters [2]. For example, NASA satellite imagery has shown that the water quality of Florida's Tampa Bay decreases in the winter months compared to the summer. More particles suspended in the water, a measure called turbidity, show up as yellow, orange, and red in December (a) than in July (b) due to seasonal freshwater discharge from nearby rivers and runoff into the bay, which carry nutrients (Figure 1). Hence, marine pollution monitoring is of great significance in terms of both theoretical and practical value [3–5]. Marine environment monitoring is an essential step in maintaining the quality of the marine environment and securing its ecology, and it is crucial for achieving sustainable marine development.

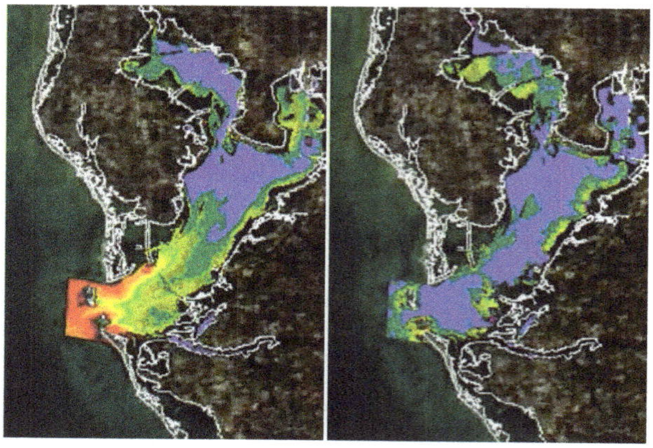

(**a**) Winter (**b**) Summer

Figure 1. Water quality of Florida's Tampa Bay decreases (Picture source NASA/USF).

Remote sensing plays a crucial role in marine monitoring, including the ecological monitoring of wetland, mangrove, coral reefs, and other key organisms (Figures 2 and 3), the environmental disaster monitoring of oil spills and red tides, the mapping of mesoscale coastlines, the measurement of chlorophyll-a, suspended sediment concentration, seawater temperature in offshore areas, and the automatic recognition of key targets such as dams, breeding areas, buildings, ports, bridges, and ships [6–8].

Based on a search of the Web of Science core collection from 2000 by using the keyword "marine remote sensing," it can be seen that marine remote sensing monitoring is a crucial aspect of marine ecological monitoring (Table 1). Remote sensing monitoring has been of interest to scholars from institutions that have published a number of important papers in this field, such as the Woods Hole Oceanographic Institution, the University of Washington, the University of California, San Diego, Oregon State University, the Plymouth Marine Laboratory, the First, Second, and Third Institutes of Oceanography, the Marine Environmental Monitoring Center, the Marine Technology Center, the Tianjin Marine Environmental Monitoring Center Station, the Ocean University of China, and Xiamen University in China. Meanwhile, government agencies around the world, such as NASA, NOAA, CNRS, and the Canadian Department of Fisheries and Oceans, invest heavily in research in this field.

Meanwhile, NASA, NOAA, and other institutes have established specific research directions in marine remote sensing technology for marine research and also collaborate with universities, whose research not only focuses on marine pollution monitoring, oil spills, dissolved organic carbon, coral issues, and thermal pollution [9–13], but also large-scale marine currents and air-sea interactions. For example, Chinese relevant institutes have specially established a research department in the field of marine remote sensing. The Second Institute of Oceanography of the Ministry of Natural Resources has taken the lead in conducting research on seawater color remote sensing in China and has achieved fruitful results in the mechanism and algorithm of seawater color remote sensing monitoring, as well as the development of domestically produced software system platforms. Relevant units such as the National Satellite Ocean Response Center and the North China Sea Administration of the Ministry of Natural Resources have all implemented operational monitoring applications for offshore oil spills. Furthermore, the Copernicus program also made a lot of contributions, whose program is the Earth observation component of the European Union's Space program, which offers information services that draw from satellite Earth observation and in-situ (non-space) data. Amounts of global data from satellites and ground-based, airborne, and seaborne measurement systems provide information to help service providers, public authorities, and other international organizations improve European citizens' quality of life and beyond. And organizations such as USEPA and US IOOS are monitoring exhaust from ships. University of Delaware professors have used remote sensing technology to monitor exhaust from ships.

Table 1. The top 10 related institutes of marine remote sensing monitoring and their research directions.

Research Institute	Number of Paper	Country/ Region	Related Laboratory	Research Direction
National Aeronautics and Space Administration	1506	United States	/	Sea level rise monitoring; Marine ecosystem research; Marine currents motion research
National Oceanic and Atmospheric Administration	1276	United States	Pacific Marine Environment Laboratory	Climate-weather research; Marine ecosystem research; Ocean and coastal evolution research
			Global Monitoring Laboratory	Greenhouse gases and carbon cycle research; Changes in clouds, aerosols and surface radiation research; Recovery of stratospheric ozone research
Chinese Academy of Sciences	1028	China	Digital Earth Key Laboratory	Frontier theory and technology of earth observation research; Earth big data science and methods research; Digital earth science and platform research; Global environmental resources spatial information system research
			Key Laboratory of Infrared Detection and Imaging Technology	High-resolution infrared imaging technology research; Hyperspectral imaging technology research; Weak target detection technology research; High quantitative remote sensing detection technology research
California Institute of Technology	853	United States	Linde Center	Earth climate change research [14]; Pollution impact research; Carbon dioxide changes research
University of Washington	617	United States	Applied Physics Laboratory- Air-sea interaction and remote sensing	Sea-air exchange research; Coastal research; Sensor research; Wave research

Table 1. *Cont.*

Research Institute	Number of Paper	Country/ Region	Related Laboratory	Research Direction
University of Maryland	576	United States	Earth System Science Interdisciplinary Center (cooperation with NASA)	Climate variability and change research; Atmospheric composition and processes research; Global carbon cycle research; Global water cycle research
University of California, San Diego	546	United States	Scripps Institution of Oceanography	Collect and process data on the Earth, oceans and atmosphere by cameras, lasers and various electromagnetic sensors
University of Colorado	476	United States	Mortenson Center in Global Engineering	Sustainable water treatment system research; Field and remote sensing research; Infrastructure resilience and disaster recovery research
			Earth Science and Observation Center, Institute for Research in Environmental Sciences	Analysis of remote sensing data, validation of data
University of Miami	445	United States	Upper Ocean Dynamics Laboratory	Experimental studies on coastal circulation processes and ocean-atmosphere interactions during the hurricane
Woods Hole Oceanographic Institution	385	United States	Claisen Laboratory	Studies on air-sea interactions and their impact on weather and climate through a wide range of measured and remote sensing data

Note: The laboratory information is compiled from the official websites of various organizations.

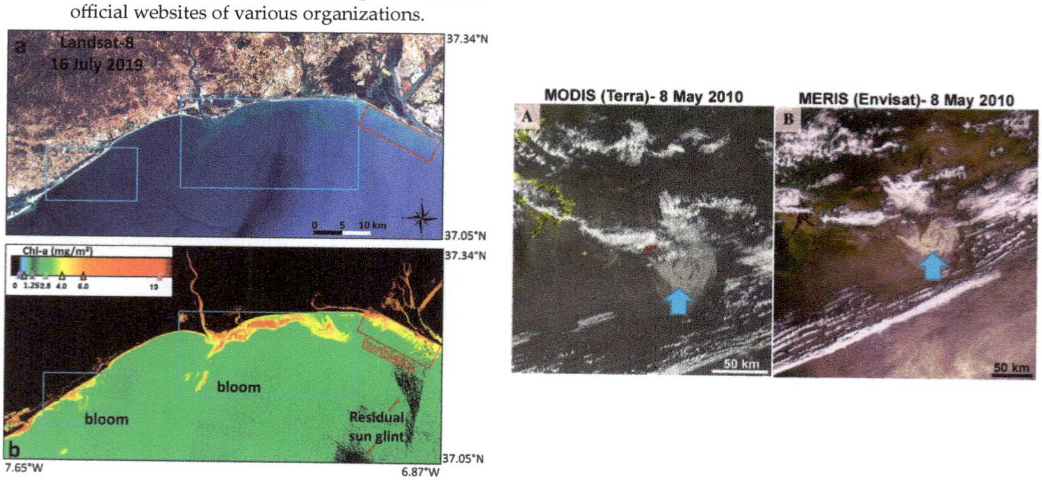

(a) (b)

Figure 2. The polyedra algal bloom and oil spill of remote sensing images. (**a**) Landsat-8 RGB (bands 4-3-2) composite in this study region on 16 July 2019; (**b**) chl-a concentration (mg/m^3) after atmospheric correction with ACOLITE for the same scene [15]. (**A**) MODIS; (**B**) MERIS showing the site of the DWH oil spill in blue arrow [16].

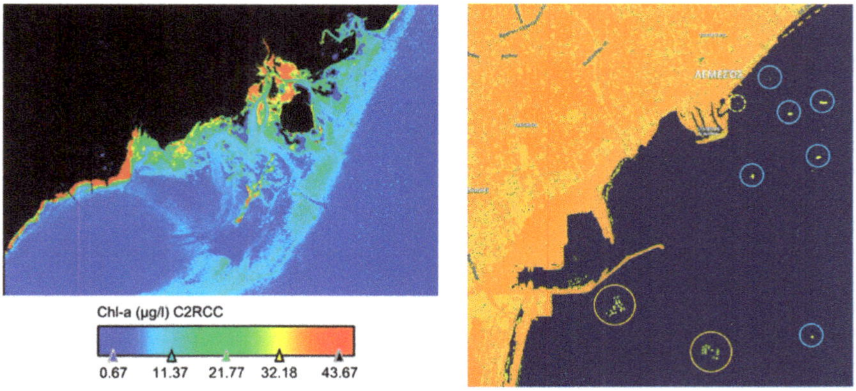

(a) (b)

Figure 3. The LiDAR and MSI remote sensing images (**a**) show the spatial distribution of Chl-a (conc_chl) C2RCC in the Kneiss Archipelago, Gulf of Gabes, Tunisia [17]. (**b**) the floating plastic litter from space using Sentinel-2 imagery [18].

Previous studies have reviewed the literature in the remote sensing field from the perspective of satellite functions and monitoring targets; however, they have not adequately reviewed the field of marine pollution monitoring. Therefore, a global review of progress in remote sensing monitoring of marine pollution is needed, which reviews the application domains of marine remote sensing technology and the progress in marine pollution monitoring (Figure 4), serving as a reference for academic research and development efforts in using remote sensing technology for marine pollution monitoring.

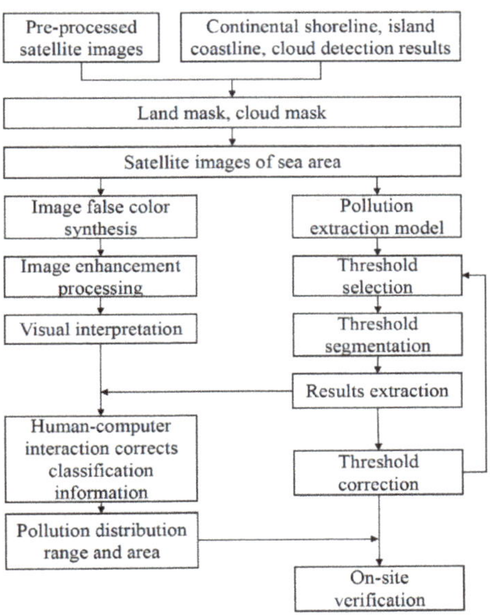

Figure 4. Framework of this study.

2. Sources and Monitoring Indices of Marine Pollution

The overexploitation of marine resources results in the degradation of the marine ecosystem and causes pollution, which damages the marine environment. The overexploitation, including major river conservation projects, land reclamation projects, coastal mining, offshore drilling, and mariculture, also leads to a shortage of marine resources and has negative impacts on coastal marine resources and environments. The utilization of marine resources varies in different regions due to differences in distribution, resulting in different types of pollution, such as land source, sea source and air source (Figure 5).

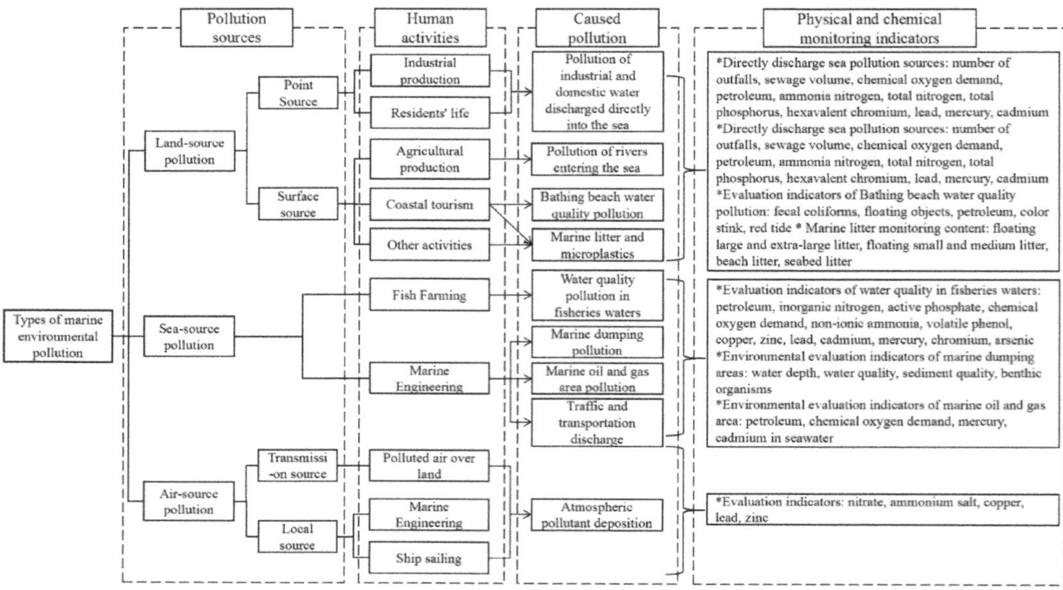

Figure 5. Source and monitoring index of marine pollution. Note: monitoring indices and methods from the Bulletin of China's Marine Ecological Environment in 2020.

Land-source pollution has the widest distribution and largest quantity, accounting for 80% of the total marine pollution and having the most serious effect. Land-source pollution can be divided into two types: point-source and area-source. Point-source pollution is primarily composed of industrial waste water and domestic sewage that is collected and treated through sewage networks and discharged into the sea through designated discharge points [19,20]. Area-source pollution comes from agricultural runoff and livestock breeding activities, which flow into the ocean through runoff [20]. The root cause of land source pollution is human economic and social activity, which contributes to the negative impacts of such activities on the marine environment [21]. This type of pollution is not limited to coastal areas but is influenced by economic and social activities and is more pronounced in offshore regions with intensive secondary industries and rapid economic development [22]. Additionally, human activities on land can cause eutrophication and biotoxicity in sea areas through river dams and water storage projects and reduce river sediment discharge, leading to coastal erosion and changes in river water and groundwater quality [23,24]. Coastal tourism, being a popular form of tourism, also contributes to marine pollution through the construction of tourist resorts, docks, and breakwaters, which fragment coastal habitats and harm biodiversity, and through the pollution of seawater quality and the flow of beach litter into the ocean from bathing beaches [25]. The water quality of direct discharge into the sea is the main evaluation index for monitoring point-source land source pollution, while the water quality of seaport rivers, beach water, marine debris, and microplastics is the main evaluation index for monitoring area-source land source pollution. Water quality is primarily evaluated according to the "Standard for Seawater Quality" (GB 3097-1997) [26], the "Standard for Surface Water Environmental Quality" (GB 3838-2002) [27] for rivers flowing into the ocean, and the "Guide for Monitoring and Evaluation of Bathing Beaches" (HY/T 0276-2019) for bathing beaches.

Sea-source pollution mainly refers to the pollution caused by human utilization of the sea. Marine pollution offshore is primarily caused by fishing and aquaculture practices. Overfishing has resulted in a reduction of fish resources, leading to the implementation of non-standard practices such as beach aquaculture, offshore cage aquaculture, and pelagic fishing, which negatively impact the marine environment by releasing nutrients and drugs into the water [28]. This can also pose a threat to coastal biodiversity, affecting beaches and mangroves [29]. Another significant source of marine pollution is marine engineering activities, such as oil and gas field exploitation, which can result in oil spills, sewage discharge, and seepage from eroded oil-bearing rocks. Additionally, the unregulated discharge of waste water from cruise ships and cargo ships in coastal tourism contributes to marine pollution [30]. Monitoring marine pollution includes monitoring the water quality of fishery waters and the discharge of waste water from marine engineering activities. The quality of fishery waters is evaluated according to the "Fishery Water Quality Standard" (GB 11607-1989) [31], while monitoring of marine engineering pollution in dumping areas and oil and gas fields should be based on the "Technical Regulations for Monitoring of Marine Dumping Areas" and "Technical Guidelines for Environmental Impact Assessment of Marine Engineering" (GB/T 17108-2006) [32].

Air-source pollution can be divided into transmission sources and local sources (Figure 6). Transmission sources refer to the pollution in the air that is transported from the land to the sea due to the influence of monsoons. Local sources refer to the pollutants that are emitted directly into the sea, such as oil and gas field exploitation, marine transportation, and fishing. The pollutants include SO_2, NOx, CO, particulate matter, and VOCs. Atmospheric pollutants fall onto the marine surface through dry deposition and wet deposition processes [33]. Dry deposition involves the physical, chemical, and biological processes by which atmospheric pollutants fall onto the marine surface, while wet deposition involves ionic pollutants and soluble pollutants from the air falling onto the marine surface through precipitation or water vapor condensation [34]. The evaluation of marine atmospheric pollution deposition is mainly conducted based on the "Technical Regulations for the Assessment of the Flux of Atmospheric Pollutants Deposition into the Sea (Trial)" and includes the observation of the elements nitrate, NH_4Cl, Cu, Pb, and Zn.

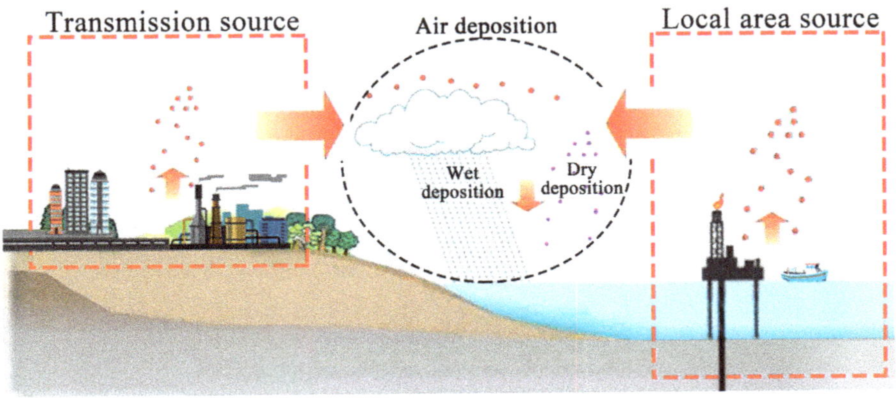

Figure 6. Mechanisms of marine atmospheric pollution transfer.

3. The Utilization of Remote Sensing Monitoring in Marine Pollution

3.1. An Overview of the Development History and Present Situation

In 1960, the United States launched the world's first meteorological satellite, TIROS-1, and used it to collect data on the marine surface temperature field from an altitude of approximately 700 km [35]. In 1978, the United States took the lead in the field of marine observation with the launch of Seasat-1, the first satellite dedicated to observing coastal seawater color [36], which marked the beginning of marine remote sensing monitoring. As the importance of marine resources continues to be recognized, more and more countries are investing in marine remote sensing technology to support monitoring efforts in the fields of shipping, environment, military, and economics [36–38].

The sensors commonly used in marine research include seawater color sensors, infrared sensors, microwave altimeters, microwave scatterometers, synthetic aperture radars, and microwave radiometers (Table 2). These sensors are capable of monitoring elements from the marine surface to the upper, lower, and bottom topography and can be used to monitor the seawater color and environment, as well as the marine dynamic environment (such as the marine surface wind field, marine surface height, effective wave height, and surface temperature) [39–41]. The sensors play a crucial role in monitoring islands, coastal zones, and other targets, including the marine wave field, storm surge floodplain, internal waves, sea ice, and oil spills. With the progress in microwave, infrared, and hyperspectral remote sensing technologies and their wider applications in the water, ecological, and air environments, the types of sensors available have become more diverse and specialized, and the spectral bands more refined. For example, China has launched a series of marine satellites, including HY-1C/D and HY-2B/C/D, equipped with new ultraviolet imagers, onboard calibration spectrometers, automatic ship identification systems (Figure 7), and other loads, providing a wealth of data for monitoring the global seawater color and dynamics [42]. The spectral bands of remote sensing images are getting finer and finer, forming multispectral, hyperspectral, and ultraspectral. At present, the spectral resolution in the 100 nm order is called multispectral, such as a remote sensor in the visible light and near-infrared spectral region only a few bands, such as the United States LandsatMSS, TM, SPOT in France, etc. The spectral resolution is in the 10 nm order, which is called HyPerspectral remote sensing. With the further improvement of the spectral resolution of remote sensing, remote sensing enters the ultraspectral stage when the spectral resolution reaches 1 nm. Hyperspectral and ultraspectral remote sensing are the frontier fields of remote sensing technology. It uses many very narrow electromagnetic wave bands to obtain relevant data from objects of interest so that substances that are not detectable in wide-band remote sensing can be detected. Chuanpeng Zhao et al. [43] proposed a new approach with solid improvements in dual-temporal image construction, misclassification processing, and tacit knowledge analysis, generated an accurate coastal salt marsh map at a national scale, and provided a classification mechanism for dual-temporal image-based coastal salt marsh identification and mapping. Remote sensing technology has huge advantages in obtaining real-time, large-scale, wide-area, and multi-period comparisons of basic coastal zone data. However, due to the indirect reflection of the characteristics of each observation object through the radiation and reflection characteristics of electromagnetic waves, the resolution is lower than that of conventional ground or ship observation methods. In addition, the presence of the phenomenon of homologous objects with different spectra and foreign objects with the same spectrum makes it difficult to recognize and interpret images.

Table 2. Satellite sensors information for marine research.

Sensor Type	Function	Representative Satellite Sensors	Satellite	Country/Region	Time	Spectral Band	Resolution	Revisit Cycle
Seawater color sensor	Monitoring of marine surface chlorophyll concentration, suspended mass concentration, marine primary productivity, diffuse attenuation coefficient, other marine optical parameters [44]	Coastal Zone Color Scanner	NIMBUS-7	United States	1978	5 visible near-infrared bands (0.443~0.750 μm), 1 thermal infrared band (11.5 μm)	0.825 km	1 day
		Sea-viewing Wide Field of View Sensor	Seastar	United States	1997	8 bands (0.402~0.885 μm)	1.1 km (IFOV)	1 day
		Moderate-resolution Imaging Spectroradiometer	EOS-AM(TERRA)/EOS-PM(AQUA)	United States	1999/2002	36 discrete spectral bands (0.4~14.4 μm)	0.25~1 km	0.5/1 day
		Chinese Ocean Color and Temperature Scanner	HY-1A	China	2002	10-band (0.402~12.50 μm)	1.1 km	1 day
			HY-1C	China	2018		1.1 km	1 day
		Medium Resolution Spectral Imager	FY-3	China	2008	20-band (0.43~12.5 μm)	0.25~1.1 km	5.5 day
		Chinese Ocean Color and Temperature Scanner	HY-1D	China	2020	10-band (0.402~12.50 μm)	1.1 km	1 day
		Ocean Colour Monitor	OceanSat-3	India	2022	3 bands	360 m	2 days
Infrared sensor	Measurement of marine surface temperature	Advanced Very High Resolution Radiometer	NOAA/TIROS	United States	1979	2 thermal infrared channels (11 μm, 12 μm)	1.1 km (IFOV)	0.5/1 day
		Along-Track Scanning Radiometer	ERS-1/2	Europe	1991/1995	2 thermal infrared channels (11 μm, 12 μm)	1 km	3 days (TIR)/6 days (SWIR)
		Advanced Along-Track Scanning Radiometer	ENVISAT	Europe	2002	11/12 μm channel during daytime and 3.7/11/12 μm channel at night	1 km (IFOV)	3 days (TIR)/6 days (SWIR)
		VistA Integration Reporting and Revenue	FY-3A	China	2008	10 channels (including visible channels, 3 infrared atmospheric window channels)	1.1 km	5.5 days

Table 2. *Cont.*

Sensor Type	Function	Representative Satellite Sensors	Satellite	Country/Region	Time	Spectral Band	Resolution	Revisit Cycle
Microwave altimeter	Measurement of mean sea level height, geoid, effective wave height, marine surface wind speed, surface laminar flow, gravity anomalies, rainfall index	Radar altimeter	ERS-1/2	Europe	1991/ 1995	C-band/5.3 GHz	20 km (IFOV)	10 days/1 month
		Dual frequency radar altimeter, Single frequency altimeter	Jason-1	United States, France	2001	NRA: Ku-band/13.6 GHz and C-band/5.3 GHz; SSALT: Ku-band/ 13.65 GHz	25 km	10 days
		Radar altimeter-2	ENVISAT	Europe	2002	Ku-band/13.5 GHz	20 km (IFOV)	10 days/1 month
		Radar altimeter	HY-2B	China	2018	Ku-band/13.58 GHz, C-band/5.25 GHz	2 km	14 days in the early stage, 168 days in the late stage
Microwave scatterom-eter	Measurement of the wind field at 10 m above sea level	Single-frame side-scan vertical transmit vertical receive (VV) radar	ERS1/2	Europe	1991/ 1995	C-band/5.3 GHz	Optimal: 50 km; Sampling: 25 km	3 days on average
		Double amplitude side scan scatterometer	ADEOS	Japan	1996	Ku-band/13.995 GHz	Optimal: 50 km; Sampling: 25 km	1.5 days
		Sea Winds scatterometer	QuikSCAT	United States	1999	Ku-band/13.4 GHz	Optimal: 50 km; Standard: 25 km; Sampling: 12.5 km	1 day
		Microwaves scatterometer	HY-2B	China	2018	Ku-band/13.256 GHz	25 km	14 days in the early stage, 168 days in the late stage
		Sector beam rotational scanning scatterometer	CFOSAT	China, France	2018	Ku-band/13.256 GHz	25/12.5 km	real time

Table 2. *Cont.*

Sensor Type	Function	Representative Satellite Sensors	Satellite	Country/Region	Time	Spectral Band	Resolution	Revisit Cycle
Synthetic aperture radar	Monitoring of wave direction spectrum, mesoscale eddy, ocean internal waves, shallow sea topography, marine surface pollution, marine surface characteristic information [45]	Active Microwave Instrument Synthetic Aperture Radar	ERS-1/2	Europe	1991/1995	L-band/1.275 GHz, S-band/3.0 GHz, C-band/5.3 GHz, X-band/9.6 GHz	30 m	1.5 month
		Ka-band Radar	SWOT	United States	2023	Ku band/C band	/	20.8 days
		Multi-polarized Advanced Synthetic Aperture Radar	ENVISAT-1	Europe	2002	C-band/5.3 GHz	30 m~1 km	5 days/3 months
		Multi-polarized C-band synthetic aperture radar	GF-3	China	2016	C-band/	1 m (highest)	1 week
			RCM-3	Canada	2019	C-band/5.405 GHz	1 m (highest)	12 days
Microwave radiometer	Measurement of marine surface temperature, marine surface wind speed, sea ice water vapor content, CO_2 and sea-air exchange	Line Leveling Passive Microwave Radiometer	DMSP	United States	1987	4 bands (19.35, 2.235, 37.0, 85.5 GHz) 7 channels	61 × 66 cm antenna diameter	1 day
		Microwave Scanning Radiometer	MOS-1	Japan	1987	2 bands (23.8 GHz, 31.4 GHz)	23 km (31.4 GHz) /32 km (23.8 GHz)	5 days
		Multi-frequency scanning microwave radiometer	Seasat-A/Nimbus-7	United States	1978	5 bands 9 channels	/22~120 km (37~6.6 GHz)	2 days
		Advanced Microwave Scanning Radiometer -E	EOS-PM(AQUA)	Europe	2002	6 bands 12 channels	Antenna diameter 1.6 m	1 day
		Advanced Microwave Scanning Radiometer	ADEOS-2	Japan	2002	8 bands 14 channels	Antenna diameter 2 m	1 day
		Scanning Microwave Radiometer Imager MWRI	HY-2B	China	2018	5 bands (6.925, 10.7, 18.7, 23.8, 37.0 GHz)	Antenna diameter 1.2 m	14 days in the early stage, 168 days in the late stage
		Calibration Microwave Radiometer	HY-2D	China	2021	6 bands (3.58, 5.25, 13.256, 18.7, 23.8, 37 GHz)	/	10 days
		Poseidon-4 radar altimeter and microwave radiometer	Sentinel-6	United States, Europe	2020	C-band/Ku-band	/	10 days

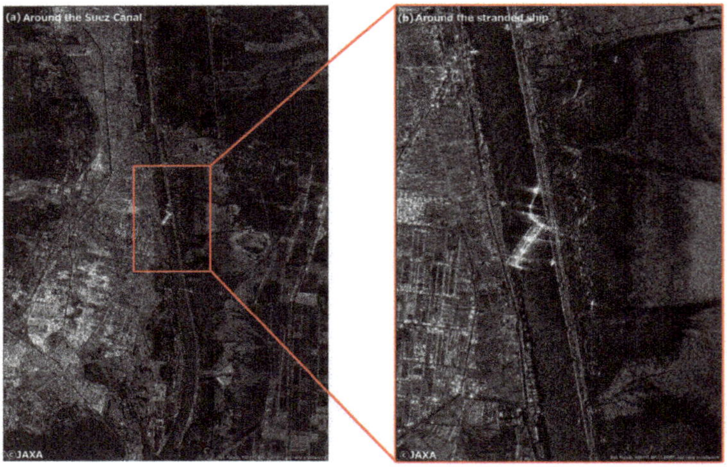

Figure 7. Image observed by PALSAR-2, synthetic aperture radar onboard ALOS-2, at 10:23 (UTC) on March 26 (**a**) around the Suez Canal; (**b**) extended image around the stranded ship (Picture source: Earth Observation Research Center).

3.2. The Monitoring Data and Methods

Marine remote sensing has the unique advantages of all-weather, wide-range, and long-term observation and is widely used in marine ecology and resources monitoring, marine disaster monitoring, marine rights and interests maintenance, marine environmental prediction and security assurance, and other fields. Therefore, marine pollution can be monitored and reversed based on marine remote sensing data. Marine pollution monitoring can be classified into three aspects: seawater quality, marine debris and microplastic pollution, and atmospheric deposition pollution. According to the application focus of remote sensing, the remote sensing data products and research methods for monitoring the pollution of these marine areas are analyzed.

3.2.1. Seawater Quality Monitoring

Marine remote sensing is widely used in a variety of fields, including marine ecology and resource monitoring and marine disaster monitoring [46]. With the help of remote sensing, it is possible to reverse and monitor marine pollution. The main environmental indices monitored by remote sensing include the suspended solids content, chlorophyll-a concentration [47], colored soluble organic matter, and other comprehensive pollution indices [46,48]. To invert the quality of water environmental indices, a spectral model must be established and combined with monitoring data. Remote sensing inversion is often performed using visible light bands (Table 3), and various methods such as empirical models, theoretical models, semi-analytic models, and others can be used (Table 4). The suspended solids content in water can be reversed by fitting a formula that correlates the measured suspended solids content with remote sensing reflectance or turbidity [49–51]. The concentration of chlorophyll-a is a direct indices of eutrophication and organic pollution, and the inversion methods of chlorophyll-a concentration include chlorophyll fluorescence, band models, nonlinear mapping, and mechanism model methods [52–54]. Chromophoric soluble organic matter is another important component of dissolved organic matter that can be monitored remotely through the development of a remote sensing inversion model [55–57]. For example, red tides have posed a huge threat to the resources and environment of coastal areas. Multi-spectral scanners can be used to monitor red tides, whose visible/infrared multispectral radiometers can provide information on the location, range, water color type, changes in phosphate concentration on the sea surface, and the direction of red tide diffusion and drift to take timely measures to control them [58].

It is difficult to find independent spectral characteristics in seawater quality indices, such as dissolved organic carbon, water temperature [59], transparency, dissolved oxygen, chemical oxygen demand, five-day biochemical oxygen demand, total nitrogen, total phosphorus, etc. As a result, indirect remote sensing analysis must rely on the correlation between different substances [60–62]. However, remote sensing monitoring also has limitations, such as difficulty in estimating pollutants in the vertical dimension of water bodies and the limitation of partial inversion to estimate only general parameters, not specific types of marine pollution. Coastal human activities and the complex dynamic environment of the sea can also affect the water quality, requiring higher performance from satellite sensors to monitor it effectively through remote sensing technology [63]. Despite the progress in remote sensing technology, the main drawback of physical-chemical sensor systems is their lack of specificity and sensitivity and their inability to assess the environmental concentration of most marine pollutants. In this regard, biosensors may offer the required specificity and selectivity [64].

Table 3. The monitoring data for seawater quality.

Seawater Quality Indices	Data	
Suspended solids content	Remote sensing data	Sentinel-3A OLCI [65], SPOT, Terra ASTER, Landsat TM/ETM, Envisat MERIS [66], COMS GOCI [67–69]
	Measured data	Seawater sampling data, official water environment monitoring samples, measured spectral data
Chlorophyll-a concentration	Remote sensing data	Non-space remote sensing data: AVIRIS, CASI, OMIS, AISA, etc.; Remote sensing data: Envisat MERIS, CBERS-2 CCD, Terra/Aqua MODIS [70,71], EO-1 Hyperion, Terra ASTER, Landsat MSS/ETM+, HJ-1 CCD, et al.
	Measured data	Ground-measured spectral data: ASD, GER, etc.
chromophoric dissolvable organic matter	Remote sensing data	Sentinel-2 [72], Landsat 8 OLI, SeaWiFS, Aqua/Terra MODIS [73], etc.
	Measured data	Water sampling data

Table 4. The monitoring methods for seawater quality.

Methods	Characteristics
Empirical models (single band model, band ratio model, multiple regression model, nonlinear model, machine learning)	Based on the relationship between water spectral information and measured water quality parameters, the algorithm is relatively mature and the process is simple, but it lacks physical basis
Theoretical model	The combination of water quality spectral characteristics and statistical analysis has a certain physical basis, but for the model requiring higher precision, the process is more complicated and the universality is poor
Semi-analytical model	Based on the relationship between apparent optical quantity and measured water quality parameters, it has a strong physical mechanism and good universality, despite the difficulty of model establishment [72]

The quality of seawater can be negatively impacted by large-scale pollutants such as red tides, green tides, and oil spills, especially due to increased human activities leading to increased inputs of nitrogen and phosphorus in the ecosystem (Table 5) [74–77]. Remote sensing technology plays a crucial role in monitoring such events by monitoring the seawater quality indices over large areas [78,79]. The research has shown that the multi-band ratio method can be used to monitor floating algae and reduce interpretation errors [80–82]. Several algorithms have been proposed, including the Floating Algae Index (FAI) [83,84], the Normalized Algae Index (NDAI) [85], the Virtual Baseline Index of Floating Algae Height (VB-FAH) [86], and the Multi-Spectral Green Tide Index (MGTI) [87,88]. However, these algorithms require complex and accurate atmospheric correction procedures, which can increase the complexity of interpretation. To address this, Zhang Hailong et al. developed the Floating Algae Index (FGTI) based on the DN values of different satellite data, and Chen Ying et al. [89] proposed the Green Tide Index (TCT-GTI) algorithm based on the Tassel Cap Transformation method, which requires no atmospheric correction.

Table 5. The Monitoring data and methods for marine green tide pollution disaster.

Author	Time	Satellite/Sensor	Waveband Range	Algorithm	Characteristics
Hu [78]	2009	Aqua/Terra MODIS	V, NIR, SWIR	Floating Algal Index	It is less sensitive to changes in environment and observational conditions (aerosol type and thickness, sun/observational geometry, and solar brilliance) and can penetrate thin clouds, providing a simple and effective means of atmospheric correction and making it easier to establish image-independent thresholds to monitor and quantify planktonic macroalgae.
Shi et al. [85]	2009	Aqua/Terra MODIS	SWIR	Normalized Algae Index	The effect of atmospheric molecular scattering is removed, making the NDAI more sensitive to the radiance signal from the marine surface [85].
Son [90]	2012	GOCI	V, NIR	GOCI floating green tide index	The muted or subtle signal of planktonic green algae is enhanced and separated from the surrounding complex water signal [90].
Xing et al. [86]	2016	HJ-1A/1B	V, NIR	Virtual Baseline Index of Floating Algal Height	Even without the use of the shortwave infrared (SWIR) band, VB-FAH appears to be comparable to the planktonic algal index (FAI).
Zhang et al. [88]	2016	GF1 WFV, HJ CCD	V	Multispectral Green Tide Index	It can effectively eliminate the influence of external interference, such as suspended sediment and thin clouds, and has low sensitivity to the environment [88].
Zhang et al. [91]	2019	GF-1 WFV1, GF-1 WFV3, HJ-1B CCD, Landsat-7 ETM+, GOCI	V, NIR, SWIR	Floating algae index	The use of tasseled cap transformation is more powerful than traditional NDVI (normalized vegetation index) in responding to perturbations from environmental conditions, observational geometry, sunlight, and thin cloud pollution [88,91].
Chen et al. [89]	2020	GOCI	V	Green Tide Index	No atmospheric correction is required [89].

Monitoring red tides can be conducted by using the spectral and temperature characteristics of the affected water bodies and through the analysis of optical satellite data (Table 6). The main methods of red tide remote sensing monitoring include the Chlorophyll Concentration Anomaly Method, the Red Tide Index (MRI), the Rrs Band Ratio Method, the Red Band Difference (RBD), the RBD_ KBBI (Karenia brevis Bloom Index), and others [92,93]. Jiang Dejuan et al. [92] emphasized that the effectiveness of red tide monitoring is dependent on the type of sensor, algorithm, and remote sensing time. Therefore, it is necessary to combine field survey data or ERGB images with the remote sensing data to verify the accuracy of red tide monitoring in different times and regions [94,95].

Table 6. The monitoring methods for the marine red tide pollution disaster.

Research Methods	Characteristics
Chlorophyll concentration anomaly method	It characterizes the most important characteristic parameters of red tide but is usually overestimated.
Red tide index	It can enhance the difference between the red tide water body and the surrounding water body; therefore, it can be used for the determination of red tide.
Band ratio method	Using the ratio of reflection and absorption bands of red tide water bodies, we can extract red tide information.
Red band difference method	The method is based on the high fluorescence property of dinolflagellate water bloom.
RBD_ KBBI	The red tide monitoring index proposed is based on RBD.

Note: Compiled from references [92].

Marine engineering encompasses a variety of activities, including marine dumping, marine oil and gas transportation, and others. The main source of seawater quality and environmental degradation in the ocean is oil spills caused by ships during berthing at ports, navigation, and accidents [96,97]. Early monitoring, tracking, and diffusion of oil spills is crucial in designing effective algorithms for remote sensing monitoring [98]. However, the current methods for oil spill monitoring often overlook the space-time characteristics and laws of the interaction between oil spills and the ocean [99]. Different remote sensing technologies, such as microwave radar, optical (multi/hyperspectral remote sensing), thermal infrared, and others, can be used to monitor marine oil spills. Microwave radar remote sensing is one of the main methods and is based on the different absorption and transmission properties of water and oil in electromagnetic waves. This method is mainly used to extract the oil spill extent by identifying the "dark pixel" feature in the microwave radar image [100,101]. Spectral response characteristics of oil spill simulation experiments can be used to identify, classify, and quantify the types of oil spills [102,103]. Optical remote sensing can also be used for the identification and classification of oil spills [104–106]. Solar reflected light can be used to monitor the oil spill thickness on the marine surface; however, the monitoring range is limited to less than 0.4 mm due to the different absorption, scattering, and reflection effects of oil spills on incident light. Thermal infrared remote sensing has a larger monitoring range; however, the ability to monitor oil spills is weaker than that of optical remote sensing due to the differences in existing spaceborne sensors [107]. In a study of oil spills caused by marine accidents, Lu Yingcheng et al. [105] used a GF-3 SAR image to delineate "suspected oil spills" and analyzed the optical anomalies of the "Sanji" oil spill incident in the East China Sea in 2018 using Sentinel-2 multi-spectral remote sensing data (MSI) from the European Space Agency. This study resulted in the optical remote sensing identification and classification of different types of oil spills. Huang Ke et al. [104] used GF-1 satellite data along with a thin oil film thickness model to calculate the oil film thickness of the oil spill area and estimate the oil spill caused by the explosion of the Huangwei oil pipeline in Qingdao in 2013. Remote sensing monitoring of oil spills at sea is influenced by marine environmental conditions, such as marine surface wind speed, and therefore, a combination of sea wave spectrum and wind wave information is necessary for further research [108,109].

3.2.2. Debris and Microplastics Monitoring

Marine debris and microplastics have a significant negative impact on marine organisms, ecosystems, fisheries, and tourism [110,111]. The monitoring of these pollutants in the ocean through remote sensing can be effectively accomplished by using sensitivity analysis, optical simulation, and satellite image spectral analysis (Table 7) [112,113]. The reflection spectra of known marine floating objects can be utilized for this purpose [110,114]. Most of the research and monitoring of marine debris is focused on plastic waste and utilizes optical sensors operating in the visible light, near-infrared, and short-wave infrared spectral ranges, as well as SAR sensors and other image data sources [115]. However, microplastics are plastic pollutants with a diameter of less than 5 mm and cannot be effectively identified through satellite images due to their small size. While there have been efforts to observe microplastics In water, through optical means such as lasers and polarization, there is still a significant gap in terms of satellite observation. SAR sensors are commonly used to monitor marine debris and microplastics by identifying surface active agents on the marine surface. The classification methods for marine debris monitoring are mainly manual classification, use of indices, spectral classification, or machine learning [115]. The use of unmanned aerial vehicles and artificial intelligence for monitoring beach litter is an emerging field [116]. Machine learning has been shown to significantly increase the speed of screening for marine debris, operating 39 times faster than human screening in cases of low sensitivity [116]. The development of a spectral characteristic model for marine debris or microplastics is also an area of exploration. Goddijn-Murphy et al. proposed a reflection model based on geometric optics and the spectral characteristics of plastics and seawater to describe the interaction between sunlight and floating microplastics on the marine surface.

The use of satellite data are useful tool in monitoring marine debris as it can provide repeated global coverage data at various scales and resolutions, which is not possible through field observations or remote sensing with ships, aircraft, or unmanned aerial vehicles. However, the limitations of optical satellite data include limited temporal coverage due to fixed time and factors such as the sun, clouds, atmospheric aerosols, sensor saturation on surfaces such as ice, sand, or snow, and a high solar zenith angle. Additionally, the spatial resolution of satellite data are typically greater than 1 m, making it difficult to monitor marine debris smaller than this resolution [117]. For the limitation, it is possible to solve this problem by improving the accuracy of remote sensing observation instruments. Furthermore, through multi-band remote sensing observation methods, we can gain a deeper understanding of the electromagnetic wave characteristics of ground objects and separate information from different observation objects.

Table 7. Remote sensing monitoring data and methods for marine debris and microplastics.

Author	Time	Band Range	Satellite/Sensor	Types of Plastic Waste	Monitoring Methods	Classification Methods
Hamilton [117]	2015	MW	RADARSAT-2	Marine surface float (marine bacteria associated with surfactants)	Observe oil slicks to monitor marine surface slicks using the association among bacteria produced by surfactants and oil slicks (Hamilton et al., 2015)	Machine learning techniques
Aoyama [118]	2016	NIR/SWIR	Worldview-2	Marine debris	Find common spectral features associated with the presence of plastic [118]	Spectral classification method
	2016	NIR/SWIR	Worldview-3	Buoys around fishing nets	Propose the spectral angle mapper (SAM) classification method [118]	Spectral classification method
Kurata [119]	2016	MW	RADARSAT-2	Marine surface float (marine bacteria associated with surfactants)	Observe oil slicks to monitor marine surface slicks using the association among bacteria produced by surfactants and oil slicks [119]	Machine learning techniques
Davaasuren [120]	2018	MW	Sentinel-1A, COSMO-SkyMed	Microplastics (surfactants, sea mud and biofilms)	Observe oil slicks to monitor marine surface slicks using the association between bacteria produced by surfactants and oil slicks [120]	Machine learning techniques
Goddijn-Murphy	2018	V~SWIR	--	Floating microplastics	Model the reflection of sunlight interacting with the marine surface of floating microplastics	Model building
Howe [121]	2018	MW	TerraSAR-X	Marine surface float (marine bacteria associated with surfactants)	Observe oil slicks to monitor marine surface slicks using the association between bacteria produced by surfactants and oil slicks [121]	Machine learning techniques
Topouzelis [115]	2019 2019	V, NIR MW	Sentinel-2A Sentinel-1	Artificial floating plastics (water bottles, LDPE plastic bags and nylon fishing nets)	Supervised classification [108]	Machine learning techniques
Biermann	2020	V, NIR, SWIR	Sentinel-2 MSI	Piece of floating large plastic	Develop novel float index (FDI) algorithms to monitor floating plastics and distinguish them from natural floats	Usage index
Kikaki [111]	2020	V, NIR NIR, SWIR NIR, SWIR	Planet P Sentinel-2 MSI Landsat-8 OLI	Floating plastic debris	Monitor and validate floating plastic debris by systematically recording and assessing the spectral characteristics of pure floating plastic and distinguishing it from other floating materials on the marine surface (e.g., Sargassum, foam)	Manual classification

3.2.3. Inversion of Atmospheric Pollution Deposition

The monitoring of atmospheric pollution deposition typically involves measured data, model simulations, and remote sensing. Atmospheric pollution deposition in the ocean can be classified into three sub-categories that are influenced by the distribution of aerosols: clean ocean, marine minerals, and marine pollution [122]. Therefore, the monitoring of atmospheric pollution deposition can be conducted through aerosols. Remote sensing by satellite is a highly effective research method for observing the global distribution of aerosols, their optical properties, and their radiation effects due to the large spatial and temporal variations of aerosol distribution [123]. Aerosol optical thickness, a key parameter of aerosols, characterizes atmospheric turbidity and is a critical factor in determining the impact of aerosols on climate [124,125]. Some of the satellite sensors used to monitor atmospheric pollution deposition include the Global Ozone Monitoring Experiment (GOME), the Atmospheric Mapping Scanning Imaging Absorption Spectrometer (SCIAMACHY), the Ozone Monitor (Aura OMI), the Interferometric Infrared Atmospheric Detector (METOP/IASI), the FY-3 Total Ozone Detector (FY-3 TOU), and the UV Ozone Vertical Detector (FY-3 SBUS) of FY-3 [126,127].

Remote sensing monitoring calculates the concentration of deposition according to the formula by reversing the density of the vertical gas column, and then calculates the dry and wet deposition flux in combination with the dry and wet deposition calculation formula. For example, Dong Haiying et al. [124] used Terra MODIS data to reverse the 10 km resolution aerosol optical thickness and analyzed that the aerosol concentration gradually decreased from the coastal waters to the outer sea due to the influence of land-based pollution components, and there was a high aerosol optical thickness in the coastal Bohai Sea, the Yellow Sea coast, and the Yangtze River estuary. SeaWiFS aerosol optical thickness products can be used to study aerosol distribution and change characteristics. Hao Zengzhou et al. [128] analyzed the aerosol distribution in China's sea area and found that the average aerosol optical thickness in the eastern China sea area has a zonal distribution centered on the middle latitude and has seasonal changes. From spring to winter, the aerosol optical thickness shifts from the high latitude sea area to the low latitude sea area, and the scope is also gradually expanding. Mao Ying [122] over the sea air will be visible infrared imaging radiometer VIIRS, medium resolution imaging spectrometer MODIS, stationary seawater color satellite imager GOCI, and geosynchronous meteorological satellite AHI H8 aerosol optical thickness product data combined with field measured data. Compared with other aerosol optical thickness remote sensing products, GOCI aerosol optical thickness products show the characteristics of high sampling frequency and high precision.

The inversion results of aerosol optical thickness remote sensing products are affected by a variety of factors, including the signal-to-noise ratio of the sensor itself, the accuracy of the calculation of the surface reflectance of the underlying surface, and the rationality of the selection of the preset aerosol model [122]. At the same time, the frequent human activities and pollutant emission and diffusion, as well as the unique physical geography and hydrological conditions of the sea area, will cause the sea area reflectivity to include many factors, not only the atmospheric impact but also the role of the seawater itself, which is more obvious in the coastal sea area.

4. Conclusions and Future Prospects

4.1. Conclusions

This paper summarizes the progress in the application of remote sensing for marine pollution monitoring. Coastal areas are subjected to various forms of pollution from land, sea, and air sources due to the difference in urbanization and industrialization, leading to a degradation in the quality of the marine environment and having significant impacts on marine ecology. Remote sensing monitoring of marine pollution encompasses various aspects, including seawater quality, marine debris and microplastic pollution, and the inversion of atmospheric pollution depositions. With the launch of various marine monitoring satellites, remote sensing provides a wealth of data sources to monitor marine hydrology and environmental parameters. Remote sensing can provide information on environmental parameters and habitats and monitor human activities, thus improving the marine environment to some extent (Figure 8). Monitoring seawater quality and temperature mainly uses remote sensing of seawater color and temperature, which is effective in monitoring seawater quality deterioration and large-scale coastal seawater quality changes during disasters. The methods of monitoring marine debris and microplastics have evolved from remote sensing inversion to machine learning. The research into remote sensing monitoring of marine atmospheric pollution deposition, which only quantifies atmospheric pollution through aerosol thickness, is inadequate. It is crucial to integrate measured data, remote sensing monitoring, and information services to achieve comprehensive monitoring and assessment of marine pollution. Moreover, there are a number of marine remote sensing products worldwide; however, there are significant differences in accuracy and other aspects of similar products in different countries, leading to technical barriers that pose challenges to marine governance decisions and environmental protection in various countries and regions. Therefore, developing three-dimensional monitoring techniques, combining different types of sensors, and establishing a comprehensive marine monitoring database is of great significance, along with data, technology, and standard sharing, to monitor the marine environment and realize the sustainable development of marine resources.

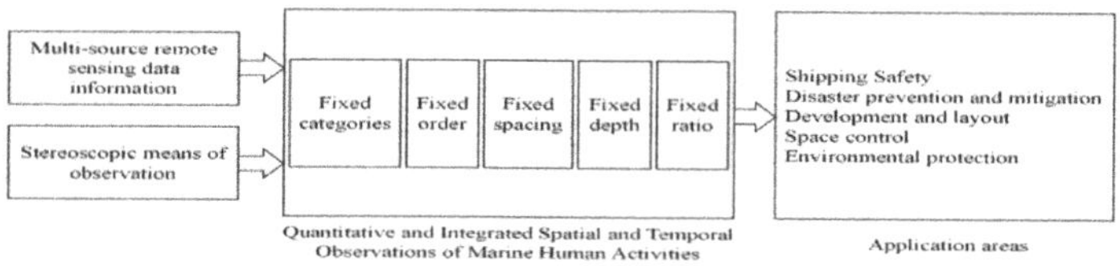

Figure 8. The extraction process of marine pollution data. Note: Redraw from reference [8].

4.2. Future Prospects

4.2.1. The Tendency of Three-Dimension in Marine Monitoring

Despite its practicality, remote sensing monitoring has limitations in its ability to identify and quantify the majority of marine pollutants. Due to the complex three-dimensional nature of the ocean, most remote sensing methods can only obtain information from the upper layer, with the exception of the altimeter used for rough sounding. The best conditions for optical sensors to penetrate the seawater surface are up to 27 m, while air sensors can only reach up to 46 m [129,130]. This resolution limitation of the image data affects the vertical dimension of marine pollution monitoring and assessment and cannot provide a quantitative method for tracing the levels of individual substances. Thus, the development of remote sensing technology that can monitor deep-sea pollution is a crucial goal for the future [117,131]. Although monitoring the deep sea may monitor less pollution than the upper sea, three-dimensional marine observation remains the key direction for the progress of remote sensing technology. Furthermore, multi-source data integration with three-dimensional ocean numerical models for marine pollution monitoring should be used. Observations obtained through remote sensing and sensors are not sufficient to project the 3D structure of the marine environment. Hence, 3D hydrodynamic and ocean circulation modeling should be developed.

4.2.2. The Tendency of Multi-Source Data Fusion in Marine Pollution Monitoring

Despite the progress in remote sensing technology and the increasing availability of data, the accuracy of remote sensing data products can still be impacted by environmental factors such as spatial-temporal variability in the sky and on earth. The diversity of human activities in the ocean also limits the validity of remote sensing data for marine pollution monitoring. Currently, marine pollution monitoring primarily relies on field observations, utilizing biochemical indices, rather than solely relying on remote sensing. To improve the accuracy and reliability of marine pollution monitoring, it is necessary to integrate remote sensing with other monitoring tools, such as ship sensors, buoy sensors, aircraft sensors, and marine ecological reserves/biosensor networks.

Remote sensing technology, such as high-resolution imaging, multi-spectral and hyperspectral, fluorescence, and Raman molecular spectroscopy, can be used to measure different types of pollution; however, it is still not enough to replace on-site observations. For example, the movement of pollutants in the ocean is influenced by complex hydrodynamics and marine circulation, making it difficult for remote sensing technology to accurately track the diffusion of marine pollution. Additionally, micropollution may not cause large-scale pollution, and it is challenging to quantify using remote sensing alone.

Therefore, multi-platform and multi-means integrated observation technology is necessary for accurate marine environmental monitoring. This includes the establishment of a marine remote sensing database, which integrates satellite, aviation, ship, and shore-based data to provide a comprehensive monitoring system for marine environmental monitoring. The development of technology for collecting, storing, processing, analyzing, and utilizing multi-source data will be crucial for the progress of marine remote sensing. By providing timely information to the government and relevant entities, this database can support disaster prevention, mitigation, and the launch of emergency plans.

In addition, the importance of ground measurements and alternative satellite calibration systems (e.g., MOBY, AERONET, and GOOS) cannot be overlooked. The Marine Optical Buoy (MOBY), a radiometric buoy stationed in the waters off Lanai, Hawaii, has been the primary basis for the on-orbit vicarious calibrations of all three US ocean color sensors and numerous international satellite sensors since late 1996 [132]. The aErosol rObotic NETwork (AERONET) program is a federation of ground-based remote sensing aerosol networks established by NASA and PHOTONS. For more than 25 years, the project has provided a long-term, continuous, and readily accessible public domain database of aerosol optical, microphysical, and radiative properties for aerosol research and characterization, validation of satellite retrievals, and synergism with other databases. The Global Ocean Observing System (GOOS) consists of a variety of observation means and is committed to obtaining and disseminating reliable assessment and forecasting information on the present and future state of the marine environment for the effective, safe, and sustainable use of the marine environment.

4.2.3. The Benefit for the Sustainable Utilization of Marine Resources

The relationship between marine resources and the environment determines the sustainable utilization of the ocean, and remote sensing technology for monitoring marine resources and the environment can enable the effective implementation of marine spatial planning. In the governance of marine resources, remote sensing methods are used to strengthen the survey of the reserves and distribution of various marine resources and monitor the impact of human activities on marine development and environmental pollution. Based on the principle of unified coordination of economic, social, and ecological benefits, by studying the intensity and methods of human utilization of the ocean, we improve marine spatial planning, ensure the rational use of marine resources, and promote sustainable development of the marine economy. In addition, comprehensive monitoring data based on remote sensing databases is also the decision-making basis for marine scientific utilization.

Author Contributions: Conceptualization, J.M. and R.M.; methodology, Q.P. and X.L.; software, Q.P. and J.W.; valida-tion, Q.P. and X.N.; formal analysis, J.M. and X.L.; investigation, Q.P. and J.W.; resources, R.M.; visualization, Q.P. and X.L.; supervision, R.M.; funding acquisition, J.M. All authors have read and agreed to the published version of the manuscript.

Funding: This research was supported by Key Laboratory of Ocean Space Resource Management Technology, MNR (Grant No. KF-2023-103), the Natural Science Foundation of Zhejiang Province, China [Grant No. LGF22D010002], the Open Funding of Zhejiang Collaborative Innovation Center for Land and Marine Spatial Utilization and Governance Research [Grant No. LHGTXT-2023-003], and Ningbo University Donghai Academy [Grant No. DH202302CBZZ03].

Data Availability Statement: Not applicable.

Conflicts of Interest: The authors declare no conflict of interest.

References

1. Sankar, T.K.; Kumar, A.; Mahto, D.K.; Das, K.C.; Narayan, P.; Fukate, M.; Awachat, P.; Padghan, D.; Mohammad, F.; Al-Lohedan, H.A.; et al. The Health Risk and Source Assessment of Polycyclic Aromatic Hydrocarbons (PAHs) in the Soil of Industrial Cities in India. *Toxics* **2023**, *11*, 515. [CrossRef] [PubMed]
2. Meshram, S.G.; Kahya, E.; Meshram, C.; Ghorbani, M.A.; Ambade, B.; Mirabbasi, R. Long-term temperature trend analysis associated with agriculture crops. *Theor. Appl. Clim.* **2020**, *140*, 1139–1159. [CrossRef]
3. Henrique, M.P.; Paul, W.L.; Vânia, P.; Rob, A.; Jörn, P.W.S.; Juan, J.F.M.; Araújo, M.B.; Patricia, B.; Reinette, B.; William, W.L.C.; et al. Scenarios for global biodiversity in the 21st century. *Science* **2010**, *330*, 1496–1501.
4. Ma, R.F.; Dou, S.M.; Gong, H.B. *China's East China Sea Sustainable Development Study: A Volume on Cross-Regional Governance of Coastal Zone Ecology and Environment*; Ocean Press: Beijing, China, 2019.
5. Sala, E.; Knowlton, N. Global Marine Biodiversity Trends. *Annu. Rev. Environ. Resour.* **2006**, *31*, 93–122. [CrossRef]
6. Lin, M.S.; Zhang, Y.G.; Yuan, X.Z. The development course and trend of ocean remote sensing satellite. *Acta Oceanol. Sin.* **2015**, *37*, 1–10.
7. Sun, F.Z. *Remote Sensing Assessment of Coastal Zones*; Science Press: Beijing, China, 2015.
8. Zou, B.; Lin, M.S.; Shi, L.J.; Zou, Y.R.; Jia, Y.J.; Zeng, T. Application of Remote Sensing Technology in Ocean Disaster. *City Disaster Reduct.* **2018**, *123*, 61–65.
9. Eck, T.F.; Holben, B.N.; Kim, J.; Beyersdorf, A.J.; Choi, M.; Lee, S.; Koo, J.-H.; Giles, D.M.; Schafer, J.S.; Sinyuk, A.; et al. Influence of cloud, fog, and high relative humidity during pollution transport events in South Korea: Aerosol properties and PM2.5 variability. *Atmos. Environ.* **2020**, *232*, 117530.
10. Fichot, C.G.; Tzortziou, M.; Mannino, A. Remote sensing of dissolved organic carbon (DOC) stocks, fluxes and transformations along the land-ocean aquatic continuum: Advances, challenges, and opportunities. *Earth-Sci. Rev.* **2023**, *242*, 104446. [CrossRef]
11. Leifer, I.; Lehr, W.J.; Simecek-Beatty, D.; Bradley, E.; Clark, R.; Dennison, P.; Hu, Y.; Matheson, S.; Jones, C.E.; Holt, B.; et al. State of the art satellite and airborne marine oil spill remote sensing: Application to the BP Deepwater Horizon oil spill. *Remote Sens. Environ.* **2012**, *124*, 185–209. [CrossRef]
12. Mace, T.H. At-sea detection of marine debris: Overview of technologies, processes, issues, and options. *Mar. Pollut. Bull.* **2012**, *65*, 23–27. [CrossRef]
13. Moreno-Madriñán, M.J.; Rickman, D.L.; Ogashawara, I.; Irwin, D.E.; Ye, J.; Al-Hamdan, M.Z. Using remote sensing to monitor the influence of river discharge on watershed outlets and adjacent coral Reefs: Magdalena River and Rosario Is-lands, Colombia. *Int. J. Appl. Earth Obs. Geoinf.* **2015**, *38*, 204–215.
14. Nellemann, C.; Hain, S.; Alde, R.J. *Dead Water: Merging of Climate Change with Pollution, Over-Harvest, and Infestations in the World's Fishing Grounds*; United Nations Environment Programme: Arendal, Norway, 2008.
15. Caballero, I.; Fernández, R.; Escalante, O.M.; Mamán, L.; Navarro, G. New capabilities of Sentinel-2A/B satellites com-bined with in situ data for monitoring small harmful algal blooms in complex coastal waters. *Sci. Rep.* **2020**, *10*, 8743.
16. Shamsudeen, T.Y. Advances in remote sensing technology, machine learning and deep learning for marine oil spill de-tection, prediction and vulnerability assessment. *Remote Sens.* **2020**, *12*, 3416.
17. Katlane, R.; Dupouy, C.; El Kilani, B.; Berges, J.C. Estimation of Chlorophyll and Turbidity Using Sentinel 2A and EO1 Data in Kneiss Archipelago Gulf of Gabes, Tunisia. *Int. J. Geosci.* **2020**, *11*, 708–728. [CrossRef]
18. Themistocleous, K.; Papoutsa, C.; Michaelides, S.; Hadjimitsis, D. Investigating Detection of Floating Plastic Litter from Space Using Sentinel-2 Imagery. *Remote Sens.* **2020**, *12*, 2648. [CrossRef]
19. Ambade, B.; Sethi, S.S.; Chintalacheruvu, M.R. Distribution, risk assessment, and source apportionment of polycyclic aromatic hydrocarbons (PAHs) using positive matrix factorization (PMF) in urban soils of East India. *Environ. Geochem. Health* **2022**, *45*, 491–505. [CrossRef]
20. Liu, H.L.; Nie, H.T.; Wang, Y.L.; Sun, X.; Wei, H. Estimation method of pollutant load into sea using statistical data—Tianjin city. *Mar. Environ. Sci.* **2019**, *38*, 968–976.
21. Kurwadkar, S.; Sankar, T.K.; Kumar, A.; Ambade, B.; Gautam, S.; Gautam, A.S.; Biswas, J.K.; Salam, M.A. Emissions of black carbon and polycyclic aromatic hydrocarbons: Potential implications of cultural practices during the Covid-19 pandemic. *Gondwana Res.* **2023**, *114*, 4–14. [CrossRef]
22. Ge, H.Q.; Lan, N. Causes and Prevention Mode on Marine Pollution from the Land-based Activities or Sources (MPLBA) in China. *China Soft Sci.* **2014**, *2*, 22–31.
23. Ma, X.D.; Mu, J.L.; Lin, Z.S.; Wang, L.J.; Yu, L.M.; Wang, Y.; Zhang, Z.; Zhang, Z.F. The Impact of Sewage from Typical Land-based Sources on the Adjacent Marine Eutrophication and Biological Toxicity. *Environ. Monitor. China* **2014**, *30*, 25–30.
24. Xu, L.H.; Li, J.L.; Li, W.F.; Zhao, S.; Yuan, Q.X.; Wang, M.Y.; Yang, L.; Lu, X.S. Progress in Impact of Human Activities on Coastal Resource and Environment. *J. Nanjing Univ. Nat. Sci. Ed.* **2014**, *37*, 124–131.

25. Davenport, J.; Davenport, J.L. The impact of tourism and personal leisure transport on coastal environments: A review. *Estuar. Coast. Shelf Sci.* **2006**, *67*, 280–292.
26. *GB 3097-1997*; Marine Water Quality Standard. State Environmental Protection Administration: Beijing, China, 1997.
27. *GB 3838-2002*; Environmental Quality Standards for Surface Water. State Environmental Protection Administration (SEPA): Beijing, China; General Administration of Quality Supervision, Inspection and Quarantine (AQSIQ) of the People's Republic of China: Beijing, China, 2002.
28. Chen, Y.B.; Song, G.B.; Zhao, W.X.; Chen, J.W. Estimating pollutant loadings from mariculture in China. *Mar. Environ. Sci.* **2016**, *35*, 1–6.
29. Liu, J.L.; Chen, X.J. Research Progress and Hotspots of Marine Biodiversity: Based on Bibliometrics and Knowledge Mapping Analysis. *Prog. Fish. Sci.* **2021**, *42*, 201–213.
30. Wang, C.Y.; He, S.J.; Li, Y.T.; Hou, X.Y.; Yang, C.Y. Study on the state and ecological effect of spilled oil pollution in Chinese Coastal Waters. *Mar. Sci.* **2009**, *33*, 57–60.
31. *GB 11607-1989*; Water Quality Standard for Fisheries. State Environmental Protection Administration: Beijing, China, 1989.
32. *GB/T 17108-2006*; Technical Directives for the Division of Marine Functional Zonation. General Administration of Quality Supervision, Inspection and Quarantine (AQSIQ): Beijing, China; Standardization Administration of the People's Republic of China (SAC): Beijing, China, 2006.
33. Kumar, A.; Ambade, B.; Sankar, T.K.; Sethi, S.S.; Kurwadkar, S. Source identification and health risk assessment of atmospheric PM2.5-bound polycyclic aromatic hydrocarbons in Jamshedpur, India. *Sustain. Cities Soc.* **2019**, *52*, 101801.
34. Ambade, B.; Kumar, A.; Latif, M. Emission sources, Characteristics and risk assessment of particulate bound Polycyclic Aromatic Hydrocarbons (PAHs) from traffic sites. 2021, *preprint*.
35. Lin, M.S.; He, X.Q.; Jia, Y.J.; Bai, Y.; Ye, X.M.; Gong, F. Advances in marine satellite remote sensing technology in China. *Acta Oceanol. Sin.* **2019**, *41*, 99–112.
36. Wang, W.J.; Jia, D.N.; Xu, J.L.; Chu, H.K.; Dong, X.R. Review of the development of global marine remote sensing satellite. *Bull. Surv. Mapp.* **2020**, *5*, 1–6.
37. Baig, M.H.A.; Zhang, L.; Shuai, T.; Tong, Q. Derivation of a tasselled cap transformation based on Landsat 8 at-satellite reflectance. *Remote Sens. Lett.* **2014**, *5*, 423–431. [CrossRef]
38. Blix, K.; Pálffy, K.; Tóth, V.R.; Eltoft, T. Remote sensing of water quality parameters over Lake Balaton by using sentinel-3 OLCI. *Water* **2018**, *10*, 1428.
39. Atlas, R.; Hoffman, R.N.; Ardizzone, J.; Leidner, S.M.; Jusem, J.C.; Smith, D.K.; Gombos, D. A cross-calibrated multi-plat-form ocean surface wind velocity product for meteorological oceanographic applications. *Bull. Am. Meteorol. Soc.* **2011**, *92*, 157–174. [CrossRef]
40. Su, H.; Li, W.; Yan, X. Retrieving temperature anomaly in the global subsurface deeper ocean from satellite observations. *J. Geophys. Res. Ocean.* **2018**, *123*, 399–410.
41. Sun, W.W.; Yang, G.; Chen, C.; Chang, M.H.; Huang, K.; Meng, X.Z.; Liu, L.Y. Development status literature analysis of China's earth observation remote sensing satellites. *J. Remote Sens.* **2020**, *24*, 479–510. [CrossRef]
42. Jiang, X.W.; Lin, M.S.; Liu, J.Q.; Zhang, Y.Q.; Xie, X.T.; Peng, H.L.; Zhou, W. The HY-2 satellite its preliminary assessment. *Int. J. Digit. Earth* **2012**, *5*, 266–281. [CrossRef]
43. Zhao, C.; Jia, M.; Wang, Z.; Mao, D.; Wang, Y. Toward a better understanding of coastal salt marsh mapping: A case from China using dual-temporal images. *Remote Sens. Environ.* **2023**, *295*, 113664. [CrossRef]
44. Dmytro, K.; Susanne, K. Evaluation of sentinel-3A OLCI products derived using the case-2 regional coast colour processor over the Baltic Sea. *Sensors* **2019**, *19*, 3609.
45. Wang, J.T.; Xu, X.G.; Meng, X.H. The modified ensemble empirical mode decomposition method and extraction of oceanic internal wave from synthetic aperture radar image. *J. Shanghai Jiaotong Univ. Sci.* **2015**, *20*, 243–250. [CrossRef]
46. Chen, C.; Yuxi, S.; Jun, T. Research on Marine Disaster Prevention and Mitigation Information Platform System Based on Big Data. *IOP Conf. Ser. Earth Environ. Sci.* **2021**, *632*, 022082. [CrossRef]
47. Yu, G.; Yang, W.; Matsushita, B.; Li, R.; Oyama, Y.; Fukushima, T. Remote Estimation of Chlorophyll-a in Inland Waters by a NIR-Red-Based Algorithm: Validation in Asian Lakes. *Remote Sens.* **2014**, *6*, 3492–3510. [CrossRef]
48. Masoud, A.A. On the Retrieval of the Water Quality Parameters from Sentinel-3/2 and Landsat-8 OLI in the Nile Delta's Coastal and Inland Waters. *Water* **2022**, *14*, 593.
49. Giorgio, D.; Anatoly, A.G. Effect of bio-optical parameter variability and uncertainties in reflectance measurements on the remote estimation of chlorophyll-a concentration in turbid productive waters: Modeling results. *Appl. Opt.* **2005**, *45*, 3577–3592.
50. Mao, Z.; Chen, J.; Pan, D.; Tao, B.; Zhu, Q. A regional remote sensing algorithm for total suspended matter in the East China Sea. *Remote Sens. Environ.* **2012**, *124*, 819–831. [CrossRef]
51. Yin, X.Q.; Yang, D.T.; Zhou, L.Z. Estimation of suspended particulate matter transport via the boundary waters of the Yellow Sea the East Sea based on satellite remote sensing. *J. Trop. Oceanogr.* **2018**, *37*, 10–16.
52. Ahn, Y.H.; Shanmugam, P. Detecting the red tide algal blooms from satellite ocean color observations in optically com-plex Northeast-Asia Coastal waters. *Remote Sens. Environ.* **2006**, *103*, 419–437. [CrossRef]
53. Hu, C.; Barnes, B.B.; Qi, L.; Corcoran, A.A. A Harmful Algal Bloom of *Karenia brevis* in the Northeastern Gulf of Mexico as Revealed by MODIS and VIIRS: A Comparison. *Sensors* **2015**, *15*, 2873–2887. [CrossRef] [PubMed]

54. Zhang, H.; Zheng, X.S. Study on red tide remote sensing monitoring in the Bohai Sea in 2014. *Mar. Sci. Bull.* **2017**, *19*, 37–51.
55. Jiang, X.W.; He, X.Q.; Lin, M.S.; Gong, F.; Ye, X.M.; Pan, D.L. Progresses on ocean satellite remote sensing application in China. *Acta Oceanol. Sin.* **2019**, *41*, 113–124.
56. Pan, D.; He, X.; Mao, T. Preliminary study on the orbit cross-calibration of CMODIS by SeaWiFS. *Prog. Nat. Sci.* **2003**, *13*, 745–749. [CrossRef]
57. Pan, Y.Y.; Guo, Q.X.; Sun, Y.H. Advances in remote sensing inversion method of chlorophyll a concentration. *Sci. Surv. Mapp.* **2017**, *42*, 43–48.
58. Zhao, J.; Ghedira, H. Monitoring red tide with satellite imagery and numerical models: A case study in the Arabian Gulf. *Mar. Pollut. Bull.* **2014**, *79*, 305–313. [CrossRef]
59. Harvey, B.P.; Gwynn-Jones, D.; Moore, P.J. Meta-analysis reveals complex marine biological responses to the interactive effects of ocean acidification and warming. *Ecol. Evol.* **2013**, *3*, 1016–1030. [CrossRef] [PubMed]
60. Caballero, I.; Stumpf, P.R.; Meredith, A. Preliminary assessment of turbidity and chlorophyll impact on bathymetry de-rived from sentinel-2A and sentinel-3A satellites in South Florida. *Remote Sens.* **2019**, *11*, 645.
61. Girolametti, F.; Fanelli, M.; Ajdini, B.; Truzzi, C.; Illuminati, S.; Susmel, S.; Celussi, M.; Šangulin, J.; Annibaldi, A. Dissolved Potentially Toxic Elements (PTEs) in Relation to Depuration Plant Outflows in Adriatic Coastal Waters: A Two Year Monitoring Survey. *Water* **2022**, *14*, 569. [CrossRef]
62. Yang, Y.P.; Wang, Q.; Wang, W.J.; Gao, S.P. Application and Advances of Remote Sensing Techniques in Determining Water Quality. *Geogr. Geo-Inf. Sci.* **2004**, *6*, 6–12.
63. Shang, S.; Lee, Z.; Wei, G. Characterization of MODIS-derived euphotic zone depth: Results for the China Sea. *Remote Sens. Environ.* **2011**, *115*, 180–186. [CrossRef]
64. Farré, M. Remote and in situ devices for the assessment of marine contaminants of emerging concern and plastic debris detection. *Curr. Opin. Environ. Sci. Health* **2020**, *18*, 79–94. [CrossRef]
65. Lu, X.M.; Su, H. Retrieving total suspended matter concentration in Fujian coastal waters using OLCI data. *Acta Sci. Circumstantiae* **2020**, *40*, 2819–2827.
66. Hou, L.L.; Ma, A.Q.; Hu, J.; Shan, G.Y.; Deng, J.M.; Han, J.Y.; Ding, Z.Y. Study on Remote Sensing Retrieval Model Op-timization of Suspended Sediment Concentration in Jiaozhou Bay. *Period. Ocean Univ. China* **2018**, *48*, 98–108.
67. Ahn, J.-H.; Park, Y.-J.; Ryu, J.-H.; Lee, B.; Oh, I.S. Development of atmospheric correction algorithm for Geostationary Ocean Color Imager (GOCI). *Ocean Sci. J.* **2012**, *47*, 247–259. [CrossRef]
68. Ryu, J.-H.; Han, H.-J.; Cho, S.; Park, Y.-J.; Ahn, Y.-H. Overview of geostationary ocean color imager (GOCI) and GOCI data processing system (GDPS). *Ocean Sci. J.* **2012**, *47*, 223–233. [CrossRef]
69. Zheng, X.S.; Zhang, Y.N.; Liu, X.H. Research on inversion method of aerosol optical depth over the Yellow Sea based on GOCI data. *Mar. Sci. Bull.* **2020**, *39*, 94–100.
70. Chen, J.; Quan, W.; Cui, T.; Song, Q. Estimation of total suspended matter concentration from MODIS data using a neural network model in the China eastern coastal zone. *Estuar. Coast. Shelf Sci.* **2015**, *155*, 104–113. [CrossRef]
71. Tian, H.Z.; Liu, Q.P.; Goes, J.I.; Gomes, H.D.R.; Yang, M.M. Temporal and spatial changes in chlorophyll-a concentrations in the Yellow Sea from 2002 to 2018 based on MODIS data. *Mar. Sci. Bull.* **2020**, *39*, 101–110.
72. Yu, G.; Fu, D.Y.; Zhong, Y.F.; Liu, D.Z.; Qi, Y.L.; Lou, Y.F. Remote Sensing Retrieval of Colored Dissolved Organic Matter in Zhanjiang Coastal Area. *J. Guangdong Ocean Univ.* **2021**, *41*, 55–62.
73. Wang, L.; Zhao, D.Z.; Yang, J.H.; Chen, Y.L. The optical properties and remote sensing retrieval model of chromophoric dissolved organic matter in the Dayang Estuary. *Acta Oceanol. Sin.* **2011**, *33*, 45–51.
74. David, O.; Anjan, D. Institutional and Policy Cocktails for Protecting Coastal and Marine Environments from Land-based Sources of Pollution. *Ocean Coast. Manag.* **2006**, *49*, 576–596.
75. Huisman, J.; Codd, G.A.; Paerl, H.W.; Paerl, H.W.; Ibelings, B.W.; Verspagen, J.M.H.; Visser, P.M. Cyanobacterial blooms. Nature reviews. *Microbiology* **2018**, *6*, 471–483.
76. Inia, M.S.; Jennifer, C.; Frank, E.M.K.; Hu, C.M.; Jennifer, W.; Dmitry, G. Evaluation and optimization of remote sensing techniques for detection of Karenia brevis blooms on the West Florida Shelf. *Remote Sens. Environ.* **2015**, *170*, 239–254.
77. Kutser, T.; Metsamaa, L.; Dekker, A.G. Influence of the vertical distribution of cyanobacteria in the water column on the remote sensing signal. *Estuar. Coast. Shelf Sci.* **2008**, *78*, 649–654. [CrossRef]
78. Hu, C. A novel ocean color index to detect floating algae in the global oceans. *Remote Sens. Environ.* **2009**, *113*, 2118–2129. [CrossRef]
79. Tao, B.; Pan, D.; Mao, Z.; Shen, Y.; Zhu, Q.; Chen, J. Optical detection of Prorocentrum donghaiense blooms based on multispectral reflectance. *Acta Oceanol. Sin.* **2013**, *32*, 48–56. [CrossRef]
80. Gower, J.; King, S.; Wei, Y.; Borstad, G.; Brown, L. Use of the 709 nm band of MERIS to detect intense plankton blooms and other conditions in coastal waters. In Proceedings of the MERIS User Workshop, Frascati, Italy, 10–13 November 2003; pp. 10–13.
81. Lee, Z.; Carder, K.L.; Arnone, R. Deriving inherent optical properties from water color: A multiband quasi-analytical algorithm for optically deep waters. *Appl. Opt.* **2002**, *41*, 5755–5772.
82. Li, X.; Liu, B.; Zheng, G.; Ren, Y.; Zhang, S.; Liu, Y.; Gao, L.; Liu, Y.; Zhang, B.; Wang, F. Deep-learning-based information mining from ocean remote-sensing imagery. *Natl. Sci. Rev.* **2020**, *7*, 1584–1605. [CrossRef] [PubMed]

83. Hu, C.; Li, X.; William, G.P.; Frank, E.M.K. Detection of natural oil slicks in the NW Gulf of Mexico using MODIS imagery. *Geophys. Res. Lett.* **2009**, *36*, L01604.
84. Hu, C.; Muller-Karger, F.E.; Taylor, C.; Carder, K.L.; Kelble, C.; Johns, E.; Heil, C.A. Red tide detection and tracing using MODIS fluorescence data: A regional example in SW Florida coastal waters. *Remote Sens. Environ.* **2005**, *97*, 311–321. [CrossRef]
85. Shi, W.; Wang, M. Green macroalgae blooms in the Yellow Sea during the spring and summer of 2008. *J. Geophys. Res. Atmos.* **2009**, *114*. [CrossRef]
86. Xing, Q.; Hu, C. Mapping macroalgal blooms in the Yellow Sea and East China Sea using HJ-1 and Landsat data: Ap-pli-cation of a virtual baseline reflectance height technique. *Remote Sens. Environ.* **2016**, *178*, 113–126.
87. Shuo, H.E.; Lou, X.L.; Shi, A.Q.; Li, D.L.; Wang, J.; Zhang, H.G. Simulation and Study of Remote Sensing Reflectance Spectra of Typical Algal Blooms in the East China Sea. *Oceanol. Limnol. Sin.* **2019**, *50*, 525–531.
88. Zhang, H.L.; Sun, D.Y.; Li, J.S.; Qiu, Z.F.; Wang, S.Q.; He, Y.J. Remote Sensing Algorithm for Detecting Green Tide in China Coastal Waters Based on GF1-WFV and H-CCD Data. *Acta Opt. Sin.* **2016**, *36*, 36–44.
89. Chen, Y.; Sun, D.Y.; Zhang, H.L.; Wang, S.Q. Remote-Sensing Monitoring of Green Tide and Its Drifting Trajectories in Yellow Sea Based on Observation Data of Geostationary Ocean Color Imager. *Acta Opt. Sin.* **2020**, *40*, 0301001. [CrossRef]
90. Son, Y.B.; Min, J.-E.; Ryu, J.-H. Detecting massive green algae (Ulva prolifera) blooms in the Yellow Sea and East China Sea using Geostationary Ocean Color Imager (GOCI) data. *Ocean Sci. J.* **2012**, *47*, 359–375. [CrossRef]
91. Zhang, H.; Qiu, Z.; Devred, E.; Sun, D.; Wang, S.; He, Y.; Yu, Y. A simple and effective method for monitoring floating green macroalgae blooms: A case study in the Yellow Sea. *Opt. Express* **2019**, *27*, 4528–4548. [CrossRef]
92. Jiang, D.J.; Wang, K.; Xia, Y. Comparative studies on remote sensing techniques for red tide monitoring in Bohai Sea. *Mar. Environ. Sci.* **2020**, *39*, 460–467.
93. Tao, B.Y.; Mao, Z.H.; Lei, H.; Pan, D.L.; Shen, Y.Z.; Bai, Y.; Zhu, Q.K.; Li, Z.E. A novel method for discriminating Prorocentrum donghaiense from diatom blooms in the East China Sea using MODIS measurements. *Remote Sens. Environ.* **2015**, *158*, 267–280.
94. Kim, Y.; Byun, Y.; Kim, Y.; Eo, Y. Detection of Cochlodinium polykrikoides red tide based on two-stage filtering using MODIS data. *Desalination* **2009**, *249*, 1171–1179. [CrossRef]
95. Liu, F.; Chen, C.; Tang, S.; Liu, D. Retrieval of chlorophyll concentration from the enveloped areas of fluorescence spectra in Pearl River estuary, China. *Proc. SPIE-Int. Soc. Opt. Eng.* **2010**, *64*, 1912–1916.
96. Commendatore, M.G.; Esteves, J.L. An Assessment of Oil Pollution in the Coastal Zone of Patagonia, Argentina. *Environ. Manag.* **2007**, *40*, 814–821. [CrossRef]
97. Zou, Y.; Zou, B.; Liang, C.; Cui, S.; Zeng, T. Multiple Index Information Extraction of Marine Oil Spills. *Geo-Inf. Sci.* **2012**, *14*, 265–269. [CrossRef]
98. Nirchio, F.; Sorgente, M.; Giancaspro, A.; Biamino, W.; Parisato, E.; Ravera, R.; Trivero, P. Automatic detection of oil spills from SAR images. *Int. J. Remote Sens.* **2005**, *26*, 1157–1174. [CrossRef]
99. Lu, Y.; Li, X.; Tian, Q.J.; Zheng, G.; Sun, S.J.; Liu, Y.X.; Yang, Q. Progress in marine oil spill optical remote sensing: De-tected targets, spectral response characteristics, and theories. *Mar. Geod.* **2013**, *36*, 334–346.
100. Yu, L.; Jonathan, L. Oil spill detection from SAR intensity imagery using a marked point process. *Remote Sens. Environ.* **2010**, *114*, 1590–1601.
101. Zhang, B.; Perrie, W.; Li, X.; Pichel, W.G. Mapping sea surface oil slicks using RADARSAT-2 quad-polarization SAR image. *Geophys. Res. Lett.* **2011**, *38*, 415–421. [CrossRef]
102. Lu, Y.; Zhan, W.; Hu, C. Detecting and quantifying oil slick thickness by thermal remote sensing: A ground-based experiment. *Remote Sens. Environ.* **2016**, *181*, 207–217.
103. Wettle, M.; Daniel, P.J.; Logan, G.A.; Thankappan, M. Assessing the effect of hydrocarbon oil type and thickness on a remote sensing signal: A sensitivity study based on the optical properties of two different oil types and the HYMAP and Quickbird sensors. *Remote Sens. Environ.* **2009**, *113*, 2000–2010. [CrossRef]
104. Huang, K.; Pan, Q.; Zhang, J.Y.; Wang, Y.L.; Yang, G.; Sun, W.W. Quantitative monitoring in oil spill incidents based on GF-1 satellite: Qingdao oil spill accident case. *Mar. Sci. Bull.* **2020**, *39*, 266–271.
105. Lu, Y.C.; Liu, J.Q.; Ding, J.; Shi, J.; Chen, J.Y.; Ye, X.M. Optical remote identification of spilled oils from the SANCHI oil tanker collision in the East China Sea. *Chin. Sci. Bull.* **2019**, *64*, 3213–3222.
106. Solberg, A.H.S.; Brekke, C.; Husoy, P.O. Oil Spill Detection in Radarsat and Envisat SAR Images. *IEEE Trans. Geosci. Remote Sens.* **2007**, *45*, 746–755. [CrossRef]
107. Jiao, J.; Lu, Y.; Hu, C.; Shi, J.; Sun, S.; Liu, Y. Quantifying ocean surface oil thickness using thermal remote sensing. *Remote Sens. Environ.* **2021**, *261*, 112513. [CrossRef]
108. Shen, Y.F.; Liu, J.Q.; Ding, J.; Jiao, J.N.; Sun, S.J.; Lu, Y.C. HY-1C COCTS and CZI observation of marine oil spills in the South China Sea. *J. Remote Sens.* **2020**, *24*, 933–944. [CrossRef]
109. Zou, Y.R.; Liang, C.; Chen, J.L.; Cui, S.X.; Lang, S.Y. An optimal parametric analysis of monitoring oil spill based on SAR. *Acta Ocean. Sin.* **2011**, *33*, 36–44.
110. Hu, C. Remote detection of marine debris using satellite observations in the visible and near infrared spectral range: Challenges and potentials. *Remote Sens. Environ.* **2021**, *259*, 112414. [CrossRef]
111. Kikaki, A.; Karantzalos, K.; Power, C.A.; Raitsos, D.E. Remotely Sensing the Source and Transport of Marine Plastic De-bris in Bay Islands of Honduras (Caribbean Sea). *Remote Sens.* **2016**, *12*, 1727.

112. Lauren, B.; Daniel, C.; Victor, M.V.; Konstantinos, T. Finding Plastic Patches in Coastal Waters using Optical Satellite Data. *Sci Rep.* **2020**, *10*, 5364.
113. Paul, M.S.H.; John, B.; Carme, A.; Montserrat, C.; Laia, R.; Salud, D. Assessment of marine litter through remote sensing: Recent approaches and future goals. *Mar. Pollut. Bull.* **2021**, *168*, 112347.
114. Lonneke, G.M.; Steef, P.; Erik, V.S.; James, N.A.; Stuart, G. Concept for a hyperspectral remote sensing algorithm for floating marine macro plastics. *Mar. Pollut. Bull.* **2018**, *126*, 255–262.
115. Topouzelis, K.; Papageorgiou, D.; Suaria, G.; Aliani, S. Floating marine litter detection algorithms and techniques using optical remote sensing data: A review. *Mar. Pollut. Bull.* **2021**, *170*, 112675. [CrossRef]
116. Martin, C.; Parkes, S.; Zhang, Q.; Zhang, X.; McCabe, M.F.; Duarte, C.M. Use of unmanned aerial vehicles for efficient beach litter monitoring. *Mar. Pollut. Bull.* **2018**, *131*, 662–673. [CrossRef]
117. Hamilton, B.; Dean, C.; Kurata, N.; Vella, K.; Soloviev, A.; Tartar, A.; Shivji, M.; Matt, S.; Perrie, W.; Lehner, S.; et al. Surfactant Associated Bacteria in the Sea Surface Microlayer: Case Studies in the Straits of Florida and the Gulf of Mexico. *Can. J. Remote Sens.* **2015**, *41*, 135–143. [CrossRef]
118. Aoyama, T. Extraction of marine debris in the Sea of Japan using high-spatial-resolution satellite images. In *Remote Sensing of the Oceans and Inland Waters: Techniques, Applications, and Challenges*; SPIE: New Delhi, India, 2016.
119. Kurata, N.; Vella, K.; Hamilton, B.; Shivji, M.; Soloviev, A.; Matt, S.; Tartar, A.; Perrie, W. Surfactant-associated bacteria in the near-surface layer of the ocean. *Sci. Rep.* **2016**, *6*, srep19123. [CrossRef]
120. Davaasuren, N.; Marino, A.; Boardman, C.P.; Alparone, M.; Nunziat, F.; Ackermann, N.; Hajnsek, I. Detecting Micro-plastics Pollution in World Oceans Using Sar Remote Sensing. In Proceedings of the IGARSS 2018-2018 IEEE International Geoscience and Remote Sensing Symposium, Valencia, Spain, 22–27 July 2018; pp. 938–941.
121. Howe, K.L.; Dean, C.W.; Kluge, J.; Soloviev, A.V.; Tartar, A.; Shivji, M.; Lehner, S.; Perrie, W. Relative abundance of *Bacillus* spp., surfactant-associated bacterium present in a natural sea slick observed by satellite SAR imagery over the Gulf of Mexico. *Elem. Sci. Anthr.* **2018**, *6*, 8. [CrossRef]
122. Mao, Y.; Zheng, J.L.; Qiu, Z.F.; Muhammad, B. Precision analysis and spatiotemporal distribution characteristics from multi source satellite aerosol optical depth data in the Yellow Sea and Bohai Sea. *Acta Sci. Circumstantiae* **2021**, *41*, 2550–2559.
123. Veefkind, J.; de Leeuw, G.; Stammes, P.; Koelemeijer, R.B. Regional Distribution of Aerosol over Land, Derived from ATSR-2 and GOME. *Remote Sens. Environ.* **2000**, *74*, 377–386. [CrossRef]
124. Dong, H.Y.; Liu, Y.; Guan, Z.Y. Validation of MODIS Aerosol Optical Depth Retrievals over East China Sea. *J. Nanjing Inst. Meteorol.* **2007**, *3*, 328–337.
125. Lee, J.; Kim, J.; Song, C.H.; Ryu, J.-H.; Ahn, Y.-H. Algorithm for retrieval of aerosol optical properties over the ocean from the Geostationary Ocean Color Imager. *Remote Sens. Environ.* **2010**, *114*, 1077–1088. [CrossRef]
126. Lin, M.S.; Ye, X.M.; Yuan, X.Z. The first quantitative joint observation of typhoon by Chinese GF-3 SAR and HY-2A microwave scatterometer. *Acta Oceanol. Sin.* **2017**, *36*, 1–3. [CrossRef]
127. Wang, S.; Feng, H.H.; Zou, B.; Yang, Z.L.; Ding, Y.; Ye, S.C.; Zhu, S.J. Research progresses on deposition monitoring of air pollution. *China Environ. Sci.* **2021**, *41*, 4961–4972.
128. Hao, Z.Z.; Pan, D.L.; Bai, Y. Characteristics of the spatial disibution monthly variation of aerosol optical thickness derived from SeaWiFS over the China Seas. *J. Mar. Sci.* **2007**, *s98*, 80–87.
129. Kachelriess, D.; Wegmann, M.; Gollock, M.; Pettorelli, N. The application of remote sensing for marine protected area management. *Ecol. Indic.* **2014**, *36*, 169–177. [CrossRef]
130. Su, H.; Liu, H.; Heyman, W.D. Automated Derivation of Bathymetric Information from Multi-Spectral Satellite Imagery Using a Non-Linear Inversion Model. *Mar. Geod.* **2008**, *31*, 281–298. [CrossRef]
131. Wu, Z.Q.; Mao, Z.H.; Wang, Z.; Qiu, Y.W.; Shen, W. Research on Remote Sensing Inversion of Shallow Water Depth Based on Extreme Learning Machine. *Hydrogr. Surv. Charting* **2019**, *39*, 11–15.
132. Franz, B.A.; Bailey, S.W.; Werdell, P.J.; McClain, C.R. Sensor-independent approach to the vicarious calibration of satellite ocean color radiometry. *Appl. Opt.* **2007**, *46*, 5068–5082. [CrossRef] [PubMed]

MDPI AG
Grosspeteranlage 5
4052 Basel
Switzerland
Tel.: +41 61 683 77 34

Water Editorial Office
E-mail: water@mdpi.com
www.mdpi.com/journal/water

www.ingramcontent.com/pod-product-compliance
Lightning Source LLC
Chambersburg PA
CBHW061437210926
51850CB00008B/1535
* 9 7 8 3 7 2 5 8 5 6 2 5 1 *